# The Variation of Animals and Plants under Domestication

## Volume 2

Charles Darwin

Alpha Editions

This edition published in 2024

ISBN : 9789362929860

Design and Setting By
**Alpha Editions**
www.alphaedis.com
Email - info@alphaedis.com

As per information held with us this book is in Public Domain. This book is a reproduction of an important historical work. Alpha Editions uses the best technology to reproduce historical work in the same manner it was first published to preserve its original nature. Any marks or number seen are left intentionally to preserve its true form.

# Contents

VOLUME II. ........................................................................- 1 -
CHAPTER 2.XIII. ..............................................................- 3 -
CHAPTER 2.XIV. .............................................................- 32 -
CHAPTER 2.XV................................................................- 53 -
CHAPTER 2.XVI. .............................................................- 66 -
CHAPTER 2.XVII. ............................................................- 79 -
CHAPTER 2.XVIII.............................................................- 111 -
CHAPTER 2.XIX. .............................................................- 140 -
CHAPTER 2.XX................................................................- 154 -
CHAPTER 2.XXI. .............................................................- 179 -
CHAPTER 2.XXII. ............................................................- 200 -
CHAPTER 2.XXIII. ...........................................................- 219 -
CHAPTER 2.XXIV. ...........................................................- 239 -
CHAPTER 2.XXV..............................................................- 263 -
CHAPTER 2.XXVI. ...........................................................- 279 -
CHAPTER 2.XXVII. ..........................................................- 293 -
CHAPTER 2.XXVIII...........................................................- 332 -

# VOLUME II.

# CHAPTER 2.XIII.

INHERITANCE continued—REVERSION OR ATAVISM.

**DIFFERENT FORMS OF REVERSION. IN PURE OR UNCROSSED BREEDS, AS IN PIGEONS, FOWLS, HORNLESS CATTLE AND SHEEP, IN CULTIVATED PLANTS. REVERSION IN FERAL ANIMALS AND PLANTS. REVERSION IN CROSSED VARIETIES AND SPECIES. REVERSION THROUGH BUD-PROPAGATION, AND BY SEGMENTS IN THE SAME FLOWER OR FRUIT. IN DIFFERENT PARTS OF THE BODY IN THE SAME ANIMAL. THE ACT OF CROSSING A DIRECT CAUSE OF REVERSION, VARIOUS CASES OF, WITH INSTINCTS. OTHER PROXIMATE CAUSES OF REVERSION. LATENT CHARACTERS. SECONDARY SEXUAL CHARACTERS. UNEQUAL DEVELOPMENT OF THE TWO SIDES OF THE BODY. APPEARANCE WITH ADVANCING AGE OF CHARACTERS DERIVED FROM A CROSS. THE GERM, WITH ALL ITS LATENT CHARACTERS, A WONDERFUL OBJECT. MONSTROSITIES. PELORIC FLOWERS DUE IN SOME CASES TO REVERSION.**

The great principle of inheritance to be discussed in this chapter has been recognised by agriculturists and authors of various nations, as shown by the scientific term ATAVISM, derived from atavus, an ancestor; by the English terms of REVERSION, or THROWING-BACK; by the French PAS-EN-ARRIERE; and by the German RUCKSCHLAG, or RUCKSCHRITT. When the child resembles either grandparent more closely than its immediate parents, our attention is not much arrested, though in truth the fact is highly remarkable; but when the child resembles some remote ancestor or some distant member in a collateral line,—and in the last case we must attribute this to the descent of all the members from a common progenitor,—we feel a just degree of astonishment. When one parent alone displays some newly-acquired and generally inheritable character, and the offspring do not inherit it, the cause may lie in the other parent having the power of prepotent transmission. But when both parents are similarly characterised, and the child does not, whatever the cause may be, inherit the character in question, but resembles its grandparents, we have one of the simplest cases of reversion. We continually see another and even more simple case of atavism, though not generally included under this head, namely, when the son more closely resembles his maternal than his paternal grand-sire in some male attribute, as in any peculiarity in the beard of man, the horns of the bull, the hackles or comb of the cock, or, as in certain diseases necessarily confined to the male sex; for as the mother cannot possess or exhibit such male attributes, the child must inherit them, through her blood, from his maternal grandsire.

The cases of reversion may be divided into two main classes which, however, in some instances, blend into one another; namely, first, those occurring in a variety or race which has not been crossed, but has lost by variation some character that it formerly possessed, and which afterwards reappears. The second class includes all cases in which an individual with some distinguishable character, a race, or species, has at some former period been crossed, and a character derived from this cross, after having disappeared during one or several generations, suddenly reappears. A third class, differing only in the manner of reproduction, might be formed to include all cases of reversion effected by means of buds, and therefore independent of true or seminal generation. Perhaps even a fourth class might be instituted, to include reversions by segments in the same individual flower or fruit, and in different parts of the body in the same individual animal as it grows old. But the two first main classes will be sufficient for our purpose.

## REVERSION TO LOST CHARACTERS BY PURE OR UNCROSSED FORMS.

Striking instances of this first class of cases were given in the sixth chapter, namely, of the occasional reappearance, in variously-coloured breeds of the pigeon, of blue birds with all the marks characteristic of the wild Columba livia. Similar cases were given in the case of the fowl. With the common ass, as the legs of the wild progenitor are almost always striped, we may feel assured that the occasional appearance of such stripes in the domestic animal is a case of simple reversion. But I shall be compelled to refer again to these cases, and therefore here pass them over.

The aboriginal species from which our domesticated cattle and sheep are descended, no doubt possessed horns; but several hornless breeds are now well established. Yet in these—for instance, in Southdown sheep—"it is not unusual to find among the male lambs some with small horns." The horns, which thus occasionally reappear in other polled breeds, either "grow to the full size," or are curiously attached to the skin alone and hang "loosely down, or drop off." (13/1. 'Youatt on Sheep' pages 20, 234. The same fact of loose horns occasionally appearing in hornless breeds has been observed in Germany; Bechstein 'Naturgesch. Deutschlands.' b. 1 s. 362.) The Galloways and Suffolk cattle have been hornless for the last 100 or 150 years, but a horned calf, with the horn often loosely attached, is occasionally produced. (13/2. 'Youatt on Cattle' pages 155, 174.)

There is reason to believe that sheep in their early domesticated condition were "brown or dingy black;" but even in the time of David certain flocks were spoken of as white as snow. During the classical period the sheep of

Spain are described by several ancient authors as being black, red, or tawny. (13/3. 'Youatt on Sheep' 1838 pages 17, 145.) At the present day, notwithstanding the great care which is taken to prevent it, particoloured lambs and some entirely black are occasionally, or even frequently, dropped by our most highly improved and valued breeds, such as the Southdowns. Since the time of the famous Bakewell, during the last century, the Leicester sheep have been bred with the most scrupulous care; yet occasionally grey-faced, or black-spotted, or wholly black lambs appear. (13/4. I have been informed of this fact through the Rev. W.D. Fox on the excellent authority of Mr. Wilmot: see also remarks on this subject in an article in the 'Quarterly Review' 1849 page 395.) This occurs still more frequently with the less improved breeds, such as the Norfolks. (13/5. Youatt pages 19, 234.) As bearing on this tendency in sheep to revert to dark colours, I may state (though in doing so I trench on the reversion of crossed breeds, and likewise on the subject of prepotency) that the Rev. W.D. Fox was informed that seven white Southdown ewes were put to a so-called Spanish ram, which had two small black spots on his sides, and they produced thirteen lambs, all perfectly black. Mr. Fox believes that this ram belonged to a breed which he has himself kept, and which is always spotted with black and white; and he finds that Leicester sheep crossed by rams of this breed always produce black lambs: he has gone on recrossing these crossed sheep with pure white Leicesters during three successive generations, but always with the same result. Mr. Fox was also told by the friend from whom the spotted breed was procured, that he likewise had gone on for six or seven generations crossing with white sheep, but still black lambs were invariably produced.

Similar facts could be given with respect to tailless breeds of various animals. For instance, Mr. Hewitt (13/6. 'The Poultry Book' by Mr. Tegetmeier 1866 page 231.) states that chickens bred from some rumpless fowls, which were reckoned so good that they won a prize at an exhibition, "in a considerable number of instances were furnished with fully developed tail-feathers." On inquiry, the original breeder of these fowls stated that, from the time when he had first kept them, they had often produced fowls furnished with tails; but that these latter would again reproduce rumpless chickens.

Analogous cases of reversion occur in the vegetable kingdom; thus "from seeds gathered from the finest cultivated varieties of Heartsease (Viola tricolor), plants perfectly wild both in their foliage and their flowers are frequently produced;" (13/7. Loudon's 'Gardener's Mag.' volume 10 1834 page 396: a nurseryman, with much experience on this subject, has likewise assured me that this sometimes occurs.) but the reversion in this instance is not to a very ancient period, for the best existing varieties of the heartsease are of comparatively modern origin. With most of our cultivated vegetables there is some tendency to reversion to what is known to be, or may be presumed to be, their aboriginal state; and this would be more evident if gardeners did not generally look over their beds of seedlings, and pull up the false plants or "rogues" as they are called. It has already been remarked, that some few seedling apples and pears generally resemble, but apparently are not identical with, the wild trees from which they are descended. In our turnip (13/8. 'Gardener's Chronicle' 1855 page 777.) and carrot-beds a few plants often "break "—that is, flower too soon; and their roots are generally hard and stringy, as in the parent-species. By the aid of a little selection, carried on during a few generations, most of our cultivated plants could probably be brought back, without any great change in their conditions of life, to a wild or nearly wild condition: Mr. Buckman has effected this with the parsnip (13/9. Ibid 1862 page 721.); and Mr. Hewett C. Watson, as he informs me, selected, during three generations, "the most diverging plants of Scotch kail, perhaps one of the least modified varieties of the cabbage; and in the third generation some of the plants came very close to the forms now established in England about old castle-walls, and called indigenous."

## REVERSION IN ANIMALS AND PLANTS WHICH HAVE RUN WILD.

In the cases hitherto considered, the reverting animals and plants have not been exposed to any great or abrupt change in their conditions of life which could have induced this tendency; but it is very different with animals and plants which have become feral or run wild. It has been repeatedly asserted in the most positive manner by various authors, that feral animals and plants invariably return to their primitive specific type. It is curious on what little evidence this belief rests. Many of our domesticated animals could not subsist in a wild state; thus, the more highly improved breeds of the pigeon will not "field" or search for their own food. Sheep have never become feral, and would be destroyed by almost every beast of prey. (13/10. Mr. Boner speaks ('Chamois-hunting' 2nd edition 1860 page 92) of sheep often running wild in the Bavarian Alps; but, on making further inquiries at my request, he found that they are not able to establish themselves; they generally perish from the frozen snow clinging to their wool, and they have lost the skill necessary to pass over steep icy slopes. On one occasion two ewes survived the winter, but their lambs perished.) In several cases we do not know the aboriginal parent- species, and cannot possibly tell whether or not there has been any close degree of reversion. It is not known in any instance what variety was first turned out; several varieties have probably in some cases run wild, and their crossing alone would tend to obliterate their proper character. Our domesticated animals and plants, when they run wild, must always be exposed to new conditions of life, for, as Mr. Wallace (13/11. See some excellent remarks on this subject by Mr. Wallace 'Journal Proc. Linn. Soc.' 1858 volume 3 page 60.) has well remarked, they have to obtain their own food, and are exposed to competition with the native productions. Under these circumstances, if our domesticated animals did not undergo change of some kind, the result would be quite opposed to the conclusions arrived at in this work. Nevertheless, I do not doubt that the simple fact of animals and plants becoming feral, does cause some tendency to reversion to the primitive state; though this tendency has been much exaggerated by some authors.

[I will briefly run through the recorded cases. With neither horses nor cattle is the primitive stock known; and it has been shown in former chapters that they have assumed different colours in different countries. Thus the horses which have run wild in South America are generally brownish-bay, and in the East dun-coloured; their heads have become larger and coarser, and this may be due to reversion. No careful description has been given of the feral goat. Dogs which have run wild in various countries have hardly anywhere assumed a uniform character; but they are probably descended from several domestic races, and aboriginally from several distinct species. Feral cats, both in Europe and La Plata, are regularly striped; in some cases they have grown to an unusually large size, but do not differ from the domestic animal in any other character. When variously-coloured tame rabbits are turned out in Europe, they generally reacquire the colouring of the wild animal; there can be no doubt that this does really occur, but we should remember that oddly-coloured and conspicuous animals would suffer much from beasts of prey and from being easily shot; this at least was the opinion of a gentleman who tried to stock his woods with a nearly white variety; if thus destroyed, they would be supplanted by, instead of being transformed into, the common rabbit. We have seen that the feral rabbits of Jamaica, and especially of Porto Santo, have assumed new colours and other new characters. The best known case of reversion, and that on which the widely spread belief in its universality apparently rests, is that of pigs. These animals have run wild in the West Indies, South America, and the Falkland Islands, and have everywhere acquired the dark colour, the thick bristles, and great tusks of the wild boar; and the young have reacquired longitudinal stripes. But even in the case of the pig, Roulin describes the half-wild animals in different parts of South America as differing in several respects. In Louisiana the pig (13/12. Dureau de la Malle 'Comptes Rendus' tome 41 1855 page 807. From the statements above given, the author concludes that the wild pigs of Louisiana are not descended from the European Sus scrofa.) has run wild, and is said to differ a little in form, and much in colour, from the domestic animal, yet does not closely resemble the wild boar of Europe. With pigeons and fowls (13/13. Capt. W. Allen, in his 'Expedition to the Niger' states that fowls have run wild on the island of Annobon, and have become modified in form and voice. The account is so meagre and vague that it did not appear to me worth copying; but I now find that Dureau de la Malle ('Comptes Rendus' tome 41

1855 page 690) advances this as a good instance of reversion to the primitive stock, and as confirmatory of a still more vague statement in classical times by Varro.), it is not known what variety was first turned out, nor what character the feral birds have assumed. The guinea-fowl in the West Indies, when feral, seems to vary more than in the domesticated state.

With respect to plants run wild, Dr. Hooker (13/14. 'Flora of Australia' 1859 Introduction page 9.) has strongly insisted on what slight evidence the common belief in their reversion to a primitive state rests. Godron (13/15. 'De l'Espece' tome 2 pages 54, 58, 60.) describes wild turnips, carrots, and celery; but these plants in their cultivated state hardly differ from their wild prototypes, except in the succulency and enlargement of certain parts,— characters which would certainly be lost by plants growing in poor soil and struggling with other plants. No cultivated plant has run wild on so enormous a scale as the cardoon (Cynara cardunculus) in La Plata. Every botanist who has seen it growing there, in vast beds, as high as a horse's back, has been struck with its peculiar appearance; but whether it differs in any important point from the cultivated Spanish form, which is said not to be prickly like its American descendant, or whether it differs from the wild Mediterranean species, which is said not to be social (though this may be due merely to the nature of the conditions), I do not know.]

## REVERSION TO CHARACTERS DERIVED FROM A CROSS, IN THE CASE OF SUB-VARIETIES, RACES, AND SPECIES.

When an individual having some recognisable peculiarity unites with another of the same sub-variety, not having the peculiarity in question, it often reappears in the descendants after an interval of several generations. Every one must have noticed, or heard from old people of children closely resembling in appearance or mental disposition, or in so small and complex a character as expression, one of their grandparents, or some more distant collateral relation. Very many anomalies of structure and diseases (13/16. Mr. Sedgwick gives many instances in the 'British and Foreign Med.-Chirurg. Review' April and July 1863 pages 448, 188.) of which instances have been given in the last chapter, have come into a family from one parent, and have reappeared in the progeny after passing over two or three generations. The following case has been communicated to me on good authority, and may, I believe, be fully trusted: a pointer-bitch produced seven puppies; four were marked with blue and white, which is so unusual a colour with pointers that she was thought to have played false with one of the greyhounds, and the whole litter was condemned; but the gamekeeper was permitted to save one as a curiosity. Two years afterwards a friend of the owner saw the young dog, and declared that he was the image of his old pointer-bitch Sappho, the only blue and white pointer of pure descent which he had ever seen. This led to close inquiry, and it was proved that he was the great-great-grandson of

Sappho; so that, according to the common expression, he had only 1/16th of her blood in his veins. I may give one other instance, on the authority of Mr. R. Walker, a large cattle- breeder in Kincardineshire. He bought a black bull, the son of a black cow with white legs, white belly and part of the tail white; and in 1870 a calf the gr.-gr.-gr.-gr.-grandchild of this cow was born coloured in the same very peculiar manner; all the intermediate offspring having been black. In these cases there can hardly be a doubt that a character derived from a cross with an individual of the same variety reappeared after passing over three generations in the one case, and five in the other.

When two distinct races are crossed, it is notorious that the tendency in the offspring to revert to one or both parent-forms is strong, and endures for many generations. I have myself seen the clearest evidence of this in crossed pigeons and with various plants. Mr. Sidney (13/17. In his edition of 'Youatt on the Pig' 1860 page 27.) states that, in a litter of Essex pigs, two young ones appeared which were the image of the Berkshire boar that had been used twenty-eight years before in giving size and constitution to the breed. I observed in the farmyard at Betley Hall some fowls showing a strong likeness to the Malay breed, and was told by Mr. Tollet that he had forty years before crossed his birds with Malays; and that, though he had at first attempted to get rid of this strain, he had subsequently given up the attempt in despair, as the Malay character would reappear.

This strong tendency in crossed breeds to revert has given rise to endless discussions in how many generations after a single cross, either with a distinct breed or merely with an inferior animal, the breed may be considered as pure, and free from all danger of reversion. No one supposes that less than three generations suffices, and most breeders think that six, seven, or eight are necessary, and some go to still greater lengths. (13/18. Dr. P. Lucas, 'Hered. Nat.' tome 2 pages 314, 892: see a good practical article on the subject in 'Gardener's Chronicle' 1856 page 620. I could add a vast number of references, but they would be superfluous.) But neither in the case of a breed which has been contaminated by a single cross, nor when, in the attempt to form an intermediate breed, half-bred animals have been matched together during many generations, can any rule be laid down how soon the tendency to reversion will be obliterated. It depends on the difference in the strength or prepotency of transmission in the two parent-forms, on their actual amount of difference, and on the nature of the conditions of life to which the crossed offspring are exposed. But we must be careful not to confound these cases of reversion to characters which were gained by a cross, with those under the first class, in which characters originally common to BOTH parents, but lost at some former period, reappear; for such characters may recur after an almost indefinite number of generations.

The law of reversion is as powerful with hybrids, when they are sufficiently fertile to breed together, or when they are repeatedly crossed with either pure parent-form, as in the case of mongrels. It is not necessary to give instances. With plants almost every one who has worked on this subject, from the time of Kolreuter to the present day, has insisted on this tendency. Gartner has recorded some good instances; but no one has given more striking ones than Naudin. (13/19. Kolreuter gives curious cases in his 'Dritte Fortsetzung' 1766 ss. 53, 59; and in his well-known 'Memoirs on Lavatera and Jalapa.' Gartner 'Bastarderzeugung' ss. 437, 441, etc. Naudin in his "Recherches sur l'Hybridite" 'Nouvelles Archives du Museum' tome 1 page 25.) The tendency differs in degree or strength in different groups, and partly depends, as we shall presently see, on whether the parent-plants have been long cultivated. Although the tendency to reversion is extremely general with nearly all mongrels and hybrids, it cannot be considered as invariably characteristic of them; it may also be mastered by long-continued selection; but these subjects will more properly be discussed in a future chapter on Crossing. From what we see of the power and scope of reversion, both in pure races, and when varieties or species are crossed, we may infer that characters of almost every kind are capable of reappearing after having been lost for a great length of time. But it does not follow from this that in each particular case certain characters will reappear; for instance, this will not occur when a race is crossed with another endowed with prepotency of transmission. Sometimes the power of reversion wholly fails, without our being able to assign any cause for the failure: thus it has been stated that in a French family in which 85 out of above 600 members, during six generations, had been subject to night-blindness, "there has not been a single example of this affection in the children of parents who were themselves free from it." (13/20. Quoted by Mr. Sedgwick in 'Med.-Chirurg. Review' April 1861 page 485. Dr. H. Dobell in 'Med.-Chirurg. Transactions' volume 46 gives an analogous case in which, in a large family, fingers with thickened joints were transmitted to several members during five generations; but when the blemish once disappeared it never reappeared.)

## REVERSION THROUGH BUD-PROPAGATION—PARTIAL REVERSION, BY SEGMENTS IN THE SAME FLOWER OR FRUIT, OR IN DIFFERENT PARTS OF THE BODY IN THE SAME INDIVIDUAL ANIMAL.

In the eleventh chapter many cases of reversion by buds, independently of seminal generation, were given—as when a leaf-bud on a variegated, a curled, or laciniated variety suddenly reassumes its proper character; or as when a Provence-rose appears on a moss-rose, or a peach on a nectarine-tree. In some of these cases only half the flower or fruit, or a smaller segment, or mere stripes, reassume their former character; and here we have reversion by segments. Vilmorin (13/21. Verlot 'Des Varietes' 1865 page 63.) has also recorded several cases with plants derived from seed, of flowers reverting by stripes or blotches to their primitive colours: he states that in all such cases a white or pale-coloured variety must first be formed, and, when this is propagated for a length of time by seed, striped seedlings occasionally make their appearance; and these can afterwards by care be multiplied by seed.

The stripes and segments just referred to are not due, as far as is known, to reversion to characters derived from a cross, but to characters lost by variation. These cases, however, as Naudin (13/22. 'Nouvelles Archives du Museum' tome 1 page 25. Alex. Braun (in his 'Rejuvenescence' Ray Soc. 1853 page 315) apparently holds a similar opinion.) insists in his discussion on disjunction of character, are closely analogous with those given in the eleventh chapter, in which crossed plants have been known to produce half-and- half or striped flowers and fruit, or distinct kinds of flowers on the same root resembling the two parent-forms. Many piebald animals probably come under this same head. Such cases, as we shall see in the chapter on Crossing, apparently result from certain characters not readily blending together, and, as a consequence of this incapacity for fusion, the offspring either perfectly resemble one of their two parents, or resemble one parent in one part, and the other parent in another part; or whilst young are intermediate in character, but with advancing age revert wholly or by segments to either parent-form, or to both. Thus, young trees of the Cytisus adami are intermediate in foliage and flowers between the two parent-forms; but when older the buds continually revert either partially or wholly to both forms. The cases given in the eleventh chapter on the changes which occurred during growth in crossed plants of Tropaeolum, Cereus, Datura, and Lathyrus are all analogous. As, however, these plants are hybrids of the first generation, and as their buds after a time come to resemble their parents and not their grandparents, these cases do not at first appear to come under the law of reversion in the ordinary sense of the word; nevertheless, as the change is effected through a succession of bud-generations on the same plant, they may be thus included.

Analogous facts have been observed in the animal kingdom, and are more remarkable, as they occur in the same individual in the strictest sense, and not as with plants through a succession of bud-generations. With animals the act of reversion, if it can be so designated, does not pass over a true generation, but merely over the early stages of growth in the same individual. For instance, I crossed several white hens with a black cock, and many of the chickens were, during the first year, perfectly white, but acquired during the second year black feathers; on the other hand, some of the chickens which were at first black, became during the second year piebald with white. A great breeder (13/23. Mr. Teebay in 'The Poultry Book' by Mr. Tegetmeier 1866 page 72.) says, that a Pencilled Brahma hen which has any of the blood of the Light Brahma in her, will "occasionally produce a pullet well pencilled during the first year, but she will most likely moult brown on the shoulders and become quite unlike her original colours in the second year." The same thing occurs with light Brahmas if of impure blood. I have observed exactly similar cases with the crossed offspring from differently coloured pigeons. But here is a more remarkable fact: I crossed a turbit, which has a frill formed by the feathers being reversed on its breast, with a trumpeter; and one of the young pigeons thus raised at first showed not a trace of the frill, but, after moulting thrice, a small yet unmistakably distinct frill appeared on its breast. According to Girou (13/24. Quoted by Hofacker 'Ueber die Eigenschaften' etc. s. 98.) calves produced from a red cow by a black bull, or from a black cow by a red bull, are not rarely born red, and subsequently become black. I possess a dog, the daughter of a white terrier by a fox- coloured bulldog; as a puppy she was quite white, but when about six months old a black spot appeared on her nose, and brown spots on her ears. When a little older she was badly wounded on the back, and the hair which grew on the cicatrix was of a brown colour, apparently derived from her father. This is the more remarkable, as with most animals having coloured hair, that which grows on a wounded surface is white.

In the foregoing cases, the characters which with advancing age reappeared, were present in the immediately preceding generations; but characters sometimes reappear in the same manner after a much longer interval of time. Thus the calves of a hornless race of cattle which originated in Corrientes, though at first quite hornless, as they become adult sometimes acquire small, crooked, and loose horns; and these in succeeding years occasionally become attached to the skull. (13/25. Azara 'Essais Hist. Nat. de Paraguay' tome 2 1801 page 372.) White and black Bantams, both of which generally breed true, sometimes assume as they grow old a saffron or red plumage. For instance, a first-rate black bantam has been described, which during three seasons was perfectly black, but then annually became more and more red; and it deserves notice that this tendency to change, whenever it occurs in a bantam, "is almost certain to prove hereditary." (13/26. These facts are given on the high authority of Mr. Hewitt in 'The Poultry Book' by Mr. Tegetmeier 1866 page 248.) The cuckoo or blue-mottled Dorking cock, when old, is liable to acquire yellow or orange hackles in place of his proper bluish-grey hackles. (13/27. 'The Poultry Book' by Tegetmeier 1866 page 97.) Now as Gallus bankiva is coloured red and orange, and as Dorking fowls and bantams are descended from this species, we can hardly doubt that the change which occasionally occurs in the plumage of these birds as their age advances, results from a tendency in the individual to revert to the primitive type.

## CROSSING AS A DIRECT CAUSE OF REVERSION.

It has long been notorious that hybrids and mongrels often revert to both or to one of their parent-forms, after an interval of from two to seven or eight, or, according to some authorities, even a greater number of generations. But that the act of crossing in itself gives an impulse towards reversion, as shown by the reappearance of long-lost characters, has never, I believe, been hitherto proved. The proof lies in certain peculiarities, which do not characterise the immediate parents, and therefore cannot have been derived from them, frequently appearing in the offspring of two breeds when crossed, which peculiarities never appear, or appear with extreme rarity, in these same breeds, as long as they are precluded from crossing. As this conclusion seems to me highly curious and novel, I will give the evidence in detail.

[My attention was first called to this subject, and I was led to make numerous experiments, by MM. Boitard and Corbie having stated that, when they crossed certain breeds of pigeons, birds coloured like the wild C. livia, or the common dovecote—namely, slaty-blue, with double black wing-bars, sometimes chequered with black, white loins, the tail barred with black, with the outer feathers edged with white,—were almost invariably produced. The breeds which I crossed, and the remarkable results attained, have been fully described in the sixth chapter. I selected pigeons belonging to true and ancient breeds, which had not a trace of blue or any of the above specified marks; but when crossed, and their mongrels recrossed, young birds were often produced, more or less plainly coloured slaty-blue, with some or all of the proper characteristic marks. I may recall to the reader's memory one case, namely, that of a pigeon, hardly distinguishable from the wild Shetland species, the grandchild of a red-spot, white fantail, and two black barbs, from any of which, when purely-bred, the production of a pigeon coloured like the wild C. livia would have been almost a prodigy.

I was thus led to make the experiments, recorded in the seventh chapter, on fowls. I selected long-established pure breeds, in which there was not a trace of red, yet in several of the mongrels feathers of this colour appeared; and one magnificent bird, the offspring of a black Spanish cock and white Silk hen, was coloured almost exactly like the wild Gallus bankiva. All who know anything of the breeding of poultry will admit that tens of thousands of pure Spanish and of pure white Silk fowls might have been reared without the appearance of a red feather. The fact, given on the authority of Mr. Tegetmeier, of the frequent appearance, in mongrel fowls, of pencilled or transversely-barred feathers, like those common to many gallinaceous birds, is likewise apparently a case of reversion to a character formerly possessed by some ancient progenitor of the family. I owe to the kindness of this excellent observer the opportunity of inspecting some neck-hackles and tail-feathers from a hybrid between the common fowl and a very distinct species, the Gallus varius; and these feathers are transversely striped in a conspicuous manner with dark metallic blue and grey, a character which could not have been derived from either immediate parent.

I have been informed by Mr. B.P. Brent, that he crossed a white Aylesbury drake and a black so-called Labrador duck, both of which are true breeds, and he obtained a young drake closely like the mallard (A. boschas). Of the musk-duck (Cairina moschata, Linn.) there are two sub-breeds, namely, white and slate-coloured; and these I am informed breed true, or nearly true. But the Rev. W.D. Fox tells me that, by putting a white drake to a slate-coloured duck, black birds, pied with white, like the wild musk-duck, were always produced. I hear from Mr. Blyth that hybrids from the canary and gold-finch almost always have streaked feathers on their backs; and this streaking must be derived from the original wild canary.

We have seen in the fourth chapter, that the so-called Himalayan rabbit, with its snow-white body, black ears, nose, tail, and feet, breeds perfectly true. This race is known to have been formed by the union of two varieties of silver-grey rabbits. Now, when a Himalayan doe was crossed by a sandy-coloured buck, a silver-grey rabbit was produced; and this is evidently a case of reversion to one of the parent varieties. The young of the Himalayan rabbit are born snow-white, and the dark marks do not appear until some time subsequently; but occasionally young Himalayan rabbits are born of a light silver-grey, which colour soon disappears; so that here we have a trace of reversion, during an early period of life, to the parent varieties, independently of any recent cross.

In the third chapter it was shown that at an ancient period some breeds of cattle in the wilder parts of Britain were white with dark ears, and that the cattle now kept half wild in certain parks, and those which have run quite wild in two distant parts of the world, are likewise thus coloured. Now, an experienced breeder, Mr. J. Beasley, of Northamptonshire (13/28. 'Gardener's Chronicle and Agricultural Gazette' 1866 page 528.), crossed some carefully selected West Highland cows with purely-bred shorthorn bulls. The bulls were red, red and white, or dark roan; and the Highland cows were all of a red colour, inclining to a light or yellow shade. But a considerable number of the offspring—and Mr. Beasley calls attention to this as a remarkable fact—were white, or white with red ears. Bearing in mind that none of the parents were white, and that they were purely-bred animals, it is highly probable that here the offspring reverted, in consequence of the cross, to the colour of some ancient and half-wild parent-breed. The following case, perhaps, comes under the same head: cows in their natural state have their udders but little developed, and do not yield nearly so much milk as our domesticated animals. Now there is some reason to believe (13/29. Ibid 1860 page 343. I am glad to find that so experienced a breeder of cattle as Mr. Willoughby Wood, 'Gardener's Chronicle' 1869 page 1216, admits my principle of a cross giving a tendency to reversion.) that cross-bred animals between two kinds, both of which are good milkers, such as Alderneys and Shorthorns, often turn out worthless in this respect.

In the chapter on the Horse reasons were assigned for believing that the primitive stock was striped and dun-coloured; and details were given, showing that in all parts of the world stripes of a dark colour frequently appear along the spine, across the legs, and on the shoulders, where they are occasionally double or treble, and even sometimes on the face and body of horses of all breeds and of all colours. But the stripes appear most frequently on the various kinds of duns. In foals they are sometimes plainly seen, and subsequently disappear. The dun-colour and the stripes are strongly transmitted when a horse thus characterised is crossed with any other; but I was not able to prove that striped duns are generally produced from the crossing of two distinct breeds, neither of which are duns, though this does sometimes occur.

The legs of the ass are often striped, and this may be considered as a reversion to the wild parent form, the Equus taeniopus of Abyssinia (13/30. Sclater in 'Proc. Zoolog. Soc.' 1862 page 163.), which is generally thus striped. In the domestic animal the stripes on the shoulder are occasionally double, or forked at the extremity, as in certain zebrine species. There is reason to believe that the foal is more frequently striped on the legs than the adult animal. As with the horse, I have not acquired any distinct evidence that the crossing of differently-coloured varieties of the ass brings out the stripes.

But now let us turn to the result of crossing the horse and ass. Although mules are not nearly so numerous in England as asses, I have seen a much greater number with striped legs, and with the stripes far more conspicuous than in either parent-form. Such mules are generally light-coloured, and might be called fallow-duns. The shoulder-stripe in one instance was deeply forked at the extremity, and in another instance was double, though united in the middle. Mr. Martin gives a figure of a Spanish mule with strong zebra-like marks on its legs (13/31. 'History of the Horse' page 212.), and remarks that mules are particularly liable to be thus striped on their legs. In South America, according to Roulin (13/32. 'Mem. presentes par divers Savans a l'Acad. Royale' tome 6 1835 page 338.), such stripes are more frequent and conspicuous in the mule than in the ass. In the United States, Mr. Gosse (13/33. 'Letters from Alabama' 1859 page 280.), speaking of these animals, says, "that in a great number, perhaps in nine out of every ten, the legs are banded with transverse dark stripes."

Many years ago I saw in the Zoological Gardens a curious triple hybrid, from a bay mare, by a hybrid from a male ass and female zebra. This animal when old had hardly any stripes; but I was assured by the superintendent, that when young it had shoulder-stripes, and faint stripes on its flanks and legs. I mention this case more especially as an instance of the stripes being much plainer during youth than in old age.

As the zebra has such a conspicuously striped body and legs, it might have been expected that the hybrids from this animal and the common ass would have had their legs in some degree striped; but it appears from the figures given in Dr. Gray's 'Knowsley Gleanings' and still more plainly from that given by Geoffroy and F. Cuvier (13/34. 'Hist. Nat. des Mammiferes' 1820 tome 1), that the legs are much more conspicuously striped than the rest of the body; and this fact is intelligible only on the belief that the ass aids in giving, through the power of reversion, this character to its hybrid offspring.

The quagga is banded over the whole front part of its body like a zebra, but has no stripes on its legs, or mere traces of them. But in the famous hybrid bred by Lord Morton (13/35. 'Philosoph. Transact.' 1821 page 20.) from a chestnut, nearly purely-bred, Arabian mare, by a male quagga, the stripes were "more strongly defined and darker than those on the legs of "the quagga." The mare was subsequently put to a black Arabian horse, and bore two colts, both of which, as formerly stated, were plainly striped on the legs, and one of them likewise had stripes on the neck and body.

The Equus indicus (13/36. Sclater in 'Proc. Zoolog. Soc.' 1862 page 163: this species is the Ghor-Khur of N.W. India, and has often been called the Hemionus of Pallas. See also Mr. Blyth's excellent paper in 'Journal of Asiatic Soc. of Bengal' volume 28 1860 page 229.) is characterised by a spinal stripe, without shoulder or leg stripes; but traces of these latter stripes may occasionally be seen even in the adult (13/37. Another species of wild ass, the true E. hemionus or Kiang, which ordinarily has no shoulder-stripes, is said occasionally to have them; and these, as with the horse and ass, are sometimes double: see Mr. Blyth in the paper just quoted and in 'Indian Sporting Review' 1856 page 320: and Col. Hamilton Smith in 'Nat. Library, Horses' page 318; and 'Dict. Class. d'Hist. Nat.' tome 3 page 563.) and Colonel S. Poole, who has had ample opportunities for observation, informs me that in the foal, when first born, the head and legs are often striped, but the shoulder-stripe is not so distinct as in the domestic ass; all these stripes, excepting that along the spine, soon disappear. Now a hybrid, raised at Knowsley (13/38. Figured in the 'Gleanings from the Knowsley Menageries' by Dr. J.E. Gray.) from a female of this species by a male domestic ass, had all four legs transversely and conspicuously striped, had three short stripes on each shoulder and had even some zebra-like stripes on its face! Dr. Gray informs me that he has seen a second hybrid of the same parentage, similarly striped.

From these facts we see that the crossing of the several equine species tends in a marked manner to cause stripes to appear on various parts of the body, especially on the legs. As we do not know whether the parent-form of the genus was striped, the appearance of the stripes can only hypothetically be attributed to reversion. But most persons, after considering the many undoubted cases of variously coloured marks reappearing by reversion in my experiments on crossed pigeons and fowls, will come to the same conclusion with respect to the horse-genus; and if so, we must admit that the progenitor of the group was striped on the legs, shoulders, face, and probably over the whole body, like a zebra.

Lastly, Professor Jaeger has given (13/39. 'Darwin'sche Theorie und ihre Stellung zu Moral und Religion' page 85.) a good case with pigs. He crossed the Japanese or masked breed with the common German breed, and the offspring were intermediate in character. He then re-crossed one of these mongrels with the pure Japanese, and in the litter thus produced one of the young resembled in all its characters a wild pig; it had a long snout and upright ears, and was striped on the back. It should be borne in mind that the young of the Japanese breed are not striped, and that they have a short muzzle and ears remarkably dependent.]

A similar tendency to the recovery of long lost characters holds good even with the instincts of crossed animals. There are some breeds of fowls which are called "everlasting layers," because they have lost the instinct of incubation; and so rare is it for them to incubate that I have seen notices published in works on poultry, when hens of such breeds have taken to sit. (13/40. Cases of both Spanish and Polish hens sitting are given in the 'Poultry Chronicle' 1855 volume 3 page 477.) Yet the aboriginal species was of course a good incubator; and with birds in a state of nature hardly any instinct is so strong as this. Now, so many cases have been recorded of the crossed offspring from two races, neither of which are incubators, becoming first-rate sitters, that the reappearance of this instinct must be attributed to reversion from crossing. One author goes so far as to say, "that a cross between two non-sitting varieties almost invariably produces a mongrel that becomes broody, and sits with remarkable steadiness." (13/41. 'The Poultry Book' by Mr. Tegetmeier 1866 pages 119, 163. The author, who remarks on the two negatives ('Journ. of Hort.' 1862 page 325), states that two broods were raised from a Spanish cock and Silver-pencilled Hamburgh hen, neither of which are incubators, and no less than seven out of eight hens in these two broods "showed a perfect obstinacy in sitting." The Rev. E.S. Dixon ('Ornamental Poultry' 1848 page 200) says that chickens reared from a cross between Golden and Black Polish fowls, are "good and steady birds to sit."

Mr. B.P. Brent informs me that he raised some good sitting hens by crossing Pencilled Hamburgh and Polish breeds. A cross-bred bird from a Spanish non-incubating cock and Cochin incubating hen is mentioned in the 'Poultry Chronicle' volume 3 page 13, as an "exemplary mother." On the other hand, an exceptional case is given in the 'Cottage Gardener' 1860 page 388 of a hen raised from a Spanish cock and black Polish hen which did not incubate.) Another author, after giving a striking example, remarks that the fact can be explained only on the principle that "two negatives make a positive." It cannot, however, be maintained that hens produced from a cross between two non-sitting breeds invariably recover their lost instinct, any more than that crossed fowls or pigeons invariably recover the red or blue plumage of their prototypes. Thus I raised several chickens from a Polish hen by a Spanish cock,—breeds which do not incubate,—and none of the young hens at first showed any tendency to sit; but one of them—the only one which was preserved—in the third year sat well on her eggs and reared a brood of chickens. So that here we have the reappearance with advancing age of a primitive instinct, in the same manner as we have seen that the red plumage of the Gallus bankiva is sometimes reacquired both by crossed and purely-bred fowls of various kinds as they grow old.

The parents of all our domesticated animals were of course aboriginally wild in disposition; and when a domesticated species is crossed with a distinct species, whether this is a domesticated or only a tamed animal, the hybrids are often wild to such a degree, that the fact is intelligible only on the principle that the cross has caused a partial return to a primitive disposition. Thus, the Earl of Powis formerly imported some thoroughly domesticated humped cattle from India, and crossed them with English breeds, which belong to a distinct species; and his agent remarked to me, without any question having been asked, how oddly wild the cross-bred animals were. The European wild boar and the Chinese domesticated pig are almost certainly specifically distinct: Sir F. Darwin crossed a sow of the latter breed with a wild Alpine boar which had become extremely tame, but the young, though having half-domesticated blood in their veins, were "extremely wild in confinement, and would not eat swill like common English pigs." Captain Hutton, in India, crossed a tame goat with a wild one from the Himalaya, and he remarked to me how surprisingly wild the offspring were. Mr. Hewitt, who has had great experience in crossing tame cock-pheasants with fowls belonging to five breeds, gives as the character of all "extraordinary wildness" (13/42. 'The Poultry Book' by Tegetmeier 1866 pages 165, 167.); but I have myself seen one exception to this rule. Mr. S. J. Salter (13/43. 'Natural History Review' 1863 April page 277.) who raised a large number of hybrids from a bantam-hen by Gallus sonneratii, states that "all were exceedingly wild." Mr. Waterton (13/44. 'Essays on Natural History' page 917.) bred some wild ducks from eggs hatched under a common duck, and the young were allowed to cross freely both amongst themselves and with the tame ducks; they were "half wild and half tame; they came to the windows to be fed, but still they had a wariness about them quite remarkable."

On the other hand, mules from the horse and ass are certainly not in the least wild, though notorious for obstinacy and vice. Mr. Brent, who has crossed canary-birds with many kinds of finches, has not observed, as he informs me, that the hybrids were in any way remarkably wild: but Mr. Jenner Weir who has had still greater experience, is of a directly opposite opinion. He remarks that the siskin is the tamest of finches, but its mules are as wild, when young, as newly caught birds, and are often lost through their continued efforts to escape. Hybrids are often raised between the common and musk duck, and I have been assured by three persons, who have kept these crossed birds, that they were not wild; but Mr. Garnett (13/45. As stated by Mr. Orton in his 'Physiology of Breeding' page 12.) observed that his hybrids were wild, and exhibited "migratory propensities" of which there is not a vestige in the common or musk duck. No case is known of this latter bird having escaped and become wild in Europe or Asia, except, according to Pallas, on the Caspian Sea; and the common domestic duck only occasionally becomes wild in districts where large lakes and fens abound. Nevertheless, a large number of cases have been recorded (13/46. M. E. de Selys-Longchamps refers ('Bulletin Acad. Roy. de Bruxelles' tome 12 No. 10) to more than seven of these hybrids shot in Switzerland and France. M. Deby asserts ('Zoologist' volume 5 1845-46 page 1254) that several have been shot in various parts of Belgium and Northern France. Audubon ('Ornitholog. Biography' volume 3 page 168), speaking of these hybrids, says that, in North America, they "now and then wander off and become quite wild.") of hybrids from these two ducks having been shot in a completely wild state, although so few are reared in comparison with purely-bred birds of either species. It is improbable that any of these hybrids could have acquired their wildness from the musk-duck having paired with a truly wild duck; and this is known not to be the case in North America; hence we must infer that they have reacquired, through reversion, their wildness, as well as renewed powers of flight.

These latter facts remind us of the statements, so frequently made by travellers in all parts of the world, on the degraded state and savage disposition of crossed races of man. That many excellent and kind-hearted mulattos have existed no one will dispute; and a more mild and gentle set of men could hardly be found than the inhabitants of the island of Chiloe, who consist of Indians commingled with Spaniards in various proportions. On the other hand, many years ago, long before I had thought of the present subject, I was struck with the fact that, in South America, men of complicated descent between Negroes, Indians, and Spaniards, seldom had, whatever the cause might be, a good expression. (13/47. 'Journal of Researches' 1845 page

71.) Livingstone—and a more unimpeachable authority cannot be quoted,—after speaking of a half-caste man on the Zambesi, described by the Portuguese as a rare monster of inhumanity, remarks, "It is unaccountable why half-castes, such as he, are so much more cruel than the Portuguese, but such is undoubtedly the case." An inhabitant remarked to Livingstone, "God made white men, and God made black men, but the Devil made halfcastes." (13/48. 'Expedition to the Zambesi' 1865 pages 25, 150.) When two races, both low in the scale, are crossed the progeny seems to be eminently bad. Thus the noble-hearted Humboldt, who felt no prejudice against the inferior races, speaks in strong terms of the bad and savage disposition of Zambos, or half-castes between Indians and Negroes; and this conclusion has been arrived at by various observers. (13/49. Dr. P. Broca on 'Hybridity in the Genus Homo' English translation 1864 page 39.) From these facts we may perhaps infer that the degraded state of so many half-castes is in part due to reversion to a primitive and savage condition, induced by the act of crossing, even if mainly due to the unfavourable moral conditions under which they are generally reared.

## SUMMARY ON THE PROXIMATE CAUSES LEADING TO REVERSION.

When purely-bred animals or plants reassume long-lost characters,—when the common ass, for instance, is born with striped legs, when a pure race of black or white pigeons throws a slaty-blue bird, or when a cultivated heartsease with large and rounded flowers produces a seedling with small and elongated flowers,—we are quite unable to assign any proximate cause. When animals run wild, the tendency to reversion, which, though it has been greatly exaggerated, no doubt exists, is sometimes to a certain extent intelligible. Thus, with feral pigs, exposure to the weather will probably favour the growth of the bristles, as is known to be the case with the hair of other domesticated animals, and through correlation the tusks will tend to be redeveloped. But the reappearance of coloured longitudinal stripes on young feral pigs cannot be attributed to the direct action of external conditions. In this case, and in many others, we can only say that any change in the habits of life apparently favour a tendency, inherent or latent in the species, to return to the primitive state.

It will be shown in a future chapter that the position of flowers on the summit of the axis, and the position of seeds within the capsule, sometimes determine a tendency towards reversion; and this apparently depends on the amount of sap or nutriment which the flower-buds and seeds receive. The position, also, of buds, either on branches or on roots, sometimes determines, as was formerly shown, the transmission of the character proper to the variety, or its reversion to a former state.

We have seen in the last section that when two races or species are crossed there is the strongest tendency to the reappearance in the offspring of long-lost characters, possessed by neither parent nor immediate progenitor. When two white, or red, or black pigeons, of well-established breeds, are united, the offspring are almost sure to inherit the same colours; but when differently-coloured birds are crossed, the opposed forces of inheritance apparently counteract each other, and the tendency which is inherent in both parents to produce slaty-blue offspring becomes predominant. So it is in several other cases. But when, for instance, the ass is crossed with E. indicus or with the horse—animals which have not striped legs—and the hybrids have conspicuous stripes on their legs and even on their faces, all that can be said is, that an inherent tendency to reversion is evolved through some disturbance in the organisation caused by the act of crossing.

Another form of reversion is far commoner, indeed is almost universal with the offspring from a cross, namely, to the characters proper to either pure parent-form. As a general rule, crossed offspring in the first generation are nearly intermediate between their parents, but the grandchildren and succeeding generations continually revert, in a greater or lesser degree, to one or both of their progenitors. Several authors have maintained that hybrids and mongrels include all the characters of both parents, not fused together, but merely mingled in different proportions in different parts of the body; or, as Naudin (13/50. 'Nouvelles Archives du Museum' tome 1 page 151.) has expressed it, a hybrid is a living mosaic-work, in which the eye cannot distinguish the discordant elements, so completely are they intermingled. We can hardly doubt that, in a certain sense, this is true, as when we behold in a hybrid the elements of both species segregating themselves into segments in the same flower or fruit, by a process of self-attraction or self-affinity; this segregation taking place either by seminal or bud-propagation. Naudin further believes that the segregation of the two specific elements or essences is eminently liable to occur in the male and female reproductive matter; and he thus explains the almost universal tendency to reversion in successive hybrid generations. For this would be the natural result of the union of pollen and ovules, in both of which the elements of the same species had been segregated by self-affinity. If, on the other hand, pollen which included the elements of one species happened to unite with ovules including the elements of the other species, the intermediate or hybrid state would still be retained, and there would be no reversion. But it would, as I suspect, be more correct to say that the elements of both parent-species exist in every hybrid in a double state, namely, blended together and completely separate. How this is possible, and what the term specific essence or element may be supposed to express, I shall attempt to show in the chapter on the hypothesis of pangenesis.

But Naudin's view, as propounded by him, is not applicable to the reappearance of characters lost long ago by variation; and it is hardly applicable to races or species which, after having been crossed at some former period with a distinct form, and having since lost all traces of the cross, nevertheless occasionally yield an individual which reverts (as in the case of the great- great-grandchild of the pointer Sappho) to the crossing form. The most simple case of reversion, namely, of a hybrid or mongrel to its grandparents, is connected by an almost perfect series with the extreme case of a purely-bred race recovering characters which had been lost during many ages; and we are thus led to infer that all the cases must be related by some common bond.

Gartner believed that only highly sterile hybrid plants exhibit any tendency to reversion to their parent-forms. This erroneous belief may perhaps be accounted for by the nature of the genera crossed by him, for he admits that the tendency differs in different genera. The statement is also directly contradicted by Naudin's observations, and by the notorious fact that perfectly fertile mongrels exhibit the tendency in a high degree,—even in a higher degree, according to Gartner himself, than hybrids. (13/51. 'Bastarderzeugung' s. 582, 438, etc.)

Gartner further states that reversions rarely occur with hybrid plants raised from species which have not been cultivated, whilst, with those which have been long cultivated, they are of frequent occurrence. This conclusion explains a curious discrepancy: Max Wichura (13/52. 'Die Bastardbefruchtung... der Weiden' 1865 s. 23. For Gartner's remarks on this head, see 'Bastarderzeugung' s. 474, 582.) who worked exclusively on willows which had not been subjected to culture, never saw an instance of reversion; and he goes so far as to suspect that the careful Gartner had not sufficiently protected his hybrids from the pollen of the parent-species: Naudin, on the other hand, who chiefly experimented on cucurbitaceous and other cultivated plants, insists more strenuously than any other author on the tendency to reversion in all hybrids. The conclusion that the condition of the parent-species, as affected by culture, is one of the proximate causes leading to reversion, agrees well with the converse case of domesticated animals and cultivated plants being liable to reversion when they become feral; for in both cases the organisation or constitution must be disturbed, though in a very different way. (13/53. Prof. Weismann in his very curious essay on the different forms produced by the same species of butterfly at different seasons ('Saison- Dimorphismus der Schmetterlinge' pages 27, 28), has come to a similar conclusion, namely, that any cause which disturbs the organisation, such as the exposure of the cocoons to heat or even to much shaking, gives a tendency to reversion.)

Finally, we have seen that characters often reappear in purely-bred races without our being able to assign any proximate cause; but when they become feral this is either indirectly or directly induced by the change in their conditions of life. With crossed breeds, the act of crossing in itself certainly leads to the recovery of long-lost characters, as well as of those derived from either parent-form. Changed conditions, consequent on cultivation, and the relative position of buds, flowers, and seeds on the plant, all apparently aid in giving this same tendency. Reversion may occur either through seminal or bud generation, generally at birth, but sometimes only with an advance of age. Segments or portions of the individual may alone be thus affected. That a being should be born resembling in certain characters an ancestor removed by two or three, and in some cases by hundreds or even thousands of generations, is assuredly a wonderful fact. In these cases the child is commonly said to inherit such characters directly from its grandparent, or more remote ancestors. But this view is hardly conceivable. If, however, we suppose that every character is derived exclusively from the father or mother, but that many characters lie latent or dormant in both parents during a long succession of generations, the foregoing facts are intelligible. In what manner characters may be conceived to lie latent, will be considered in a future chapter to which I have lately alluded.

**LATENT CHARACTERS.**

But I must explain what is meant by characters lying latent. The most obvious illustration is afforded by secondary sexual characters. In every female all the secondary male characters, and in every male all the secondary female characters, apparently exist in a latent state, ready to be evolved under certain conditions. It is well known that a large number of female birds, such as fowls, various pheasants, partridges, peahens, ducks, etc., when old or diseased, or when operated on, assume many or all of the secondary male characters of their species. In the case of the hen-pheasant this has been observed to occur far more frequently during certain years than during others. (13/54. Yarrell 'Phil. Transact.' 1827 page 268; Dr. Hamilton in 'Proc. Zoolog. Soc.' 1862 page 23.) A duck ten years old has been known to assume both the perfect winter and summer plumage of the drake. (13/55. 'Archiv. Skand. Beitrage zur Naturgesch.' 8 s. 397-413.) Waterton (13/56. In his 'Essays on Nat. Hist.' 1838 Mr. Hewitt gives analogous cases with hen-pheasants in 'Journal of Horticulture' July 12, 1864 page 37. Isidore Geoffroy Saint-Hilaire in his 'Essais de Zoolog. Gen.' ('Suites a Buffon' 1842 pages 496-513), has collected such cases in ten different kinds of birds. It appears that Aristotle was well aware of the change in mental disposition in old hens. The case of the female deer acquiring horns is given at page 513.) gives a curious case of a hen which had ceased laying, and had assumed the plumage, voice, spurs, and warlike disposition of the cock; when opposed to an enemy she would erect her hackles and show fight. Thus every character, even to the instinct and manner of fighting, must have lain dormant in this hen as long as her ovaria continued to act. The females of two kinds of deer, when old, have been known to acquire horns; and, as Hunter has remarked, we see something of an analogous nature in the human species.

On the other hand, with male animals, it is notorious that the secondary sexual characters are more or less completely lost when they are subjected to castration. Thus, if the operation be performed on a young cock, he never, as Yarrell states, crows again; the comb, wattles, and spurs do not grow to their full size, and the hackles assume an intermediate appearance between true hackles and the feathers of the hen. Cases are recorded of confinement, which often affects the reproductive system, causing analogous results. But characters properly confined to the female are likewise acquired by the male; the capon takes to sitting on eggs, and will bring up chickens; and what is more curious, the utterly sterile male hybrids from the pheasant and the fowl act in the same manner, "their delight being to watch when the hens leave their nests, and to take on themselves the office of a sitter." (13/57. 'Cottage Gardener' 1860 page 379.) That admirable observer Reaumur (13/58. 'Art de faire Eclore' etc. 1749 tome 2 page 8.) asserts that a cock, by being long confined in solitude and darkness, can be taught to take charge of young chickens; he then utters a peculiar cry, and retains during his whole life this newly acquired maternal instinct. The many well-ascertained cases of various male mammals giving milk shows that their rudimentary mammary glands retain this capacity in a latent condition.

We thus see that in many, probably in all cases, the secondary characters of each sex lie dormant or latent in the opposite sex, ready to be evolved under peculiar circumstances. We can thus understand how, for instance, it is possible for a good milking cow to transmit her good qualities through her male offspring to future generations; for we may confidently believe that these qualities are present, though latent, in the males of each generation. So it is with the game-cock, who can transmit his superiority in courage and vigour through his female to his male offspring; and with man it is known (13/59. Sir H. Holland 'Medical Notes and Reflections' 3rd edition 1855 page 31.) that diseases, such as hydrocele, necessarily confined to the male sex, can be transmitted through the female to the grandson. Such cases as these offer, as was remarked at the commencement of this chapter, the simplest possible examples of reversion; and they are intelligible on the belief that characters common to the grandparent and grandchild of the same sex are present, though latent, in the intermediate parent of the opposite sex.

The subject of latent characters is so important, as we shall see in a future chapter, that I will give another illustration. Many animals have the right and left sides of their body unequally developed: this is well known to be the case with flat-fish, in which the one side differs in thickness and colour and in the shape of the fins, from the other, and during the growth of the young fish one eye is gradually twisted from the lower to the upper surface. (13/60. See Steenstrup on the 'Obliquity of Flounders' in Annals and Mag. of Nat. Hist.' May 1865 page 361. I have given an abstract of Malm's explanation of this wonderful phenomenon in the 'Origin of Species' 6th Edition page 186.) In most flat-fishes the left is the blind side, but in some it is the right; though in both cases reversed or "wrong fishes," are occasionally developed; and in Platessa flesus the right or left side is indifferently the upper one. With gasteropods or shell-fish, the right and left sides are extremely unlike; the far greater number of species are dextral, with rare and occasional reversals of development; and some few are normally sinistral; but certain species of Bulimus, and many Achatinellae (13/61. Dr. E. von Martens in 'Annals and Mag. of Nat. Hist.' March 1866 page 209.) are as often sinistral as dextral. I will give an analogous case in the great articulate kingdom: the two sides of Verruca (13/62. Darwin 'Balanidae' Ray Soc. 1854 page 499: see also the appended remarks on the apparently capricious development of the thoracic limbs on the right and left sides in the higher crustaceans.) are so wonderfully unlike, that without careful dissection it is extremely difficult to recognise the corresponding parts on the opposite sides of the body; yet it is apparently a mere matter of chance whether it be the right or the left side that undergoes so singular amount of change. One plant is known to me (13/63. Mormodes ignea: Darwin 'Fertilisation of Orchids' 1862 page 251.) in which the flower, according as it stands on the one or other side of the spike, is unequally developed. In all the foregoing cases the two sides are perfectly symmetrical at an early period of growth. Now, whenever a species is as liable to be unequally developed on the one as on the other side, we may infer that the capacity for such development is present, though latent, in the undeveloped side. And as a reversal of development occasionally occurs in animals of many kinds, this latent capacity is probably very common.

The best yet simplest cases of characters lying dormant are, perhaps, those previously given, in which chickens and young pigeons, raised from a cross between differently coloured birds, are at first of one colour, but in a year or two acquire feathers of the colour of the other parent; for in this case the tendency to a change of plumage is clearly latent in the young bird. So it is with hornless breeds of cattle, some of which acquire small horns as they grow old. Purely bred black and white bantams, and some other fowls, occasionally assume, with advancing years, the red feathers of the parent-species. I will here add a somewhat different case, as it connects in a striking manner latent characters of two classes. Mr. Hewitt (13/64. 'Journal of Horticulture' July 1864 page 38. I have had the opportunity of examining these remarkable feathers through the kindness of Mr. Tegetmeier.) possessed an excellent Sebright gold-laced bantam hen, which, as she became old, grew diseased in her ovaria, and assumed male characters. In this breed the males resemble the females in all respects except in their combs, wattles, spurs, and instincts; hence it might have been expected that the diseased hen would have assumed only those masculine characters which are proper to the breed, but she acquired, in addition, well-arched tail sickle-feathers quite a foot in length, saddle-feathers on the loins, and hackles on the neck,—ornaments which, as Mr. Hewitt remarks, "would be held as abominable in this breed." The Sebright bantam is known (13/65. 'The Poultry Book' by Mr. Tegetmeier 1866 page 241.) to have originated about the year 1800 from a cross between a common bantam and a Polish fowl, recrossed by a hen-tailed bantam, and carefully selected; hence there can hardly be a doubt that the sickle-feathers and hackles which appeared in the old hen were derived from the Polish fowl or common bantam; and we thus see that not only certain masculine characters proper to the Sebright bantam, but other masculine characters derived from the first progenitors of the breed, removed by a period of above sixty years, were lying latent in this henbird, ready to be evolved as soon as her ovaria became diseased.

From these several facts it must be admitted that certain characters, capacities, and instincts, may lie latent in an individual, and even in a succession of individuals, without our being able to detect the least sign of their presence. When fowls, pigeons, or cattle of different colours are crossed, and their offspring change colour as they grow old, or when the crossed turbit acquired the characteristic frill after its third moult, or when rarely-bred bantams partially assume the red plumage of their prototype, we cannot doubt that these qualities were from the first present, though latent, in the individual animal, like the characters of a moth in the caterpillar. Now, if these animals had produced offspring before they had acquired with advancing age their new characters, nothing is more probable than that they would have transmitted them to some of their offspring, who in this case would in appearance have received such characters from their grand-parents or more distant progenitors. We should then have had a case of reversion, that is, of the reappearance in the child of an ancestral character, actually present, though during youth completely latent, in the parent; and this we may safely conclude is what occurs in all reversions to progenitors, however remote.

This view of the latency in each generation of all the characters which appear through reversion, is also supported by their actual presence in some cases during early youth alone, or by their more frequent appearance and greater distinctness at this age than during maturity. We have seen that this is often the case with the stripes on the legs and faces of the several species of the horse genus. The Himalayan rabbit, when crossed, sometimes produces offspring which revert to the parent silver-grey breed, and we have seen that in purely bred animals pale-grey fur occasionally reappears during early youth. Black cats, we may feel assured, would occasionally produce by reversion tabbies; and on young black kittens, with a pedigree (13/66. Carl Vogt 'Lectures on Man' English translation 1864 page 411.) known to have been long pure, faint traces of stripes may almost always be seen which afterwards disappear. Hornless Suffolk cattle occasionally produce by reversion horned animals; and Youatt (13/67. 'On Cattle' page 174.) asserts that even in hornless individuals "the rudiment of a horn may be often felt at an early age."

No doubt it appears at first sight in the highest degree improbable that in every horse of every generation there should be a latent capacity and tendency to produce stripes, though these may not appear once in a thousand generations; that in every white, black, or other coloured pigeon, which may have transmitted its proper colour during centuries, there should be a latent capacity in the plumage to become blue and to be marked with certain characteristic bars; that in every child in a six-fingered family there should be the capacity for the production of an additional digit; and so in other cases. Nevertheless, there is no more inherent improbability in this being the case than in a useless and rudimentary organ, or even in only a tendency to the production of a rudimentary organ, being inherited during millions of generations, as is well known to occur with a multitude of organic beings. There is no more inherent improbability in each domestic pig, during a thousand generations, retaining the capacity and tendency to develop great tusks under fitting conditions, than in the young calf having retained, for an indefinite number of generations rudimentary incisor teeth, which never protrude through the gums.

I shall give at the end of the next chapter a summary of the three preceding chapters; but as isolated and striking cases of reversion have here been chiefly insisted on, I wish to guard the reader against supposing that reversion is due to some rare or accidental combination of circumstances. When a character, lost during hundreds of generations, suddenly reappears, no doubt some such combination must occur; but reversions, to the immediately preceding generations may be constantly observed, at least, in the offspring of most unions. This has been universally recognised in the case of hybrids and mongrels, but it has been recognised simply from the difference between the united forms rendering the resemblance of the offspring to their grandparents or more remote progenitors of easy detection. Reversion is likewise almost invariably the rule, as Mr. Sedgwick has shown, with certain diseases. Hence we must conclude that a tendency to this peculiar form of transmission is an integral part of the general law of inheritance.

## MONSTROSITIES.

A large number of monstrous growths and of lesser anomalies are admitted by every one to be due to an arrest of development, that is, to the persistence of an embryonic condition. But many monstrosities cannot be thus explained; for parts of which no trace can be detected in the embryo, but which occur in other members of the same class of animals occasionally appear, and these may probably with truth be attributed to reversion. As, however, I have treated this subject as fully as I could in my 'Descent of Man' (ch. 1 2nd edition), I will not here recur to it.

[When flowers which have normally an irregular structure become regular or peloric, the change is generally looked at by botanists as a return to the primitive state. But Dr. Maxwell Masters (13/68. 'Natural Hist. Review' April 1863 page 258. See also his Lecture, Royal Institution, March 16, 1860. On same subject see Moquin-Tandon 'Elements de Teratologie' 1841 pages 184, 352. Dr. Peyritsch has collected a large number of very interesting cases 'Sitzb. d. k. Akad. d. Wissensch.' Wien b. 60 and especially b. 66 1872 page 125.), who has ably discussed this subject, remarks that when, for instance, all the sepals of a Tropaeolum become green and of the same shape, instead of being coloured with one prolonged into a spur, or when all the petals of a Linaria become simple and regular, such cases may be due merely to an arrest of development; for in these flowers all the organs during their earliest condition are symmetrical, and, if arrested at this stage of growth, they would not become irregular. If, moreover, the arrest were to take place at a still earlier period of development, the result would be a simple tuft of green leaves; and no one probably would call this a case of reversion. Dr. Masters designates the cases first alluded to as regular peloria; and others, in which all the corresponding parts assume a similar form of irregularity, as when all the petals in a Linaria become spurred, as irregular peloria. We have no right to attribute these latter cases to reversion, until it can be shown that the parent-form, for instance, of the genus Linaria had had all its petals spurred; for a chance of this nature might result from the spreading of an anomalous structure, in accordance with the law, to be discussed in a future chapter, of homologous parts tending to vary in the same manner. But as both forms of peloria frequently occur on the same individual plant of the Linaria (13/69. Verlot 'Des Varietes' 1865 page 89; Naudin 'Nouvelles Archives du Museum' tome 1 page 137.), they probably stand in some close relation to one another. On the doctrine that peloria is simply the result of an arrest of development, it is difficult to understand how an organ arrested at a very early period of growth should acquire its full functional perfection;—how a petal, supposed to be thus arrested, should acquire its brilliant colours, and serve as an envelope to the flower, or a stamen produce efficient pollen; yet this occurs with many peloric flowers. That pelorism is not due to mere chance variability, but either to an arrest of development or to reversion, we may infer from an observation made by Ch. Morren (13/70. In his discussion on some curious peloric Calceolarias quoted in 'Journal of Horticulture' February 24, 1863 page 152.) namely, that families which have irregular flowers often "return by these monstrous growths to their regular form; whilst we never see a regular flower realise the structure of an irregular one."

Some flowers have almost certainly become more or less completely peloric through reversion, as the following interesting case shows. Corydalis tuberosa properly has one of its two nectaries colourless, destitute of nectar, only half the size of the other, and therefore, to a certain extent, in a rudimentary state; the pistil is curved towards the perfect nectary, and the hood, formed of the inner petals, slips off the pistil and stamen in one direction alone, so that, when a bee sucks the perfect nectary, the stigma and stamens are exposed and rubbed against the insect's body. In several closely allied genera, as in Dielytra, etc., there are two perfect nectaries, the pistil is straight, and the hood slips off on either side, according as the bee sucks either nectary. Now, I have examined several flowers of Corydalis tuberosa, in which both nectaries were equally developed and contained nectar; in this we see only the redevelopment of a partially aborted organ; but with this redevelopment the pistil becomes straight, and the hood slips off in either direction, so that these flowers have acquired the perfect structure, so well adapted for insect agency, of Dielytra and its allies. We cannot attribute these coadapted modifications to chance, or to correlated variability; we must attribute them to reversion to a primordial condition of the species.

The peloric flowers of Pelargonium have their five petals in all respects alike, and there is no nectary so that they resemble the symmetrical flowers of the closely allied genus Geranium; but the alternate stamens are also sometimes destitute of anthers, the shortened filaments being left as rudiments, and in this respect they resemble the symmetrical flowers of the closely allied genus Erodium. Hence we may look at the peloric flowers of Pelargonium as having reverted to the state of some primordial form, the progenitor of the three closely related genera of Pelargonium, Geranium, and Erodium.

In the peloric form of Antirrhinum majus, appropriately called the" Wonder," the tubular and elongated flowers differ wonderfully from those of the common snapdragon; the calyx and the mouth of the corolla consist of six equal lobes, and include six equal instead of four unequal stamens. One of the two additional stamens is manifestly formed by the development of a microscopically minute papilla, which may be found at the base of the upper lip of the flower of the common snapdragons in the nineteen plants examined by me. That this papilla is a rudiment of a stamen was well shown by its various degrees of development in crossed plants between the common and the peloric Antirrhinum. Again, a peloric Galeobdolon luteum, growing in my garden, had five equal petals, all striped like the ordinary lower lip, and included five equal instead of four unequal stamens; but Mr. R. Keeley, who sent me this plant, informs me that the flowers vary greatly, having from four to six lobes to the corolla, and from three to six stamens. (13/71. For other cases of six divisions in peloric flowers of the Labiatae and Scrophulariaceae see Moquin-Tandon 'Teratologie' page 192.) Now, as the members of the two great families to which the Antirrhinum and Galeobdolon belong are properly pentamerous, with some of the parts confluent and others suppressed, we ought not to look at the sixth stamen and the sixth lobe to the corolla in either case as due to reversion, any more than the additional petals in double flowers in these same two families. But the case is different with the fifth stamen in the peloric Antirrhinum, which is produced by the redevelopment of a rudiment always present, and which probably reveals to us the state of the flower, as far as the stamens are concerned, at some ancient epoch. It is also difficult to believe that the other four stamens and the petals, after an arrest of development at a very early embryonic age, would have come to full perfection in colour, structure, and function, unless these organs had at some former period normally passed through a similar course of growth. Hence it appears to me probable that the progenitor of the genus Antirrhinum must at some remote epoch have included five stamens and borne flowers in some degree resembling those now produced by the peloric form. The conclusion that peloria is not a mere monstrosity, irrespective of any former state of the species, is supported by the fact that this structure is often strongly inherited, as in the case of the peloric Antirrhinum and Gloxinia and sometimes in that of the peloric Corydalis solida. (13/72. Godron reprinted from the 'Memoires de l'Acad. de Stanislas' 1868.)

Lastly I may add that many instances have been recorded of flowers, not generally considered as peloric, in which certain organs are abnormally augmented in number. As an increase of parts cannot be looked at as an arrest of development, nor as due to the redevelopment of rudiments, for no rudiments are present, and as these additional parts bring the plant into closer relationship with its natural allies, they ought probably to be viewed as reversions to a primordial condition.]

These several facts show us in an interesting manner how intimately certain abnormal states are connected together; namely, arrests of development causing parts to become rudimentary or to be wholly suppressed,—the redevelopment of parts now in a more or less rudimentary condition,—the reappearance of organs of which not a vestige can be detected,—and to these may be added, in the case of animals, the presence during youth, and subsequent disappearance, of certain characters which occasionally are retained throughout life. Some naturalists look at all such abnormal structures as a return to the ideal state of the group to which the affected being belongs; but it is difficult to conceive what is meant to be conveyed by this expression. Other naturalists maintain, with greater probability and distinctness of view, that the common bond of connection between the several foregoing cases is an actual, though partial, return to the structure of the ancient progenitor of the group. If this view be correct, we must believe that a vast number of characters, capable of evolution, lie hidden in every organic being. But it would be a mistake to suppose that the number is equally great in all beings. We know, for instance, that plants of many orders occasionally become peloric; but many more cases have been observed in the Labiatae and Scrophulariaceae than in any other order; and in one genus of the Scrophulariaceae, namely Linaria, no less than thirteen species have been described in this condition (13/73. Moquin- Tandon 'Teratologie' page 186.) On this view of the nature of peloric flowers, and bearing in mind certain monstrosities in the animal kingdom, we must conclude that the progenitors of most plants and animals have left an impression, capable of redevelopment, on the germs of their descendants, although these have since been profoundly modified.

The fertilised germ of one of the higher animals, subjected as it is to so vast a series of changes from the germinal cell to old age,—incessantly agitated by what Quatrefages well calls the tourbillon vital,—is perhaps the most wonderful object in nature. It is probable that hardly a change of any kind affects either parent, without some mark being left on the germ. But on the doctrine of reversion, as given in this chapter, the germ becomes a far more marvellous object, for, besides the visible changes which it undergoes, we must believe that it is crowded with invisible characters, proper to both sexes, to both the right and left side of the body, and to a long line of male and female ancestors separated by hundreds or even thousands of generations from the present time: and these characters, like those written on paper with invisible ink, lie ready to be evolved whenever the organisation is disturbed by certain known or unknown conditions.

# CHAPTER 2.XIV.

INHERITANCE continued.—FIXEDNESS OF CHARACTER—PREPOTENCY—SEXUAL LIMITATION—CORRESPONDENCE OF AGE.

**FIXEDNESS OF CHARACTER APPARENTLY NOT DUE TO ANTIQUITY OF INHERITANCE. PREPOTENCY OF TRANSMISSION IN INDIVIDUALS OF THE SAME FAMILY, IN CROSSED BREEDS AND SPECIES; OFTEN STRONGER IN ONE SEX THAN THE OTHER; SOMETIMES DUE TO THE SAME CHARACTER BEING PRESENT AND VISIBLE IN ONE BREED AND LATENT IN THE OTHER. INHERITANCE AS LIMITED BY SEX. NEWLY-ACQUIRED CHARACTERS IN OUR DOMESTICATED ANIMALS OFTEN TRANSMITTED BY ONE SEX ALONE, SOMETIMES LOST BY ONE SEX ALONE. INHERITANCE AT CORRESPONDING PERIODS OF LIFE. THE IMPORTANCE OF THE PRINCIPLE WITH RESPECT TO EMBRYOLOGY; AS EXHIBITED IN DOMESTICATED ANIMALS: AS EXHIBITED IN THE APPEARANCE AND DISAPPEARANCE OF INHERITED DISEASES; SOMETIMES SUPERVENING EARLIER IN THE CHILD THAN IN THE PARENT. SUMMARY OF THE THREE PRECEDING CHAPTERS.**

In the last two chapters the nature and force of Inheritance, the circumstances which interfere with its power, and the tendency to Reversion, with its many remarkable contingencies, were discussed. In the present chapter some other related phenomena will be treated of, as fully as my materials permit.

**FIXEDNESS OF CHARACTER.**

It is a general belief amongst breeders that the longer any character has been transmitted by a breed, the more fully it will continue to be transmitted. I do not wish to dispute the truth of the proposition that inheritance gains strength simply through long continuance, but I doubt whether it can be proved. In one sense the proposition is little better than a truism; if any character has remained constant during many generations, it will be likely to continue so, if the conditions of life remain the same. So, again, in improving a breed, if care be taken for a length of time to exclude all inferior individuals, the breed will obviously tend to become truer, as it will not have been crossed during many generations by an inferior animal. We have previously seen, but without being able to assign any cause, that, when a new character appears, it is occasionally from the first constant, or fluctuates much, or wholly fails to be transmitted. So it is with the aggregate of slight differences which characterise a new variety, for some propagate their kind from the first much truer than others. Even with plants multiplied by bulbs, layers, etc., which may in one sense be said to form parts of the same individual, it is well known that certain varieties retain and transmit through successive bud-generations their newly-acquired characters more truly than others. In none of these, nor in the following cases, does there appear to be any relation between the force with which a character is transmitted and the length of time during which it has been transmitted. Some varieties, such as white and yellow hyacinths and white sweet-peas, transmit their colours more faithfully than do the varieties which have retained their natural colour. In the Irish family, mentioned in the twelfth chapter, the peculiar tortoiseshell-like colouring of the eyes was transmitted far more faithfully than any ordinary colour. Ancon and Mauchamp sheep and niata cattle, which are all comparatively modern breeds, exhibit remarkably strong powers of inheritance. Many similar cases could be adduced.

As all domesticated animals and cultivated plants have varied, and yet are descended from aboriginally wild forms, which no doubt had retained the same character from an immensely remote epoch, we see that scarcely any degree of antiquity ensures a character being transmitted perfectly true. In this case, however, it may be said that changed conditions of life induce certain modifications, and not that the power of inheritance fails; but in every case of failure, some cause, either internal or external, must interfere. It will generally be found that the organs or parts which in our domesticated productions have varied, or which still continue to vary,—that is, which fail to retain their former state,—are the same with the parts which differ in the natural species of the same genus. As, on the theory of descent with modification, the species of the same genus have been modified since they branched off from a common progenitor, it follows that the characters by which they differ from one another have varied, whilst other parts of the organisation have remained unchanged; and it might be argued that these same characters now vary under domestication, or fail to be inherited, from their lesser antiquity. But variation in a state of nature seems to stand in some close relation with changed conditions of life, and characters which have already varied under such conditions would be apt to vary under the still greater changes consequent on domestication, independently of their greater or less antiquity.

Fixedness of character, or the strength of inheritance, has often been judged of by the preponderance of certain characters in the crossed offspring between distinct races; but prepotency of transmission here comes into play, and this, as we shall immediately see, is a very different consideration from the strength or weakness of inheritance. (14/1. See 'Youatt on Cattle' pages

92, 69, 78, 88, 163; and 'Youatt on Sheep' page 325. Also Dr. Lucas 'L'Hered. Nat.' tome 2 page 310.) It has often been observed that breeds of animals inhabiting wild and mountainous countries cannot be permanently modified by our improved breeds; and as these latter are of modern origin, it has been thought that the greater antiquity of the wilder breeds has been the cause of their resistance to improvement by crossing; but it is more probably due to their structure and constitution being better adapted to the surrounding conditions. When plants are first subjected to culture, it has been found that, during several generations, they transmit their characters truly, that is, do not vary, and this has been attributed to ancient characters being strongly inherited: but it may with equal or greater probability be consequent on changed conditions of life requiring a long time for their cumulative action. Notwithstanding these considerations, it would perhaps be rash to deny that characters become more strongly fixed the longer they are transmitted; but I believe that the proposition resolves itself into this,—that characters of all kinds, whether new or old, tend to be inherited, and that those which have already withstood all counteracting influences and been truly transmitted, will, as a general rule, continue to withstand them, and consequently be faithfully inherited.

## PREPOTENCY IN THE TRANSMISSION OF CHARACTER.

When individuals, belonging to the same family, but distinct enough to be recognised, or when two well-marked races, or two species, are crossed, the usual result, as stated in the previous chapter, is, that the offspring in the first generation are intermediate between their parents, or resemble one parent in one part and the other parent in another part. But this is by no means the invariable rule; for in many cases it is found that certain individuals, races, and species, are prepotent in transmitting their likeness. This subject has been ably discussed by Prosper Lucas (14/2. 'Hered. Nat.' tome 2 pages 112-120.), but is rendered extremely complex by the prepotency sometimes running equally in both sexes, and sometimes more strongly in one sex than in the other; it is likewise complicated by the presence of secondary sexual characters, which render the comparison of crossed breeds with their parents difficult.

It would appear that in certain families some one ancestor, and after him others in the same family, have had great power in transmitting their likeness through the male line; for we cannot otherwise understand how the same features should so often be transmitted after marriages with many females, as in the case of the Austrian Emperors; and so it was, according to Niebuhr, with the mental qualities of certain Roman families. (14/3. Sir H. Holland 'Chapters on Mental Physiology' 1852 page 234.) The famous bull Favourite is believed (14/4. 'Gardener's Chronicle' 1860 page 270.) to have had a prepotent influence on the shorthorn race. It has also been observed (14/5.

Mr. N.H. Smith 'Observations on Breeding' quoted in 'Encyclop. of Rural Sports' page 278.) with English racehorses that certain mares have generally transmitted their own character, whilst other mares of equally pure blood have allowed the character of the sire to prevail. A famous black greyhound, Bedlamite, as I hear from Mr. C.M. Brown "invariably got all his puppies black, no matter what was the colour of the bitch;" but then Bedlamite "had a preponderance of black in his blood, both on the sire and dam side."

[The truth of the principle of prepotency comes out more clearly when distinct races are crossed. The improved Shorthorns, notwithstanding that the breed is comparatively modern, are generally acknowledged to possess great power in impressing their likeness on all other breeds; and it is chiefly in consequence of this power that they are so highly valued for exportation. (14/6. Quoted by Bronn 'Geshichte der Natur' b. 2 s. 170. See Sturm 'Ueber Racen' 1825 s. 104-107. For the niata cattle see my 'Journal of Researches' 1845 page 146.) Godine has given a curious case of a ram of a goat-like breed of sheep from the Cape of Good Hope, which produced offspring hardly to be distinguished from himself, when crossed with ewes of twelve other breeds. But two of these half-bred ewes, when put to a merino ram, produced lambs closely resembling the merino breed. Girou de Buzareingues (14/7. Lucas 'L'Heredite Nat.' tome 2 page 112.) found that of two races of French sheep the ewes of one, when crossed during successive generations with merino rams, yielded up their character far sooner than the ewes of the other race. Sturm and Girou have given analogous cases with other breeds of sheep and with cattle, the prepotency running in these cases through the male side; but I was assured on good authority in South America, that when niata cattle are crossed with common cattle, though the niata breed is prepotent whether males or females are used, yet that the prepotency is strongest through the female line. The Manx cat is tailless and has long hind legs; Dr. Wilson crossed a male Manx with common cats, and, out of twenty-three kittens, seventeen were destitute of tails; but when the female Manx was crossed by common male cats all the kittens had tails, though they were generally short and imperfect. (14/8. Mr. Orton 'Physiology of Breeding' 1855 page 9.)

In making reciprocal crosses between pouter and fantail pigeons, the pouter-race seemed to be prepotent through both sexes over the fantail. But this is probably due to weak power in the fantail rather than to any unusually strong power in the pouter, for I have observed that barbs also preponderate over fantails. This weakness of transmission in the fantail, though the breed is an ancient one, is said (14/9. Boitard and Corbie 'Les Pigeons' 1824 page 224.) to be general; but I have observed one exception to the rule, namely, in a cross between a fantail and laugher. The most curious instance known to me of weak power in both sexes is in the trumpeter pigeon. This breed has been well known for at least 130 years: it breeds perfectly true, as I have been assured by those who have long kept many birds: it is characterised by a peculiar tuft of feathers over the beak, by a crest on the head, by a singular coo quite unlike that of any other breed, and by much-feathered feet. I have crossed both sexes with turbits of two sub-breeds, with almond tumblers, spots, and runts, and reared many mongrels and recrossed them; and though the crest on the head and feathered feet were inherited (as is generally the case with most breeds), I have never seen a vestige of the tuft over the beak or heard the peculiar coo. Boitard and Corbie (14/10. 'Les Pigeons' pages 168, 198.) assert that this is the invariable result of crossing trumpeters with other breeds: Neumeister (14/11. 'Das Ganze' etc. 1837 s. 39.), however, states that in Germany mongrels have been obtained, though very rarely, which were furnished with the tuft and would trumpet: but a pair of these mongrels with a tuft, which I imported, never trumpeted. Mr. Brent states (14/12. 'The Pigeon Book' page 46.) that the crossed offspring of a trumpeter were crossed with trumpeters for three generations, by which time the mongrels had 7/8ths of this blood in their veins, yet the tuft over the beak did not appear. At the fourth generation the tuft appeared, but the birds though now having 15-16ths trumpeter's blood still did not trumpet. This case well shows the wide difference between inheritance and prepotency; for here we have a well- established old race which transmits its characters faithfully, but which, when crossed with any other race, has the feeblest power of transmitting its two chief characteristic qualities.

I will give one other instance with fowls and pigeons of weakness and strength in the transmission of the same character to their crossed offspring. The Silk fowl breeds true, and there is reason to believe is a very ancient race; but when I reared a large number of mongrels from a Silk hen by a Spanish cock, not one exhibited even a trace of the so-called silkiness. Mr. Hewitt also asserts that in no instance are the silky feathers transmitted by this breed when crossed with any other variety. But three birds out of many raised by Mr. Orton from a cross between a silk cock and a bantam hen had silky feathers. (14/13. 'Physiology of Breeding' page 22; Mr. Hewitt in 'The Poultry Book' by Tegetmeier 1866 page 224.) So that it is certain that this breed very seldom has the power of transmitting its peculiar plumage to its crossed progeny. On the other hand, there is a silk sub-variety of the fantail pigeon, which has its feathers in nearly the same state as in the Silk fowl: now we have already seen that fantails, when crossed, possess singularly weak power in transmitting their general qualities; but the silk sub-variety when crossed with any other small-sized race invariably transmits its silky feathers! (14/14. Boitard and Corbie 'Les Pigeons' 1824 page 226.)

The well-known horticulturist, Mr. Paul, informs me that he fertilised the Black Prince hollyhock with pollen of the White Globe and the Lemonade and Black Prince hollyhocks reciprocally; but not one seedling from these three crosses inherited the black colour of the Black Prince. So, again, Mr. Laxton, who has had such great experience in crossing peas, writes to me that "whenever a cross has been effected between a white-blossomed and a purple- blossomed pea, or between a white-seeded and a purple-spotted, brown or maple- seeded pea, the offspring seems to lose nearly all the characteristics of the white-flowered and white-seeded varieties; and this result follows whether these varieties have been used as the pollen-bearing or seed-producing parents."

The law of prepotency comes into action when species are crossed, as with races and individuals. Gartner has unequivocally shown (14/15. 'Bastarderzeugung' s. 256, 290, etc. Naudin 'Nouvelles Archives du Museum' tome 1 page 149 gives a striking instance of prepotency in Datura stramonium when crossed with two other species.) that this is the case with plants. To give one instance: when Nicotiana paniculata and vincaeflora are crossed, the character of N. paniculata is almost completely lost in the hybrid; but if N. quadrivalvis be crossed with N. vincaeflora, this latter species, which was before so prepotent, now in its turn almost disappears under the power of N. quadrivalvis. It is remarkable that the prepotency of one species over another in transmission is quite independent, as shown by Gartner, of the greater or less facility with which the one fertilises the other.

With animals, the jackal is prepotent over the dog, as is stated by Flourens, who made many crosses between these animals; and this was likewise the case with a hybrid which I once saw between a jackal and a terrier. I cannot doubt, from the observations of Colin and others, that the ass is prepotent over the horse; the prepotency in this instance running more strongly through the male than through the female ass; so that the mule resembles the ass more closely than does the hinny. (14/16. Flourens 'Longevite Humaine' page 144 on crossed jackals. With respect to the difference between the mule and the hinny I am aware that this has generally been attributed to the sire and dam transmitting their characters differently; but Colin, who has given in his 'Traite Phys. Comp.' tome 2 pages 537-539, the fullest description which I have met with of these reciprocal hybrids, is strongly of opinion that the ass preponderates in both crosses, but in an unequal degree. This is likewise the conclusion of Flourens, and of Bechstein in his 'Naturgeschichte Deutschlands' b. 1 s. 294. The tail of the hinny is much more like that of the horse than is the tail of the mule, and this is generally accounted for by the males of both species transmitting with greater power this part of their structure; but a compound hybrid which I saw in the Zoological Gardens, from a mare by a hybrid ass- zebra, closely resembled its mother in its tail.) The male pheasant, judging from Mr. Hewitt's descriptions (14/17. Mr. Hewitt who has had such great experience in raising these hybrids says ('Poultry Book' by Mr. Tegetmeier 1866 pages 165-167) that in all, the head was destitute of wattles, comb, and ear-lappets; and all closely resembled the pheasant in the shape of the tail and general contour of the body. These hybrids were raised from hens of several breeds by a cock-pheasant; but another hybrid, described by Mr. Hewitt, was raised from a hen-pheasant, by a silver-laced Bantam cock, and this possessed a rudimental comb and wattles.), and from the hybrids which I have seen, preponderates over the domestic fowl; but the latter, as far as colour is concerned, has considerable power of transmission, for hybrids raised from five differently coloured hens differed greatly in plumage. I formerly examined some curious hybrids in the Zoological Gardens, between the Penguin variety of the common duck and the Egyptian goose (Anser aegyptiacus); and although I will not assert that the domesticated variety preponderated over the natural species, yet it had strongly impressed its unnatural upright figure on these hybrids.

I am aware that such cases as the foregoing have been ascribed by various authors, not to one species, race, or individual being prepotent over the other in impressing its character on its crossed offspring, but to such rules as that the father influences the external characters and the mother the internal or vital organs. But the great diversity of the rules given by various authors almost proves their falseness. Dr. Prosper Lucas has fully discussed this point, and has shown (14/18. 'L'Hered. Nat.' tome 2 book 2 chapter 1.) that none of the rules (and I could add others to those quoted by him) apply to all animals. Similar rules have been announced for plants, and have been proved by Gartner (14/19. 'Bastarderzeugung' s. 264-266. Naudin 'Nouvelles Archives du Museum' tome 1 page 148 has arrived at a similar conclusion.) to be all erroneous. If we confine our view to the domesticated races of a single species, or perhaps even to the species of the same genus, some such rules may hold good; for instance, it seems that in reciprocally crossing various breeds of fowls the male generally gives colour (14/20. 'Cottage Gardener' 1856 pages 101, 137.); but conspicuous exceptions have passed under my own eyes. It seems that the ram usually gives its peculiar horns and fleece to its crossed offspring, and the bull the presence or absence of horns.

In the following chapter on Crossing I shall have occasion to show that certain characters are rarely or never blended by crossing, but are transmitted in an unmodified state from either parent-form; I refer to this fact here because it is sometimes accompanied on the one side by prepotency, which thus acquires the false appearance of unusual strength. In the same chapter I shall show that the rate at which a species or breed absorbs and obliterates another by repeated crosses, depends in chief part on prepotency in transmission.]

In conclusion, some of the cases above given,—for instance, that of the trumpeter pigeon,—prove that there is a wide difference between mere inheritance and prepotency. This latter power seems to us, in our ignorance, to act in most cases quite capriciously. The very same character, even though it be an abnormal or monstrous one, such as silky feathers, may be transmitted by different species, when crossed, either with prepotent force or singular feebleness. It is obvious, that a purely-bred form of either sex, in all cases in which prepotency does not run more strongly in one sex than the other, will transmit its character with prepotent force over a mongrelised and already variable form. (14/21. See some remarks on this head with respect to sheep by Mr. Wilson in 'Gardener's Chronicle' 1863 page 15. Many striking instances of this result are given by M. Malingie-Nouel 'Journ. R. Agricult. Soc.' volume 14 1853 page 220 with respect to crosses between English and French sheep. He found that he obtained the desired influence of the English breeds by crossing intentionally mongrelised French breeds with pure English breeds.) From several of the above-given cases we may conclude that mere antiquity of character does not by any means necessarily make it prepotent. In some cases prepotency apparently depends on the same character being present and visible in one of the two breeds which are crossed, and latent or invisible in the other breed; and in this case it is natural that the character which is potentially present in both breeds should be prepotent. Thus, we have reason to believe that there is a latent tendency in all horses to be dun-coloured and striped; and when a horse of this kind is crossed with one of any other colour, it is said that the offspring are almost sure to be striped. Sheep have a similar latent tendency to become dark-coloured, and we have seen with what prepotent force a ram with a few black spots, when crossed with white sheep of various breeds, coloured its offspring. All pigeons have a latent tendency to become slaty-blue, with certain characteristic marks, and it is known that, when a bird thus coloured is crossed with one of any other colour, it is most difficult afterwards to eradicate the blue tint. A nearly parallel case is offered by those black bantams which, as they grow old, develop a latent tendency to acquire red feathers. But there are exceptions to the rule: hornless breeds of cattle possess a latent capacity to reproduce horns, yet when crossed with horned breeds they do not invariably produce offspring bearing horns.

We meet with analogous cases with plants. Striped flowers, though they can be propagated truly by seed, have a latent tendency to become uniformly coloured, but when once crossed by a uniformly coloured variety, they ever afterwards fail to produce striped seedlings. (14/22. Verlot 'Des Varietes' 1865 page 66.) Another case is in some respects more curious: plants bearing peloric flowers have so strong a latent tendency to reproduce their normally irregular flowers, that this often occurs by buds when a plant is transplanted into poorer or richer soil. (14/23. Moquin-Tandon 'Teratologie' page 191.) Now I crossed the peloric snapdragon (Antirrhinum majus), described in the last chapter, with pollen of the common form; and the latter, reciprocally, with peloric pollen. I thus raised two great beds of seedlings, and not one was peloric. Naudin (14/24. 'Nouvelles Archives du Museum' tome 1 page 137.) obtained the same result from crossing a peloric Linaria with the common form. I carefully examined the flowers of ninety plants of the crossed Antirrhinum in the two beds, and their structure had not been in the least affected by the cross, except that in a few instances the minute rudiment of the fifth stamen, which is always present, was more fully or even completely developed. It must not be supposed that this entire obliteration of the peloric structure in the crossed plants can be accounted for by any incapacity of transmission; for I raised a large bed of plants from the peloric Antirrhinum, artificially fertilised by its own pollen, and sixteen plants, which alone survived the winter, were all as perfectly peloric as the parent-plant. Here we have a good instance of the wide difference between the inheritance of a character and the power of transmitting it to crossed offspring. The crossed plants, which perfectly resembled the common snapdragon, were allowed to sow themselves, and out of a hundred and twenty-seven seedlings, eighty-eight proved to be common snapdragons, two were in an intermediate condition between the peloric and normal state, and thirty-seven were perfectly peloric, having reverted to the structure of their one grand-parent. This case seems at first sight to offer an exception to the rule just given, namely, that a character which is present in one form and latent in the other is generally transmitted with prepotent force when the two forms are crossed. For in all the Scrophulariaceae, and especially in the genera Antirrhinum and Linaria, there is, as was shown in the last chapter, a strong latent tendency to become peloric; but there is also, as we have seen, a still stronger tendency in all peloric plants to reacquire their normal irregular structure. So that we have two opposed latent tendencies in the same plants. Now, with the crossed Antirrhinums the tendency to produce normal or irregular flowers, like those of the common Snapdragon, prevailed in the first generation; whilst the tendency to pelorism, appearing to gain strength by the intermission of a generation, prevailed to a large extent in the second set of seedlings. How it is possible for a character to gain strength by the intermission of a generation, will be considered in the chapter on pangenesis.

On the whole, the subject of prepotency is extremely intricate,—from its varying so much in strength, even in regard to the same character, in different animals,—from its running either equally in both sexes, or, as frequently is the case with animals, but not with plants, much stronger in one sex than the other,—from the existence of secondary sexual characters,—from the transmission of certain characters being limited, as we shall immediately see, by sex,—from certain characters not blending together,—and, perhaps, occasionally from the effects of a previous fertilisation on the mother. It is therefore not surprising that no one has hitherto succeeded in drawing up general rules on the subject of prepotency.

## INHERITANCE AS LIMITED BY SEX.

New characters often appear in one sex, and are afterwards transmitted to the same sex, either exclusively or in a much greater degree than to the other. This subject is important, because with animals of many kinds in a state of nature, both high and low in the scale, secondary sexual characters, not directly connected with the organs of reproduction, are conspicuously present. With our domesticated animals, characters of this kind often differ widely from those distinguishing the two sexes of the parent species; and the principle of inheritance, as limited by sex, explains how this is possible.

[Dr. P. Lucas has shown (14/25. 'L'Hered. Nat.' tome 2 pages 137-165. See also Mr. Sedgwick's four memoirs, immediately to be referred to.) that when a peculiarity, in no manner connected with the reproductive organs, appears in either parent, it is often transmitted exclusively to the offspring of the same sex, or to a much greater number of them than of the opposite sex. Thus, in the family of Lambert, the horn-like projections on the skin were transmitted from the father to his sons and grandsons alone; so it has been with other cases of ichthyosis, with supernumerary digits, with a deficiency of digits and phalanges, and in a lesser degree with various diseases, especially with colour-blindness and the haemorrhagic diathesis, that is, an extreme liability to profuse and uncontrollable bleeding from trifling wounds. On the other hand, mothers have transmitted, during several generations, to their daughters alone, supernumerary and deficient digits, colour-blindness and other peculiarities. So that the very same peculiarity may become attached to either sex, and be long inherited by that sex alone; but the attachment in certain cases is much more frequent to one than the other sex. The same peculiarities also may be promiscuously transmitted to either sex. Dr. Lucas gives other cases, showing that the male occasionally transmits his peculiarities to his daughters alone, and the mother to her sons alone; but even in this case we see that inheritance is to a certain extent, though inversely, regulated by sex. Dr. Lucas, after weighing the whole evidence, comes to the conclusion that every peculiarity tends to be transmitted in a greater or lesser degree to that sex in which it first appears. But a more definite rule, as I have elsewhere shown (14/26. 'Descent of Man' 2nd edition page 32.) generally holds good, namely, that variations which first appear in either sex at a late period of life, when the reproductive functions are active, tend to be developed in that sex alone; whilst variations which first appear early in life in either sex are commonly transmitted to both sexes. I am, however, far from supposing that this is the sole determining cause.

A few details from the many cases collected by Mr. Sedgwick (14/27. On Sexual Limitation in Hereditary Diseases 'Brit. and For. Med.-Chirurg. Review' April 1861 page 477; July page 198; April 1863 page 445; and July page 159. Also in 1867 'On the influence of Age in Hereditary Disease.'), may be here given. Colour-blindness, from some unknown cause, shows itself much oftener in males than in females; in upwards of two hundred cases collected by Mr. Sedgwick, nine-tenths related to men; but it is eminently liable to be transmitted through women. In the case given by Dr. Earle, members of eight related families were affected during five generations: these families consisted of sixty-one individuals, namely, of thirty-two males, of whom nine-sixteenths were incapable of distinguishing colour, and of twenty-nine females, of whom only one-fifteenth were thus affected. Although colour-blindness thus generally clings to the male sex, nevertheless, in one instance in which it first appeared in a female, it was transmitted during five generations to thirteen individuals, all of whom were females. The haemorrhagic diathesis, often accompanied by rheumatism, has been known to affect the males alone during five generations, being transmitted, however, through the females. It is said that deficient phalanges in the fingers have been inherited by the females alone during ten generations. In another case, a man thus deficient in both hands and feet, transmitted the peculiarity to his two sons and one daughter; but in the third generation,—out of nineteen grandchildren, twelve sons had the family defect, whilst the seven daughters were free. In ordinary cases of sexual limitation, the sons or daughters inherit the peculiarity, whatever it may be, from their father or mother, and transmit it to their children of the same sex; but generally with the haemorrhagic diathesis, and often with colour-blindness, and in some other cases, the sons never inherit the peculiarity directly from their fathers, but the daughters alone transmit the latent tendency, so that the sons of the daughters alone exhibit it. Thus the father, grandson, and great-great-grandson will exhibit a peculiarity,— the grandmother, daughter, and great-grand-daughter having transmitted it in a latent state. Hence we have, as Mr. Sedgwick remarks, a double kind of atavism or reversion; each grandson apparently receiving and developing the peculiarity from his grandfather, and each daughter apparently receiving the latent tendency from her grandmother.

From the various facts recorded by Dr. Prosper Lucas, Mr. Sedgwick, and others, there can be no doubt that peculiarities first appearing in either sex, though not in any way necessarily or invariably connected with that sex, strongly tend to be inherited by the offspring of the same sex, but are often transmitted in a latent state through the opposite sex.

Turning now to domesticated animals, we find that certain characters not proper to the parent species are often confined to, and inherited by, one sex alone; but we do not know the history of the first appearance of such characters. In the chapter on Sheep, we have seen that the males of certain races differ greatly from the females in the shape of their horns, these being absent in the ewes of some breeds; they differ also in the development of fat in the tail and in the outline of the forehead. These differences, judging from the character of the allied wild species, cannot be accounted for by supposing that they have been derived from distinct parent forms. There is, also, a great difference between the horns of the two sexes in one Indian breed of goats. The bull zebu is said to have a larger hump than the cow. In the Scotch deer-hound the two sexes differ in size more than in any other variety of the dog (14/28. W. Scrope 'Art of Deer Stalking' page 354.) and, judging from analogy, more than in the aboriginal parent-species. The peculiar colour called tortoise-shell is very rarely seen in a male cat; the males of this variety being of a rusty tint.

In various breeds of the fowl the males and females often differ greatly; and these differences are far from being the same with those which distinguish the two sexes of the parent-species, the Gallus bankiva; and consequently have originated under domestication. In certain sub-varieties of the Game race we have the unusual case of the hens differing from each other more than the cocks. In an Indian breed of a white colour shaded with black, the hens invariably have black skins, and their bones are covered by a black periosteum, whilst the cocks are never or most rarely thus characterised. Pigeons offer a more interesting case; for throughout the whole great family the two sexes do not often differ much; and the males and females of the parent-form, the C. livia, are undistinguishable: yet we have seen that with pouters the male has the characteristic quality of pouting more strongly developed than the female; and in certain sub-varieties the males alone are spotted or striated with black, or otherwise differ in colour. When male and female English carrier-pigeons are exhibited in separate pens, the difference in the development of the wattle over the beak and round the eyes is conspicuous. So that here we have instances of the appearance of secondary sexual characters in the domesticated races of a species in which such differences are naturally quite absent.]

On the other hand, secondary sexual characters which belong to the species in a state of nature are sometimes quite lost, or greatly diminished, under domestication. We see this in the small size of the tusks in our improved breeds of the pig, in comparison with those of the wild boar. There are sub-breeds of fowls, in which the males have lost the fine-flowing tail-feathers and hackles; and others in which there is no difference in colour between the two sexes. In some cases the barred plumage, which in gallinaceous birds is commonly the attribute of the hen, has been transferred to the cock, as in the cuckoo sub-breeds. In other cases masculine characters have been partly transferred to the female, as with the splendid plumage of the golden-spangled Hamburgh hen, the enlarged comb of the Spanish hen, the pugnacious disposition of the Game hen, and as in the well-developed spurs which occasionally appear in the hens of various breeds. In Polish fowls both sexes are ornamented with a topknot, that of the male being formed of hackle-like feathers, and this is a new male character in the genus Gallus. On the whole, as far as I can judge, new characters are more apt to appear in the males of our domesticated animals than in the females (14/29. I have given in my 'Descent of Man' 2nd edition page 223 sufficient evidence that male animals are usually more variable than the females.), and afterwards to be inherited exclusively or more strongly by the males. Finally, in accordance with the principle of inheritance as limited by sex, the preservation and augmentation of secondary sexual characters in natural species offers no especial difficulty, as this would follow through that form of selection which I have called sexual selection.

## INHERITANCE AT CORRESPONDING PERIODS OF LIFE.

This is an important subject. Since the publication of my 'Origin of Species' I have seen no reason to doubt the truth of the explanation there given of one of the most remarkable facts in biology, namely, the difference between the embryo and the adult animal. The explanation is, that variations do not necessarily or generally occur at a very early period of embryonic growth, and that such variations are inherited at a corresponding age. As a consequence of this the embryo, even after the parent-form has undergone great modification, is left only slightly modified; and the embryos of widely-different animals which are descended from a common progenitor remain in many important respects like one another and probably like their common progenitor. We can thus understand why embryology throws a flood of light on the natural system of classification, as this ought to be as far as possible genealogical. When the embryo leads an independent life, that is, becomes a larva, it has to be adapted to the surrounding conditions in its structure and instincts, independently of those of its parents; and the principle of inheritance at corresponding periods of life renders this possible.

This principle is, indeed, in one way so obvious that it escapes attention. We possess a number of races of animals and plants, which, when compared with one another and with their parent-forms, present conspicuous differences, both in their immature and mature states. Look at the seeds of the several kinds of peas, beans, maize, which can be propagated truly, and see how they differ in size, colour, and shape, whilst the full-grown plants differ but little. Cabbages, on the other hand, differ greatly in foliage and manner of growth, but hardly at all in their seeds; and generally it will be found that the differences between cultivated plants at different periods of growth are not necessarily closely connected together, for plants may differ much in their seeds and little when full-grown, and conversely may yield seeds hardly distinguishable, yet differ much when full-grown. In the several breeds of poultry, descended from a single species, differences in the eggs and chickens whilst covered with down, in the plumage at the first and subsequent moults, as well as in the comb and wattles, are all inherited. With man peculiarities in the milk and second teeth (of which I have received the details) are inheritable, and longevity is often transmitted. So again with our improved breeds of cattle and sheep, early maturity, including the early development of the teeth, and with certain breeds of fowl the early appearance of secondary sexual characters, all come under the same head of inheritance at corresponding periods.

Numerous analogous facts could be given. The silk-moth, perhaps, offers the best instance; for in the breeds which transmit their characters truly, the eggs differ in size, colour, and shape: the caterpillars differ, in moulting three or four times, in colour, even in having a dark-coloured mark like an eyebrow, and in the loss of certain instincts;—the cocoons differ in size, shape, and in the colour and quality of the silk; these several differences being followed by slight or barely distinguishable differences in the mature moth.

But it may be said that, if in the above cases a new peculiarity is inherited, it must be at the corresponding stage of development; for an egg or seed can resemble only an egg or seed, and the horn in a full-grown ox can resemble only a horn. The following cases show inheritance at corresponding periods more plainly, because they refer to peculiarities which might have supervened, as far as we can see, earlier or later in life, yet are inherited at the same period at which they first appeared.

[In the Lambert family the porcupine-like excrescences appeared in the father and sons at the same age, namely, about nine weeks after birth. (14/30. Prichard 'Phys. Hist. of Mankind' 1851 volume 1 page 349.) In the extraordinary hairy family described by Mr. Crawfurd (14/31. 'Embassy to the Court of Ava' volume 1 page 320. The third generation is described by Capt. Yule in his 'Narrative of the Mission to the Court of Ava' 1855 page 94.), children were produced during three generations with hairy ears; in the father the hair began to grow over his body at six years old; in his daughter somewhat earlier, namely, at one year; and in both generations the milk teeth appeared late in life, the permanent teeth being afterwards singularly deficient. Greyness of hair at an unusually early age has been transmitted in some families. These cases border on diseases inherited at corresponding periods of life, to which I shall immediately refer.

It is a well-known peculiarity with almond-tumbler pigeons, that the full beauty and peculiar character of the plumage does not appear until the bird has moulted two or three times. Neumeister describes and figures a brace of pigeons in which the whole body is white except the breast, neck, and head; but in their first plumage all the white feathers have coloured edges. Another breed is more remarkable: its first plumage is black, with rusty-red wing-bars and a crescent-shaped mark on the breast; these marks then become white, and remain so during three or four moults; but after this period the white spreads over the body, and the bird loses its beauty. (14/32. 'Das Ganze der Taubenzucht' 1837 s. 24 tab. 4 figure 2 s. 21 tab. 1 figure 4.) Prize canary-birds have their wings and tail black: "this colour, however, is only retained until the first moult, so that they must be exhibited ere the change takes place. Once moulted, the peculiarity has ceased. Of course all the birds emanating from this stock have black wings and tails the first year." (14/33. Kidd 'Treatise on the Canary' page 18.) A curious and somewhat analogous account has been given (14/34. Charlesworth 'Mag. of Nat. Hist.' volume 1 1837 page 167.) of a family of wild pied rooks which were first observed in 1798, near Chalfont, and which every year from that date up to the period of the published notice, viz., 1837 "have several of their brood particoloured, black and white. This variegation of the plumage, however, disappears with the first moult; but among the next young families there are always a few pied ones." These changes of plumage, which are inherited at various corresponding periods of life in the pigeon, canary-bird, and rook, are remarkable, because the parent-species passes through no such change.

Inherited diseases afford evidence in some respects of less value than the foregoing cases, because diseases are not necessarily connected with any change in structure; but in other respects of more value, because the periods have been more carefully observed. Certain diseases are communicated to the child apparently by a process like inoculation, and the child is from the first affected; such cases may be here passed over. Large classes of diseases usually appear at certain ages, such as St. Vitus's dance in youth, consumption in early mid-life, gout later, and apoplexy still later; and these are naturally inherited at the same period. But even in diseases of this class, instances have been recorded, as with St. Vitus's dance, showing that an unusually early or late tendency to the disease is inheritable. (14/35. Dr. Prosper Lucas 'Hered. Nat.' tome 2 page 713.) In most cases the appearance of any inherited disease is largely determined by certain critical periods in each person's life, as well as by unfavourable conditions. There are many other diseases, which are not attached to any particular period, but which certainly tend to appear in the child at about the same age at which the parent was first attacked. An array of high authorities, ancient and modern, could be given in support of this proposition. The illustrious Hunter believed in it; and Piorry (14/36. 'L'Hered. dans les Maladies' 1840 page 135. For Hunter see Harlan 'Med. Researches' page 530.) cautions the physician to look closely to the child at the period when any grave inheritable disease attacked the parent. Dr. Prosper Lucas (14/37. 'L'Hered. Nat.' tome 2 page 850.), after collecting facts from every source, asserts that affections of all kinds, though not related to any particular period of life, tend to reappear in the offspring at whatever period of life they first appeared in the progenitor.

As the subject is important, it may be well to give a few instances, simply as illustrations, not as proof; for proof, recourse must be had to the authorities above quoted. Some of the following cases have been selected for the sake of showing that, when a slight departure from the rule occurs, the child is affected somewhat earlier in life than the parent. In the family of Le Compte blindness was inherited through three generations, and no less than twenty-seven children and grandchildren were all affected at about the same age; their blindness in general began to advance about the fifteenth or sixteenth year, and ended in total deprivation of sight at the age of about twenty-two. (14/38. Sedgwick 'Brit. and For. Med.-Chirurg. Review' April 1861 page 485. In some accounts the number of children and grandchildren is given as 37; but this seems to be an error judging from the paper first published in the 'Baltimore Med. and Phys. Reg.' 1809 of which Mr. Sedgwick has been so kind as to send me a copy.) In another case a father and his four children all became blind at twenty-one years old; in another, a grandmother grew blind at thirty-five, her daughter at nineteen, and three grandchildren at the ages of thirteen and eleven. (14/39. Prosper Lucas 'Hered. Nat.' tome 1 page 400.) So with deafness, two brothers, their father and paternal grandfather, all became deaf at the age of forty. (14/40. Sedgwick ibid July 1861 page 202.)

Esquirol gives several striking instances of insanity coming on at the same age, as that of a grandfather, father, and son, who all committed suicide near their fiftieth year. Many other cases could be given, as of a whole family who became insane at the age of forty. (14/41. Piorry page 109; Prosper Lucas tome 2 page 759.) Other cerebral affections sometimes follow the same rule,—for instance, epilepsy and apoplexy. A woman died of the latter disease when sixty-three years old; one of her daughters at forty-three, and the other at sixty-seven: the latter had twelve children, who all died from tubercular meningitis. (14/42. Prosper Lucas tome 2 page 748.) I mention this latter case because it illustrates a frequent occurrence, namely, a change in the precise nature of an inherited disease, though still affecting the same organ.

Asthma has attacked several members of the same family when forty years old, and other families during infancy. The most different diseases, such as angina pectoris, stone in the bladder, and various affections of the skin, have appeared in successive generations at nearly the same age. The little finger of a man began from some unknown cause to grow inwards, and the same finger in his two sons began at the same age to bend inwards in a similar manner. Strange and inexplicable neuralgic affections have caused parents and children to suffer agonies at about the same period of life. (14/43. Prosper Lucas tome 3 pages 678, 700, 702; Sedgwick ibid April 1863 page 449 and July 1863 page 162. Dr. J. Steinan 'Essay on Hereditary Disease' 1843 pages 27, 34.)

I will give only two other cases, which are interesting as illustrating the disappearance as well as the appearance of disease at the same age. Two brothers, their father, their paternal uncles, seven cousins, and their paternal grandfather, were all similarly affected by a skin-disease, called pityriasis versicolor; "the disease, strictly limited to the males of the family (though transmitted through the females), usually appeared at puberty, and disappeared at about the age of forty or forty-five years." The second case is that of four brothers, who when about twelve years old suffered almost every week from severe headaches, which were relieved only by a recumbent position in a dark room. Their father, paternal uncles, paternal grandfather, and granduncles all suffered in the same way from headaches, which ceased at the age of fifty-four or fifty-five in all those who lived so long. None of the females of the family were affected. (14/44. These cases are given by Mr. Sedgwick on the authority of Dr. H. Stewart in 'Med.-Chirurg. Review' April 1863 pages 449, 477.)]

It is impossible to read the foregoing accounts, and the many others which have been recorded, of diseases coming on during three or even more generations in several members of the same family at the same age, especially in the case of rare affections in which the coincidence cannot be attributed to chance, and to doubt that there is a strong tendency to inheritance in disease at corresponding periods of life. When the rule fails, the disease is apt to come on earlier in the child than in the parent; the exceptions in the other direction being very much rarer. Dr. Lucas (14/45. 'Hered. Nat.' tome 2 page 852.) alludes to several cases of inherited diseases coming on at an earlier period. I have already given one striking instance with blindness during three generations; and Mr. Bowman remarks that this frequently occurs with cataract. With cancer there seems to be a peculiar liability to earlier inheritance: Sir J. Paget, who has particularly attended to this subject, and tabulated a large number of cases, informs me that he believes that in nine cases out of ten the later generation suffers from the disease at an earlier period than the previous generation. He adds, "In the instances in which the opposite relation holds, and the members of later generations have cancer at a later age than their predecessors, I think it will be found that the non-cancerous parents have lived to extreme old ages." So that the longevity of a non-affected parent seems to have the power of influencing the fatal period in the offspring; and we thus apparently get another element of complexity in inheritance.

The facts, showing that with certain diseases the period of inheritance occasionally or even frequently advances, are important with respect to the general descent-theory, for they render it probable that the same thing would occur with ordinary modifications of structure. The final result of a long series of such advances would be the gradual obliteration of characters proper to the embryo and larva, which would thus come to resemble more and more closely the mature parent-form. But any structure which was of service to the embryo or larva would be preserved by the destruction at this stage of growth of each individual which manifested any tendency to lose its proper character at too early an age.

Finally, from the numerous races of cultivated plants and domestic animals, in which the seeds or eggs, the young or old, differ from one another and from those of the parent-species;—from the cases in which new characters have appeared at a particular period, and afterwards been inherited at the same period;—and from what we know with respect to disease, we must believe in the truth of the great principle of inheritance at corresponding periods of life.

## SUMMARY OF THE THREE PRECEDING CHAPTERS.

Strong as is the force of inheritance, it allows the incessant appearance of new characters. These, whether beneficial or injurious,—of the most trifling importance, such as a shade of colour in a flower, a coloured lock of hair, or a mere gesture,—or of the highest importance, as when affecting the brain, or an organ so perfect and complex as the eye,—or of so grave a nature as to deserve to be called a monstrosity,—or so peculiar as not to occur normally in any member of the same natural class,—are often inherited by man, by the lower animals, and plants. In numberless cases it suffices for the inheritance of a peculiarity that one parent alone should be thus characterised. Inequalities in the two sides of the body, though opposed to the law of symmetry, may be transmitted. There is ample evidence that the effects of mutilations and of accidents, especially or perhaps exclusively when followed by disease, are occasionally inherited. There can be no doubt that the evil effects of the long-continued exposure of the parent to injurious conditions are sometimes transmitted to the offspring. So it is, as we shall see in a future chapter, with the effects of the use and disuse of parts, and of mental habits. Periodical habits are likewise transmitted, but generally, as it would appear, with little force.

Hence we are led to look at inheritance as the rule, and non-inheritance as the anomaly. But this power often appears to us in our ignorance to act capriciously, transmitting a character with inexplicable strength or feebleness. The very same peculiarity, as the weeping habit of trees, silky feathers, etc., may be inherited either firmly or not at all by different members of the same group, and even by different individuals of the same species, though treated in the same manner. In this latter case we see that the power of transmission is a quality which is merely individual in its attachment. As with single characters, so it is with the several concurrent slight differences which distinguish sub-varieties or races; for of these, some can be propagated almost as truly as species, whilst others cannot be relied on. The same rule holds good with plants, when propagated by bulbs, offsets, etc., which in one sense still form parts of the same individual, for some varieties retain or inherit through successive bud-generations their character far more truly than others.

Some characters not proper to the parent-species have certainly been inherited from an extremely remote epoch, and may therefore be considered as firmly fixed. But it is doubtful whether length of inheritance in itself gives fixedness of character; though the chances are obviously in favour of any character which has long been transmitted true or unaltered still being transmitted true as long as the conditions of life remain the same. We know that many species, after having retained the same character for countless ages, whilst living under their natural conditions, when domesticated have varied in the most diversified manner, that is, have failed to transmit their original form; so that no character appears to be absolutely fixed. We can sometimes account for the failure of inheritance by the conditions of life being opposed to the development of certain characters; and still oftener, as with plants cultivated by grafts and buds, by the conditions causing new and slight modifications incessantly to appear. In this latter case it is not that inheritance wholly fails, but that new characters are continually superadded. In some few cases, in which both parents are similarly characterised, inheritance seems to gain so much force by the combined action of the two parents, that it counteracts its own power, and a new modification is the result.

In many cases the failure of the parents to transmit their likeness is due to the breed having been at some former period crossed; and the child takes after his grandparent or more remote ancestor of foreign blood. In other cases, in which the breed has not been crossed, but some ancient character has been lost through variation, it occasionally reappears through reversion, so that the parents apparently fail to transmit their own likeness. In all cases, however, we may safely conclude that the child inherits all its characters from its parents, in whom certain characters are latent, like the secondary sexual characters of one sex in the other. When, after a long succession of bud-generations, a flower or fruit becomes separated into distinct segments, having the colours or other attributes of both parent-forms, we cannot doubt that these characters were latent in the earlier buds, though they could not then be detected, or could be detected only in an intimately commingled state. So it is with animals of crossed parentage, which with advancing years occasionally exhibit characters derived from one of their two parents, of which not a trace could at first be perceived. Certain monstrosities, which resemble what naturalists call the typical form of the group in question, apparently come under the same law of reversion. It is assuredly an astonishing fact that the male and female sexual elements, that buds, and even full-grown animals, should retain characters, during several generations in the case of crossed breeds, and during thousands of generations in the case of pure breeds, written as it were in invisible ink, yet ready at any time to be evolved under certain conditions.

What these conditions precisely are, we do not know. But any cause which disturbs the organisation or constitution seems to be sufficient. A cross certainly gives a strong tendency to the reappearance of long-lost characters, both corporeal and mental. In the case of plants, this tendency is much stronger with those species which have been crossed after long cultivation and which therefore have had their constitutions disturbed by this cause as well as by crossing, than with species which have always lived under their natural conditions and have then been crossed. A return, also, of domesticated animals and cultivated plants to a wild state favours reversion; but the tendency under these circumstances has been much exaggerated.

When individuals of the same family which differ somewhat, and when races or species are crossed, the one is often prepotent over the other in transmitting its character. A race may possess a strong power of inheritance, and yet when crossed, as we have seen with trumpeter-pigeons, yield to the prepotency of every other race. Prepotency of transmission may be equal in the two sexes of the same species, but often runs more strongly in one sex. It plays an important part in determining the rate at which one race can be modified or wholly absorbed by repeated crosses with another. We can seldom tell what makes one race or species prepotent over another; but it sometimes depends on the same character being present and visible in one parent, and latent or potentially present in the other.

Characters may first appear in either sex, but oftener in the male than in the female, and afterwards be transmitted to the offspring of the same sex. In this case we may feel confident that the peculiarity in question is really present though latent in the opposite sex! hence the father may transmit through his daughter any character to his grandson; and the mother conversely to her granddaughter. We thus learn, and the fact is an important one, that transmission and development are distinct powers. Occasionally these two powers seem to be antagonistic, or incapable of combination in the same individual; for several cases have been recorded in which the son has not directly inherited a character from his father, or directly transmitted it to his son, but has received it by transmission through his non-affected mother, and transmitted it through his non-affected daughter. Owing to inheritance being limited by sex, we see how secondary sexual characters may have arisen under nature; their preservation and accumulation being dependent on their service to either sex.

At whatever period of life a new character first appears, it generally remains latent in the offspring until a corresponding age is attained, and then is developed. When this rule fails, the child generally exhibits the character at an earlier period than the parent. On this principle of inheritance at corresponding periods, we can understand how it is that most animals display from the germ to maturity such a marvellous succession of characters.

Finally, though much remains obscure with respect to Inheritance, we may look at the following laws as fairly well established.

FIRSTLY, a tendency in every character, new and old, to be transmitted by seminal and bud generation, though often counteracted by various known and unknown causes.

SECONDLY, reversion or atavism, which depends on transmission and development being distinct powers: it acts in various degrees and manners through both seminal and bud generation.

THIRDLY, prepotency of transmission, which may be confined to one sex, or be common to both sexes.

FOURTHLY, transmission, as limited by sex, generally to the same sex in which the inherited character first appeared; and this in many, probably most cases, depends on the new character having first appeared at a rather late period of life.

FIFTHLY, inheritance at corresponding periods of life, with some tendency to the earlier development of the inherited character.

In these laws of Inheritance, as displayed under domestication, we see an ample provision for the production, through variability and natural selection, of new specific forms.

# CHAPTER 2.XV.

ON CROSSING.

FREE INTERCROSSING OBLITERATES THE DIFFERENCES BETWEEN ALLIED BREEDS. WHEN THE NUMBERS OF TWO COMMINGLING BREEDS ARE UNEQUAL, ONE ABSORBS THE OTHER. THE RATE OF ABSORPTION DETERMINED BY PREPOTENCY OF TRANSMISSION, BY THE CONDITIONS OF LIFE, AND BY NATURAL SELECTION. ALL ORGANIC BEINGS OCCASIONALLY INTERCROSS; APPARENT EXCEPTIONS. ON CERTAIN CHARACTERS INCAPABLE OF FUSION; CHIEFLY OR EXCLUSIVELY THOSE WHICH HAVE SUDDENLY APPEARED IN THE INDIVIDUAL. ON THE MODIFICATION OF OLD RACES, AND THE FORMATION OF NEW RACES BY CROSSING. SOME CROSSED RACES HAVE BRED TRUE FROM THEIR FIRST PRODUCTION. ON THE CROSSING OF DISTINCT SPECIES IN RELATION TO THE FORMATION OF DOMESTIC RACES.

In the two previous chapters, when discussing reversion and prepotency, I was necessarily led to give many facts on crossing. In the present chapter I shall consider the part which crossing plays in two opposed directions,—firstly, in obliterating characters, and consequently in preventing the formation of new races; and secondly, in the modification of old races, or in the formation of new and intermediate races, by a combination of characters. I shall also show that certain characters are incapable of fusion.

The effects of free or uncontrolled breeding between the members of the same variety or of closely allied varieties are important; but are so obvious that they need not be discussed at much length. It is free intercrossing which chiefly gives uniformity, both under nature and under domestication, to the individuals of the same species or variety, when they live mingled together and are not exposed to any cause inducing excessive variability. The prevention of free crossing, and the intentional matching of individual animals, are the corner-stones of the breeder's art. No man in his senses would expect to improve or modify a breed in any particular manner, or keep an old breed true and distinct, unless he separated his animals. The killing of inferior animals in each generation comes to the same thing as their separation. In savage and semi-civilised countries, where the inhabitants have not the means of separating their animals, more than a single breed of the same species rarely or never exists. In former times, even in the United States, there were no distinct races of sheep, for all had been mingled together. (15/1. 'Communications to the Board of Agriculture' volume 1 page 367.)

The celebrated agriculturist Marshall (15/2. 'Review of Reports, North of England' 1808 page 200.) remarks that "sheep that are kept within fences, as well as shepherded flocks in open countries, have generally a similarity, if not a uniformity, of character in the individuals of each flock;" for they breed freely together, and are prevented from crossing with other kinds; whereas in the unenclosed parts of England the unshepherded sheep, even of the same flock, are far from true or uniform, owing to various breeds having mingled and crossed. We have seen that the half-wild cattle in each of the several British parks are nearly uniform in character; but in the different parks, from not having mingled and crossed during many generations, they differ to a certain small extent.

We cannot doubt that the extraordinary number of varieties and sub-varieties of the pigeon, amounting to at least one hundred and fifty, is partly due to their remaining, differently from other domesticated birds, paired for life once matched. On the other hand, breeds of cats imported into this country soon disappear, for their nocturnal and rambling habits render it hardly possible to prevent free crossing. Rengger (15/3. 'Saugethiere von Paraguay' 1830 s. 212.) gives an interesting case with respect to the cat in Paraguay: in all the distant parts of the kingdom it has assumed, apparently from the effects of the climate, a peculiar character, but near the capital this change has been prevented, owing, as he asserts, to the native animal frequently crossing with cats imported from Europe. In all cases like the foregoing, the effects of an occasional cross will be augmented by the increased vigour and fertility of the crossed offspring, of which fact evidence will hereafter be given; for this will lead to the mongrels increasing more rapidly than the pure parent-breeds.

When distinct breeds are allowed to cross freely, the result will be a heterogeneous body; for instance, the dogs in Paraguay are far from uniform, and can no longer be affiliated to their parent-races. (15/4. Rengger 'Saugethiere' etc. s. 154.) The character which a crossed body of animals will ultimately assume must depend on several contingencies,—namely, on the relative members of the individuals belonging to the two or more races which are allowed to mingle; on the prepotency of one race over the other in the transmission of character; and on the conditions of life to which they are exposed. When two commingled breeds exist at first in nearly equal numbers, the whole will sooner or later become intimately blended, but not so soon, both breeds being equally favoured in all respects, as might have been expected. The following calculation (15/5. White 'Regular Gradation in Man' page 146.) shows that this is the case: if a colony with an equal number of black and white men were founded, and we assume that they marry indiscriminately, are equally prolific, and that one in thirty annually dies and is born; then "in 65 years the number of blacks, whites, and mulattoes would be equal. In 91 years the whites would be 1-10th, the blacks 1-10th, and the mulattoes, or people of intermediate degrees of colour, 8-10ths of the whole number. In three centuries not 1-100th part of the whites would exist."

When one of two mingled races exceed the other greatly in number, the latter will soon be wholly, or almost wholly, absorbed and lost. (15/6. Dr. W.F. Edwards in his 'Caracteres Physiolog. des Races Humaines' page 24 first called attention to this subject, and ably discussed it.) Thus European pigs and dogs have been largely introduced in the islands of the Pacific Ocean, and the native races have been absorbed and lost in the course of about fifty or sixty years (15/7. Rev. D. Tyerman and Bennett 'Journal of Voyages' 1821-1829 volume 1 page 300.); but the imported races no doubt were favoured. Rats may be considered as semi-domesticated animals. Some snake-rats (Mus alexandrinus) escaped in the Zoological Gardens of London "and for a long time afterwards the keepers frequently caught cross-bred rats, at first half-breds, afterwards with less of the character of the snake-rat, till at length all traces of it disappeared." (15/8. Mr. S.J. Salter 'Journal Linn. Soc.' volume 6 1862 page 71.) On the other hand, in some parts of London, especially near the docks, where fresh rats are frequently imported, an endless variety of intermediate forms may be found between the brown, black, and snake rat, which are all three usually ranked as distinct species.

How many generations are necessary for one species or race to absorb another by repeated crosses has often been discussed (15/9. Sturm 'Ueber Racen, etc.' 1825 s. 107. Bronn 'Geschichte der Natur' b. 2 s. 170 gives a table of the proportions of blood after successive crosses. Dr. P. Lucas 'L'Heredite Nat.' tome 2 page 308.); and the requisite number has probably been much exaggerated. Some writers have maintained that a dozen or score, or even more generations, are necessary; but this in itself is improbable, for in the tenth generation there would be only 1-1024th part of foreign blood in the offspring. Gartner found (15/10. 'Bastarderzeugung' s. 463, 470.), that with plants, one species could be made to absorb another in from three to five generations, and he believes that this could always be effected in from six to seven generations. In one instance, however, Kolreuter (15/11. 'Nova Acta Petrop.' 1794 page 393: see also previous volume.) speaks of the offspring of Mirabilis vulgaris, crossed during eight successive generations by M. longiflora, as resembling this latter species so closely, that the most scrupulous observer could detect "vix aliquam notabilem differentiam" or, as he says, he succeeded, "ad plenariam fere transmutationem." But this expression shows that the act of absorption was not even then absolutely complete, though these crossed plants contained only the 1-256th part of M. vulgaris. The conclusions of such accurate observers as Gartner and Kolreuter are of far higher worth than those made without scientific aim by breeders. The most precise account which I have met with is given by

Stonehenge (15/12. 'The Dog' 1867 pages 179-184.) and is illustrated by photographs. Mr. Hanley crossed a greyhound bitch with a bulldog; the offspring in each succeeding generation being recrossed with first-rate greyhounds. As Stonehenge remarks, it might naturally be supposed that it would take several crosses to get rid of the heavy form of the bulldog; but Hysterics, the gr-gr-granddaughter of a bulldog, showed no trace whatever of this breed in external form. She and all of the same litter, however, were "remarkably deficient in stoutness, though fast as well as clever." I believe clever refers to skill in turning. Hysterics was put to a son of Bedlamite, "but the result of the fifth cross is not as yet, I believe, more satisfactory than that of the fourth." On the other hand, with sheep, Fleischmann (15/13. As quoted in the 'True Principles of Breeding' by C.H. Macknight and Dr. H. Madden 1865 page 11.) shows how persistent the effects of a single cross may be: he says "that the original coarse sheep (of Germany) have 5500 fibres of wool on a square inch; grades of the third or fourth Merino cross produced about 8000, the twentieth cross 27,000, the perfect pure Merino blood 40,000 to 48,000." So that common German sheep crossed twenty times successively with Merino did not by any means acquire wool as fine as that of the pure breed. But in all cases, the rate of absorption will depend largely on the conditions of life being favourable to any particular character; and we may suspect that there would be a constant tendency to degeneration in the wool of Merinos under the climate of Germany, unless prevented by careful selection; and thus perhaps the foregoing remarkable case may be explained. The rate of absorption must also depend on the amount of distinguishable difference between the two forms which are crossed, and especially, as Gartner insists, on prepotency of transmission in the one form over the other. We have seen in the last chapter that one of two French breeds of sheep yielded up its character, when crossed with Merinos, very much more slowly than the other; and the common German sheep referred to by Fleischmann may be in this respect analogous. In all cases there will be more or less liability to reversion during many subsequent generations, and it is this fact which has probably led authors to maintain that a score or more of generations are requisite for one race to absorb another. In considering the final result of the commingling of two or more breeds, we must not forget that the act of crossing in itself tends to bring back long-lost characters not proper to the immediate parent-forms.

With respect to the influence of the conditions of life on any two breeds which are allowed to cross freely, unless both are indigenous and have long been accustomed to the country where they live, they will, in all probability, be unequally affected by the conditions, and this will modify the result. Even with indigenous breeds, it will rarely or never occur that both are equally well adapted to the surrounding circumstances; more especially when permitted to roam freely, and not carefully tended, as is generally the case with breeds allowed to cross. As a consequence of this, natural selection will to a certain extent come into action, and the best fitted will survive, and this will aid in determining the ultimate character of the commingled body.

How long a time it would require before such a crossed body of animals would assume a uniform character within a limited area, no one can say; that they would ultimately become uniform from free intercrossing, and from the survival of the fittest, we may feel assured; but the characters thus acquired would rarely or never, as may be inferred from the previous considerations, be exactly intermediate between those of the two parent-breeds. With respect to the very slight differences by which the individuals of the same sub-variety, or even of allied varieties, are characterised, it is obvious that free crossing would soon obliterate such small distinctions. The formation of new varieties, independently of selection, would also thus be prevented; except when the same variation continually recurred from the action of some strongly predisposing cause. We may therefore conclude that free crossing has in all cases played an important part in giving uniformity of character to all the members of the same domestic race and of the same natural species, though largely governed by natural selection and by the direct action of the surrounding conditions.

## ON THE POSSIBILITY OF ALL ORGANIC BEINGS OCCASIONALLY INTERCROSSING.

But it may be asked, can free crossing occur with hermaphrodite animals and plants? All the higher animals, and the few insects which have been domesticated, have separate sexes, and must inevitably unite for each birth. With respect to the crossing of hermaphrodites, the subject is too large for the present volume, but in the 'Origin of Species' I have given a short abstract of the reasons which induce me to believe that all organic beings occasionally cross, though perhaps in some cases only at long intervals of time. (15/14. With respect to plants, an admirable essay on this subject (Die Geschlechter-Vertheilung bei den Pflanzen: 1867) has been published by Dr. Hildebrand who arrives at the same general conclusions as I have done. Various other treatises have since appeared on the same subject, more especially by Hermann Muller and Delpino.) I will merely recall the fact that many plants, though hermaphrodite in structure, are unisexual in function;—such as those called by C.K. Sprengel DICHOGAMOUS, in which the pollen and stigma of the same flower are matured at different periods; or those called by me RECIPROCALLY DIMORPHIC, in which the flower's own pollen is not fitted to fertilise its own stigma; or again, the many kinds in which curious mechanical contrivances exist, effectually preventing self-fertilisation. There are, however, many hermaphrodite plants which are not in any way specially constructed to favour intercrossing, but which nevertheless commingle almost as freely as animals with separated sexes. This is the case with cabbages, radishes, and onions, as I know from having experimented on them: even the peasants of Liguria say that cabbages must be prevented "from falling in love" with each other. In the orange tribe, Gallesio (15/15. 'Teoria della Riproduzione Vegetal' 1816 page 12.) remarks that the amelioration of the various kinds is checked by their continual and almost regular crossing. So it is with numerous other plants.

On the other hand, some cultivated plants rarely or never intercross, for instance, the common pea and sweet-pea (Lathyrus odoratus); yet their flowers are certainly adapted for cross fertilisation. The varieties of the tomato and aubergine (Solanum) and the pimenta (Pimenta vulgaris?) are said (15/16. Verlot 'Des Varietes' 1865 page 72.) never to cross, even when growing alongside one another. But it should be observed that these are all exotic plants, and we do not know how they would behave in their native country when visited by the proper insects. With respect to the common pea, I have ascertained that it is rarely crossed in this country owing to premature fertilisation. There exist, however, some plants which under their natural conditions appear to be always self-fertilised, such as the Bee Ophrys (Ophrys apifera) and a few other Orchids; yet these plants exhibit the plainest adaptations for cross-fertilisation. Again, some few plants are believed to produce only closed flowers, called cleistogene, which cannot possibly be crossed. This was long thought to be the case with the Leersia oryzoides (15/17. Duval Jouve 'Bull. Soc. Bot. de France' tome 10 1863 page 194. With respect to the perfect flowers setting seed see Dr. Ascherson in 'Bot. Zeitung' 1864 page 350.), but this grass is now known occasionally to produce perfect flowers, which set seed.

Although some plants, both indigenous and naturalised, rarely or never produce flowers, or if they flower never produce seeds, yet no one doubts that phanerogamic plants are adapted to produce flowers, and the flowers to produce seed. When they fail, we believe that such plants under different conditions would perform their proper function, or that they formerly did so, and will do so again. On analogous grounds, I believe that the flowers in the above specified anomalous cases which do not now intercross, either would do so occasionally under different conditions, or that they formerly did so—the means for affecting this being generally still retained—and will again intercross at some future period, unless indeed they become extinct. On this view alone, many points in the structure and action of the reproductive organs in hermaphrodite plants and animals are intelligible,— for instance, the fact of the male and female organs never being so completely enclosed as to render access from without impossible. Hence we may conclude that the most important of all the means for giving uniformity to the individuals of the same species, namely, the capacity of occasionally intercrossing, is present, or has been formerly present, with all organic beings, except, perhaps, some of the lowest.

## [ON CERTAIN CHARACTERS NOT BLENDING.

When two breeds are crossed their characters usually become intimately fused together; but some characters refuse to blend, and are transmitted in an unmodified state either from both parents or from one. When grey and white mice are paired, the young are piebald, or pure white or grey, but not of an intermediate tint; so it is when white and common collared turtle-doves are paired. In breeding Game fowls, a great authority, Mr. J. Douglas, remarks, "I may here state a strange fact: if you cross a black with a white game, you get birds of both breeds of the clearest colour." Sir R. Heron crossed during many years white, black, brown, and fawn-coloured Angora rabbits, and never once got these colours mingled in the same animal, but often all four colours in the same litter. (15/18. Extract of a letter from Sir R. Heron 1838 given me by Mr. Yarrell. With respect to mice see 'Annal. des Sc. Nat.' tome 1 page 180; and I have heard of other similar cases. For turtle-doves Boitard and Corbie 'Les Pigeons' etc. page 238. For the Game fowl 'The Poultry Book' 1866 page 128. For crosses of tailless fowls see Bechstein 'Naturges. Deutsch.' b. 3 s. 403. Bronn 'Geschichte der Natur' b. 2 s. 170 gives analogous facts with horses. On the hairless condition of crossed South American dogs see Rengger 'Saugethiere von Paraguay' s. 152; but I saw in the Zoological Gardens mongrels, from a similar cross, which were hairless, quite hairy, or hairy in patches, that is, piebald with hair. For crosses of Dorking and other fowls see 'Poultry Chronicle' volume 2 page 355. About the crossed pigs, extract of letter from Sir R. Heron to Mr. Yarrell. For other cases see P. Lucas 'L'Hered. Nat.' tome 1 page 212.) From cases like these, in which the colours of the two parents are transmitted quite separately to the offspring, we have all sorts of gradations, leading to complete fusion. I will give an instance: a gentleman with a fair complexion, light hair but dark eyes, married a lady with dark hair and complexion: their three children have very light hair, but on careful search about a dozen black hairs were found scattered in the midst of the light hair on the heads of all three.

When turnspit dogs and ancon sheep, both of which have dwarfed limbs, are crossed with common breeds, the offspring are not intermediate in structure, but take after either parent. When tailless or hornless animals are crossed with perfect animals, it frequently, but by no means invariably, happens that the offspring are either furnished with these organs in a perfect state, or are quite destitute of them. According to Rengger, the hairless condition of the Paraguay dog is either perfectly or not at all transmitted to its mongrel offspring; but I have seen one partial exception in a dog of this parentage which had part of its skin hairy, and part naked, the parts being distinctly separated as in a piebald animal. When Dorking fowls with five toes are crossed with other breeds, the chickens often have five toes on one foot and four on the other. Some crossed pigs raised by Sir R. Heron between the solid- hoofed and common pig had not all four feet in an intermediate condition, but two feet were furnished with properly divided, and two with united hoofs.

Analogous facts have been observed with plants: Major Trevor Clarke crossed the little, glabrous-leaved, annual stock (Matthiola), with pollen of a large, red-flowered, rough-leaved, biennial stock, called cocardeau by the French, and the result was that half the seedlings had glabrous and the other half rough leaves, but none had leaves in an intermediate state. That the glabrous seedlings were the product of the rough-leaved variety, and not accidentally of the mother-plant's own pollen, was shown by their tall and strong habit of growth. (15/19. 'Internat. Hort. and Bot. Congress of London' 1866.) in the succeeding generations raised from the rough-leaved crossed seedlings, some glabrous plants appeared, showing that the glabrous character, though incapable of blending with and modifying the rough leaves, was all the time latent in this family of plants. The numerous plants formerly referred to, which I raised from reciprocal crosses between the peloric and common Antirrhinum, offer a nearly parallel case; for in the first generation all the plants resembled the common form, and in the next generation, out of one hundred and thirty-seven plants, two alone were in an intermediate condition, the others perfectly resembling either the peloric or common form. Major Trevor Clarke also fertilised the above-mentioned red-flowered stock with pollen from the purple Queen stock, and about half the seedlings scarcely differed in habit, and not at all in the red colour of the flower, from the mother-plant, the other half bearing blossoms of a rich purple, closely like those of the paternal plant. Gartner crossed many white and yellow-flowered species and varieties of Verbascum; and these colours were never blended, but the offspring bore either pure white or pure yellow blossoms; the former in the larger proportion. (15/20. 'Bastarderzeugung' s. 307. Kolreuter 'Dritte Fortsetszung' s. 34, 39 however, obtained intermediate tints from similar crosses in the genus Verbascum. With respect to the turnips see Herbert 'Amaryllidaceae' 1837 page 370.) Dr. Herbert raised many seedlings, as he informed me, from Swedish turnips crossed by two other varieties, and these never produced flowers of an intermediate tint, but always like one of their parents. I fertilised the purple sweet-pea (Lathyrus odoratus), which has a dark reddish-purple standard-petal and violet-coloured wings and keel, with pollen of the painted lady sweet-pea, which has a pale cherry-coloured standard, and almost white wings and keel; and from the same pod I twice raised plants perfectly resembling both sorts; the greater number resembling the father. So perfect was the resemblance, that I should have thought there had been some mistake, if the plants which were at first identical with the paternal variety, namely, the painted-lady, had not later in the season produced, as mentioned in a former chapter, flowers blotched and streaked with dark purple. I raised grandchildren and great-grandchildren from these crossed plants, and they continued to resemble the painted-lady, but during later generations became rather more blotched with purple, yet none reverted completely to the original mother-plant, the purple sweet-pea. The following case is slightly different, but still shows the same principle: Naudin (15/21. 'Nouvelles Archives du Museum' tome 1 page 100.) raised numerous hybrids between the yellow Linaria vulgaris and the purple L. purpurea, and during three successive generations the colours kept distinct in different parts of the same flower.

From cases such as the foregoing, in which the offspring of the first generation perfectly resemble either parent, we come by a small step to those cases in which differently coloured flowers borne on the same root resemble both parents, and by another step to those in which the same flower or fruit is striped or blotched with the two parental colours, or bears a single stripe of the colour or other characteristic quality of one of the parent-forms. With hybrids and mongrels it frequently or even generally happens that one part of the body resembles more or less closely one parent and another part the other parent; and here again some resistence to fusion, or, what comes to the same thing, some mutual affinity between the organic atoms of the same nature, apparently comes into play, for otherwise all parts of the body would be equally intermediate in character. So again, when the offspring of hybrids or mongrels, which are themselves nearly intermediate in character, revert either wholly or by segments to their ancestors, the principle of the affinity of similar, or the repulsion of dissimilar atoms, must come into action. To this principle, which seems to be extremely general, we shall recur in the chapter on pangenesis.

It is remarkable, as has been strongly insisted upon by Isidore Geoffroy St. Hilaire in regard to animals, that the transmission of characters without fusion occurs very rarely when species are crossed; I know of one exception alone, namely, with the hybrids naturally produced between the common and hooded crow (Corvus corone and cornix), which, however, are closely allied species, differing in nothing except colour. Nor have I met with any well-ascertained cases of transmission of this kind, even when one form is strongly prepotent over another, when two races are crossed which have been slowly formed by man's selection, and therefore resemble to a certain extent natural species. Such cases as puppies in the same litter closely resembling two distinct breeds, are probably due to superfoetation,—that is, to the influence of two fathers. All the characters above enumerated, which are transmitted in a perfect state to some of the offspring and not to others,— such as distinct colours, nakedness of skin, smoothness of leaves, absence of horns or tail, additional toes, pelorism, dwarfed structure, etc.,—have all been known to appear suddenly in individual animals and plants. From this fact, and from the several slight, aggregated differences which distinguish domestic races and species from one another, not being liable to this peculiar form of transmission, we may conclude that it is in some way connected with the sudden appearance of the characters in question.]

## ON THE MODIFICATION OF OLD RACES AND THE FORMATION OF NEW RACES BY CROSSING.

We have hitherto chiefly considered the effects of crossing in giving uniformity of character; we must now look to an opposite result. There can be no doubt that crossing, with the aid of rigorous selection during several generations, has been a potent means in modifying old races, and in forming new ones. Lord Orford crossed his famous stud of greyhounds once with the bulldog, in order to give them courage and perseverance. Certain pointers have been crossed, as I hear from the Rev. W.D. Fox, with the foxhound, to give them dash and speed. Certain strains of Dorking fowls have had a slight infusion of Game blood; and I have known a great fancier who on a single occasion crossed his turbit-pigeons with barbs, for the sake of gaining greater breadth of beak.

In the foregoing cases breeds have been crossed once, for the sake of modifying some particular character; but with most of the improved races of the pig, which now breed true, there have been repeated crosses,—for instance, the improved Essex owes its excellence to repeated crosses with the Neapolitan, together probably with some infusion of Chinese blood. (15/22. Richardson 'Pigs' 1847 pages 37, 42; S. Sidney's edition of 'Youatt on the Pig' 1860 page 3.) So with our British sheep: almost all the races, except the Southdown, have been largely crossed; "this, in fact, has been the history of our principal breeds." (15/23. See Mr. W.C. Spooner's excellent paper on Cross-Breeding 'Journal Royal Agricult. Soc.' volume 20 part 2: see also an equally good article by Mr. Ch. Howard in 'Gardener's Chronicle' 1860 page 320.) To give an example, the "Oxfordshire Downs" now rank as an established breed. (15/24. 'Gardener's Chronicle' 1857 pages 649, 652.) They were produced about the year 1830 by crossing "Hampshire and in some instances Southdown ewes with Cotswold rams:" now the Hampshire ram was itself produced by repeated crosses between the native Hampshire sheep and Southdowns; and the long-woolled Cotswold were improved by crosses with the Leicester, which latter again is believed to have been a cross between several long-woolled sheep. Mr. Spooner, after considering the various cases which have been carefully recorded, concludes, "that from a judicious pairing of cross-bred animals it is practicable to establish a new breed." On the continent the history of several crossed races of cattle and of other animals has been well ascertained. To give one instance: the King of Wurtemburg, after twenty-five years' careful breeding, that is, after six or seven generations, made a new breed of cattle from a cross between a Dutch and a Swiss breed, combined with other breeds. (15/25. 'Bulletin de La Soc. d'Acclimat.' 1862 tome 9 page 463. See also for other cases MM. Moll and Gayot 'Du Boeuf' 1860 page 32.) The Sebright bantam, which breeds as true as any other kind of fowl, was formed about sixty years ago by a complicated cross. (15/26. 'Poultry Chronicle' volume 2 1854 page 36.) Dark Brahmas, which are believed by some fanciers to constitute a distinct species, were undoubtedly formed (15/27. 'The Poultry Book' by W.B. Tegetmeier 1866 page 58.) in the United States, within a recent period, by a cross between Chittagongs and Cochins. With plants there is little doubt that the Swede-turnip originated from a cross; and the history of a variety of wheat, raised from two very distinct varieties, and which after six years' culture presented an even sample, has been recorded on good authority. (15/28. 'Gardener's Chronicle' 1852 page 765.)

Until lately, cautious and experienced breeders, though not averse to a single infusion of foreign blood, were almost universally convinced that the attempt to establish a new race, intermediate between two widely distinct races, was hopeless "they clung with superstitious tenacity to the doctrine of purity of blood, believing it to be the ark in which alone true safety could be found." (15/29. Spooner in 'Journal Royal Agricult. Soc.' volume 20 part 2) Nor was this conviction unreasonable: when two distinct races are crossed, the offspring of the first generation are generally nearly uniform in character; but even this sometimes fails to be the case, especially with crossed dogs and fowls, the young of which from the first are sometimes much diversified. As cross-bred animals are generally of large size and vigorous, they have been raised in great numbers for immediate consumption. But for breeding they are found utterly useless; for though they may themselves be uniform in character, they yield during many generations astonishingly diversified offspring. The breeder is driven to despair, and concludes that he will never form an intermediate race. But from the cases already given, and from others which have been recorded, it appears that patience alone is necessary; as Mr. Spooner remarks, "nature opposes no barrier to successful admixture; in the course of time, by the aid of selection and careful weeding, it is practicable to establish a new breed." After six or seven generations the hoped-for result will in most cases be obtained; but even then an occasional reversion, or failure to keep true, may be expected. The attempt, however, will assuredly fail if the conditions of life be decidedly unfavourable to the characters of either parent-breed. (15/30. See Colin 'Traite de Phys. Comp. des Animaux Domestiques' tome 2 page 536, where this subject is well treated.)

Although the grandchildren and succeeding generations of cross-bred animals are generally variable in an extreme degree, some curious exceptions to the rule have been observed both with crossed races and species. Thus

Boitard and Corbie (15/31. 'Les Pigeons' page 37.) assert that from a Pouter and a Runt "a Cavalier will appear, which we have classed amongst pigeons of pure race, because it transmits all its qualities to its posterity." The editor of the 'Poultry Chronicle' (15/32. Volume 1 1854 page 101.) bred some bluish fowls from a black Spanish cock and a Malay hen; and these remained true to colour "generation after generation." The Himalayan breed of rabbits was certainly formed by crossing two sub-varieties of the silver-grey rabbit; although it suddenly assumed its present character, which differs much from that of either parent-breed, yet it has ever since been easily and truly propagated. I crossed some Labrador and Penguin ducks, and recrossed the mongrels with Penguins; afterwards most of the ducks reared during three generations were nearly uniform in character, being brown with a white crescentic mark on the lower part of the breast, and with some white spots at the base of the beak; so that by the aid of a little selection a new breed might easily have been formed. With regard to crossed varieties of plants, Mr. Beaton (15/33. 'Cottage Gardener' 1856 page 110.) remarks that "Melville's extraordinary cross between the Scotch kale and an early cabbage is as true and genuine as any on record;" but in this case no doubt selection was practised. Gartner (15/34. 'Bastarderzeugung' s. 553.) has given five cases of hybrids, in which the progeny kept constant; and hybrids between Dianthus armeria and deltoides remained true and uniform to the tenth generation. Dr. Herbert likewise showed me a hybrid from two species of Loasa which from its first production had kept constant during several generations.

We have seen in the first chapter, that the several kinds of dogs are almost certainly descended from more than one species, and so it is with cattle, pigs and some other domesticated animals. Hence the crossing of aboriginally distinct species probably came into play at an early period in the formation of our present races. From Rutimeyer's observations there can be little doubt that this occurred with cattle; but in most cases one form will probably have absorbed and obliterated the other, for it is not likely that semi-civilised men would have taken the necessary pains to modify by selection their commingled, crossed, and fluctuating stock. Nevertheless, those animals which were best adapted to their conditions of life would have survived through natural selection; and by this means crossing will often have indirectly aided in the formation of primeval domesticated breeds. Within recent times, as far as animals are concerned, the crossing of distinct species has done little or nothing towards the formation or modification of our races. It is not yet known whether the several species of silk-moth which have been recently crossed in France will yield permanent races. With plants which can be multiplied by buds and cuttings, hybridisation has done wonders, as with many kinds of Roses, Rhododendrons, Pelargoniums, Calceolarias, and

Petunias. Nearly all these plants can be propagated by seed, most of them freely; but extremely few or none come true by seed.

Some authors believe that crossing is the chief cause of variability,—that is, of the appearance of absolutely new characters. Some have gone so far as to look at it as the sole cause; but this conclusion is disproved by the facts given in the chapter on Bud-variation. The belief that characters not present in either parent or in their ancestors frequently originate from crossing is doubtful; that they occasionally do so is probable; but this subject will be more conveniently discussed in a future chapter on the causes of Variability.

A condensed summary of this and of the three following chapters, together with some remarks on Hybridism, will be given in the nineteenth chapter.

# CHAPTER 2.XVI.

CAUSES WHICH INTERFERE WITH THE FREE CROSSING OF VARIETIES—INFLUENCE OF DOMESTICATION ON FERTILITY.

DIFFICULTIES IN JUDGING OF THE FERTILITY OF VARIETIES WHEN CROSSED. VARIOUS CAUSES WHICH KEEP VARIETIES DISTINCT, AS THE PERIOD OF BREEDING AND SEXUAL PREFERENCE. VARIETIES OF WHEAT SAID TO BE STERILE WHEN CROSSED. VARIETIES OF MAIZE, VERBASCUM, HOLLYHOCK, GOURDS, MELONS, AND TOBACCO, RENDERED IN SOME DEGREE MUTUALLY STERILE. DOMESTICATION ELIMINATES THE TENDENCY TO STERILITY NATURAL TO SPECIES WHEN CROSSED. ON THE INCREASED FERTILITY OF UNCROSSED ANIMALS AND PLANTS FROM DOMESTICATION AND CULTIVATION.

The domesticated races of both animals and plants, when crossed, are, with extremely few exceptions, quite prolific,—in some cases even more so than the purely-bred parent-races. The offspring, also, raised from such crosses are likewise, as we shall see in the following chapter, generally more vigorous and fertile than their parents. On the other hand, species when crossed, and their hybrid offspring, are almost invariably in some degree sterile; and here there seems to exist a broad and insuperable distinction between races and species. The importance of this subject as bearing on the origin of species is obvious; and we shall hereafter recur to it.

It is unfortunate how few precise observations have been made on the fertility of mongrel animals and plants during several successive generations. Dr. Broca (16/1. 'Journal de Physiolog.' tome 2 1859 page 385.) has remarked that no one has observed whether, for instance, mongrel dogs, bred inter se, are indefinitely fertile; yet, if a shade of infertility be detected by careful observation in the offspring of natural forms when crossed, it is thought that their specific distinction is proved. But so many breeds of sheep, cattle, pigs, dogs, and poultry, have been crossed and recrossed in various ways, that any sterility, if it had existed, would from being injurious almost certainly have been observed. In investigating the fertility of crossed varieties many sources of doubt occur. Whenever the least trace of sterility between two plants, however closely allied, was observed by Kolreuter, and more especially by Gartner, who counted the exact number of seed in each capsule, the two forms were at once ranked as distinct species; and if this rule be followed, assuredly it will never be proved that varieties when crossed are in any degree sterile. We have formerly seen that certain breeds of dogs do not readily pair together; but no observations have been made whether, when paired, they produce the full number of young, and whether the latter are perfectly fertile inter se; but, supposing that some degree of sterility were found to exist, naturalists would simply infer that these breeds were descended from aboriginally distinct species; and it would be scarcely possible to ascertain whether or not this explanation was the true one.

The Sebright Bantam is much less prolific than any other breed of fowls, and is descended from a cross between two very distinct breeds, recrossed by a third sub-variety. But it would be extremely rash to infer that the loss of fertility was in any manner connected with its crossed origin, for it may with more probability be attributed either to long-continued close interbreeding, or to an innate tendency to sterility correlated with the absence of hackles and sickle tail-feathers.

Before giving the few recorded cases of forms, which must be ranked as varieties, being in some degree sterile when crossed, I may remark that other causes sometimes interfere with varieties freely intercrossing. Thus they may differ too greatly in size, as with some kinds of dogs and fowls: for instance, the editor of the 'Journal of Horticulture, etc.' (16/2. December 1863 page 484.) says that he can keep Bantams with the larger breeds without much danger of their crossing, but not with the smaller breeds, such as Games, Hamburghs, etc. With plants a difference in the period of flowering serves to keep varieties distinct, as with the various kinds of maize and wheat: thus Colonel Le Couteur (16/3. On 'The Varieties of Wheat' page 66.) remarks, "the Talavera wheat, from flowering much earlier than any other kind, is sure to continue pure." In different parts of the Falkland Islands the cattle are breaking up into herds of different colours; and those on the higher ground, which are generally white, usually breed, as I am informed by Sir J. Sulivan, three months earlier than those on the lowland; and this would manifestly tend to keep the herds from blending.

Certain domestic races seem to prefer breeding with their own kind; and this is a fact of some importance, for it is a step towards that instinctive feeling which helps to keep closely allied species in a state of nature distinct. We have now abundant evidence that, if it were not for this feeling, many more hybrids would be naturally produced than in this case. We have seen in the first chapter that the alco dog of Mexico dislikes dogs of other breeds; and the hairless dog of Paraguay mixes less readily with the European races, than the latter do with each other. In Germany the female Spitz-dog is said to receive the fox more readily than will other dogs; a female Australian Dingo in England attracted the wild male foxes. But these differences in the sexual instinct and attractive power of the various breeds may be wholly due to their descent from distinct species. In Paraguay the horses have much freedom, and an excellent observer (16/4. Rengger 'Saugethiere von Paraguay' s. 336.)

believes that the native horses of the same colour and size prefer associating with each other, and that the horses which have been imported from Entre Rios and Banda Oriental into Paraguay likewise prefer associating together. In Circassia six sub-races of the horse have received distinct names; and a native proprietor of rank (16/5. See a memoir by MM. Lherbette and De Quatrefages in 'Bull. Soc. d'Acclimat.' tome 8 July 1861 page 312.) asserts that horses of three of these races, whilst living a free life, almost always refuse to mingle and cross, and will even attack one another.

It has been observed, in a district stocked with heavy Lincolnshire and light Norfolk sheep, that both kinds; though bred together, when turned out, "in a short time separate to a sheep;" the Lincolnshires drawing off to the rich soil, and the Norfolks to their own dry light soil; and as long as there is plenty of grass, "the two breeds keep themselves as distinct as rooks and pigeons." In this case different habits of life tend to keep the races distinct. On one of the Faroe islands, not more than half a mile in diameter, the half-wild native black sheep are said not to have readily mixed with the imported white sheep. It is a more curious fact that the semi-monstrous ancon sheep of modern origin "have been observed to keep together, separating themselves from the rest of the flock, when put into enclosures with other sheep." (16/6. For the Norfolk sheep see Marshall 'Rural Economy of Norfolk' volume 2 page 136. See Rev. L. Landt 'Description of Faroe' page 66. For the ancon sheep see 'Phil. Transact.' 1813 page 90.) With respect to fallow-deer, which live in a semi-domesticated condition, Mr. Bennett (16/7. White 'Nat. Hist. of Selbourne' edited by Bennett page 39. With respect to the origin of the dark-coloured deer see 'Some Account of English Deer Parks' by E.P. Shirley, Esq.) states that the dark and pale coloured herds, which have long been kept together in the Forest of Dean, in High Meadow Woods, and in the New Forest, have never been known to mingle: the dark-coloured deer, it may be added, are believed to have been first brought by James I. from Norway, on account of their greater hardiness. I imported from the island of Porto Santo two of the feral rabbits, which differ, as described in the fourth chapter, from common rabbits; both proved to be males, and, though they lived during some years in the Zoological Gardens, the superintendent, Mr. Bartlett, in vain endeavoured to make them breed with various tame kinds; but whether this refusal to breed was due to any change in the instinct, or simply to their extreme wildness, or whether confinement had rendered them sterile, as often occurs, cannot be determined.

Whilst matching for the sake of experiment many of the most distinct breeds of pigeons, it frequently appeared to me that the birds, though faithful to their marriage vow, retained some desire after their own kind. Accordingly I asked Mr. Wicking, who has kept a larger stock of various breeds together than any man in England, whether he thought that they would prefer pairing with their own kind, supposing that there were males and females enough of each; and he without hesitation answered that he was convinced that this was the case. It has often been noticed that the dovecote pigeon seems to have an actual aversion towards the several fancy breeds (16/8. 'The Dovecote' by the Rev. E.S. Dixon page 155; Bechstein 'Naturgesch. Deutschlands' b. 4 1795 page 17.) yet all have certainly sprung from a common progenitor. The Rev. W.D. Fox informs me that his flocks of white and common Chinese geese kept distinct.

These facts and statements, though some of them are incapable of proof, resting only on the opinion of experienced observers, show that some domestic races are led by different habits of life to keep to a certain extent separate, and that others prefer coupling with their own kind, in the same manner as species in a state of nature, though in a much less degree.

[With respect to sterility from the crossing of domestic races, I know of no well-ascertained case with animals. This fact, seeing the great difference in structure between some breeds of pigeons, fowls, pigs, dogs, etc., is extraordinary, in contrast with the sterility of many closely allied natural species when crossed; but we shall hereafter attempt to show that it is not so extraordinary as it at first appears. And it may be well here to recall to mind that the amount of external difference between two species is not a safe guide for predicting whether or not they will breed together,—some closely allied species when crossed being utterly sterile, and others which are extremely unlike being moderately fertile. I have said that no case of sterility in crossed races rests on satisfactory evidence; but here is one which at first seems trustworthy. Mr. Youatt (16/9. 'Cattle' page 202.) and a better authority cannot be quoted, states, that formerly in Lancashire crosses were frequently made between longhorn and shorthorn cattle; the first cross was excellent, but the produce was uncertain; in the third or fourth generation the cows were bad milkers; "in addition to which, there was much uncertainty whether the cows would conceive; and full one-third of the cows among some of these half-breds failed to be in calf." This at first seems a good case: but Mr. Wilkinson states (16/10. Mr. J. Wilkinson in 'Remarks addressed to Sir J. Sebright' 1820 page 38.), that a breed derived from this same cross was actually established in another part of England; and if it had failed in fertility, the fact would surely have been noticed. Moreover, supposing that Mr. Youatt had proved his case, it might be argued that the sterility was wholly due to the two parent-breeds being descended from primordially distinct species.

In the case of plants Gartner states that he fertilised thirteen heads (and subsequently nine others) on a dwarf maize bearing yellow seed (16/11. 'Bastarderzeugung' s. 87, 169. See also the Table at the end of volume.) with pollen of a tall maize having red seed; and one head alone produced good seed, but only five in number. Though these plants are monoecious, and therefore do not require castration, yet I should have suspected some accident in the manipulation, had not Gartner expressly stated that he had during many years grown these two varieties together, and they did not spontaneously cross; and this, considering that the plants are monoecious and abound with pollen, and are well known generally to cross freely, seems explicable only on the belief that these two varieties are in some degree mutually infertile. The hybrid plants raised from the above five seeds were intermediate in structure, extremely variable, and perfectly fertile. (16/12. 'Bastarderzeugung' s. 87, 577.) In like manner Prof. Hildebrand (16/13. 'Bot. Zeitung' 1868 page 327.) could not succeed in fertilising the female flowers of a plant bearing brown grains with pollen from a certain kind bearing yellow grains; although other flowers on the same plant, which were fertilised with their own pollen, yielded good seed. No one, I believe, even suspects that these varieties of maize are distinct species; but had the hybrids been in the least sterile, no doubt Gartner would at once have so classed them. I may here remark, that with undoubted species there is not necessarily any close relation between the sterility of a first cross and that of the hybrid offspring. Some species can be crossed with facility, but produce utterly sterile hybrids; others can be crossed with extreme difficulty, but the hybrids when produced are moderately fertile. I am not aware, however, of any instance quite like this of the maize, namely, of a first cross made with difficulty, but yielding perfectly fertile hybrids. (16/14. Mr. Shirreff formerly thought ('Gardener's Chronicle' 1858 page 771) that the offspring from a cross between certain varieties of wheat became sterile in the fourth generation; but he now admits ('Improvement of the Cereals' 1873) that this was an error.)

The following case is much more remarkable, and evidently perplexed Gartner, whose strong wish it was to draw a broad line of distinction between species and varieties. In the genus Verbascum, he made, during eighteen years, a vast number of experiments, and crossed no less than 1085 flowers and counted their seeds. Many of these experiments consisted in crossing white and yellow varieties of both V. lychnitis and V. blattaria with nine other species and their hybrids. That the white and yellow flowered plants of these two species are really varieties, no one has doubted; and Gartner actually raised in the case of both species one variety from the seed of the other. Now in two of his works (16/15. 'Kenntniss der Befruchtung' s. 137; 'Bastarderzeugung' s. 92, 181. On raising the two varieties from seed see s. 307.) he distinctly asserts that crosses between similarly-coloured flowers yield more seed than between dissimilarly-coloured; so that the yellow-flowered variety of either species (and conversely with the white-flowered variety), when crossed with pollen of its own kind, yields more seed than when crossed with that of the white variety; and so it is when differently coloured species are crossed. The general results may be seen in the Table at the end of his volume. In one instance he gives (16/16. 'Bastarderzeugung' s. 216.) the following details; but I must premise that Gartner, to avoid exaggerating the degree of sterility in his crosses, always compares the MAXIMUM number obtained from a cross with the AVERAGE number naturally given by the pure mother-plant. The white variety of V. lychnitis, naturally fertilised by its own pollen, gave from an AVERAGE of twelve capsules ninety-six good seeds in each; whilst twenty flowers fertilised with pollen from the yellow variety of this same species, gave as the MAXIMUM only eighty-nine good seeds; so that we have the proportion of 1000 to 908, according to Gartner's usual scale. I should have thought it possible that so small a difference in fertility might have been accounted for by the evil effects of the necessary castration; but Gartner shows that the white variety of V. lychnitis, when fertilised first by the white variety of V. blattaria, and then by the yellow variety of this species, yielded seed in the proportion of 622 to 438; and in both these cases castration was performed. Now the sterility which results from the crossing of the differently coloured varieties of the same species, is fully as great as that which occurs in many cases when distinct species are crossed. Unfortunately Gartner compared the results of the first unions alone, and not the sterility of the two sets of hybrids produced from the white variety of V. lychnitis when fertilised by the white and yellow varieties of V. blattaria, for it is probable that they would have differed in this respect.

Mr. J. Scott has given me the results of a series of experiments on Verbascum, made by him in the Botanic Gardens of Edinburgh. (16/17. The results have since been published in 'Journ. Asiatic Soc. of Bengal' 1867 page 145.) He repeated some of Gartner's experiments on distinct species, but obtained only fluctuating results, some confirmatory, the greater number contradictory; nevertheless these seem hardly sufficient to overthrow the conclusion arrived at by Gartner from experiments tried on a larger scale. Mr. Scott also experimented on the relative fertility of unions between similarly and dissimilarly-coloured varieties of the same species. Thus he fertilised six flowers of the yellow variety of V. lychnitis by its own pollen, and obtained six capsules; and calling, for the sake of comparison, the average number of good seed in each of their capsules one hundred, he found that this same yellow variety, when fertilised by the white variety, yielded from seven capsules an average of ninety-four seed. On the same principle, the white variety of V. lychnitis by its own pollen (from six capsules), and by the pollen of the yellow variety (eight capsules), yielded seed in the proportion of 100 to 82. The yellow variety of V. thapsus by its own pollen (eight capsules), and by that of the white variety (only two capsules), yielded seed in the proportion of 100 to 94. Lastly, the white variety of V. blattaria by its own pollen (eight capsules), and by that of the yellow variety (five capsules), yielded seed in the proportion of 100 to 79. So that in every case the unions of similarly-coloured varieties of the same species were more fertile than the unions of dissimilarly-coloured varieties; when all the cases are grouped together, the difference of fertility is as 100 to 86. Some additional trials were made, and altogether thirty-six similarly-coloured unions yielded thirty-five good capsules; whilst thirty-five dissimilarly- coloured unions yielded only twenty-six good capsules. Besides the foregoing experiments, the purple V. phoeniceum was crossed by a rose-coloured and a white variety of the same species; these two varieties were also crossed together, and these several unions yielded less seed than V. phoeniceum by its own pollen. Hence it follows from Mr. Scott's experiments, that in the genus Verbascum the similarly and dissimilarly-coloured varieties of the same species behave, when crossed, like closely allied but distinct species. (16/18. The following facts, given by Kolreuter in his 'Dritte Fortsetzung' ss. 34, 39, appear at first sight strongly to confirm Mr. Scott's and Gartner's statements; and to a certain limited extent they do so. Kolreuter asserts, from innumerable observations, that insects incessantly carry pollen from one species and variety of Verbascum to another; and I can confirm this assertion; yet he found that the white and yellow varieties of Verbascum lychnitis often grew wild mingled together: moreover, he cultivated these two varieties in considerable numbers during four years in his garden, and they kept true by seed; but when he crossed them, they produced flowers of an intermediate tint. Hence it might have been thought that both varieties must have a stronger elective affinity for the pollen of their own variety than for that of the other; this elective affinity, I may add of each species for its own pollen (Kolreuter 'Dritte Forts.' s. 39 and Gartner 'Bastarderz.' passim) being a perfectly well-ascertained power. But the force of the foregoing facts is much lessened by Gartner's numerous experiments, for, differently from Kolreuter, he never once got ('Bastarderz.' s. 307) an intermediate tint when he crossed the yellow and white flowered varieties of Verbascum. So that the fact of the white and yellow varieties keeping true to their colour by seed does not prove that they were not mutually fertilised by the pollen carried by insects from one to the other.)

This remarkable fact of the sexual affinity of similarly-coloured varieties, as observed by Gartner and Mr. Scott, may not be of very rare occurrence; for the subject has not been attended to by others. The following case is worth giving, partly to show how difficult it is to avoid error. Dr. Herbert (16/19. 'Amaryllidaceae' 1837 page 366. Gartner has made a similar observation.) has remarked that variously-coloured double varieties of the Hollyhock (Althea rosea) may be raised with certainty by seed from plants growing close together. I have been informed that nurserymen who raise seed for sale do not separate their plants; accordingly I procured seed of eighteen named varieties; of these, eleven varieties produced sixty-two plants all perfectly true to their kind; and seven produced forty-nine plants, half of which were true and half false. Mr. Masters of Canterbury has given me a more striking case; he saved seed from a great bed of twenty-four named varieties planted in closely adjoining rows, and each variety reproduced itself truly with only sometimes a shade of difference in tint. Now in the hollyhock the pollen, which is abundant, is matured and nearly all shed before the stigma of the same flower is ready to receive it (16/20. Kolreuter first observed this fact, 'Mem. de l'Acad. de St. Petersburg' volume 3 page 127. See also C.K. Sprengel 'Das Entdeckte Geheimniss' s. 345.); and as bees covered with pollen incessantly fly from plant to plant, it would appear that adjoining varieties could not escape being crossed. As, however, this does not occur, it appeared to me probable that the pollen of each variety was prepotent on its own stigma over that of all other varieties, but I have no evidence on this point. Mr. C. Turner of Slough, well known for his success in the cultivation of this plant, informs me that it is the doubleness of the flowers which prevents the bees gaining access to the pollen and stigma; and he finds that it is difficult even to cross them artificially. Whether this explanation will fully account for varieties in close proximity propagating themselves so truly by seed, I do not know.

The following cases are worth giving, as they relate to monoecious forms, which do not require, and consequently cannot have been injured by, castration. Girou de Buzareingues crossed what he designates three varieties of gourd (16/21. Namely Barbarines, Pastissons, Giraumous: 'Annal. des Sc. Nat.' tome 30 1833 pages 398 and 405.), and asserts that their mutual fertilisation is less easy in proportion to the difference which they present. I am aware how imperfectly the forms in this group were until recently known; but Sageret (16/22. 'Memoire sur les Cucurbitaceae' 1826 pages 46, 55.), who ranked them according to their mutual fertility, considers the three forms above alluded to as varieties, as does a far higher authority, namely, M. Naudin. (16/23. 'Annales des Sc. Nat.' 4th series tome 6. M. Naudin considers these forms as undoubtedly varieties of Cucurbita pepo.) Sageret (16/24. 'Mem. Cucurb.' page 8.) has observed that certain melons have a greater tendency, whatever the cause may be, to keep true than others; and M. Naudin, who has had such immense experience in this group, informs me that he believes that certain varieties intercross more readily than others of the same species; but he has not proved the truth of this conclusion; the frequent abortion of the pollen near Paris being one great difficulty. Nevertheless, he has grown close together, during seven years, certain forms of Citrullus, which, as they could be artificially crossed with perfect facility and produced fertile offspring, are ranked as varieties; but these forms when not artificially crossed kept true. Many other varieties, on the other hand, in the same group cross with such facility, as M. Naudin repeatedly insists, that without being grown far apart they cannot be kept in the least true.

Another case, though somewhat different, may be here given, as it is highly remarkable, and is established on excellent evidence. Kolreuter minutely describes five varieties of the common tobacco (16/25. 'Zweite Forts.' s. 53 namely Nicotiana major vulgaris; (2) perennis; (3) transylvanica; (4) a sub-var. of the last; (5) major latifol. fl. alb.) which were reciprocally crossed, and the offspring were intermediate in character and as fertile as their parents: from this fact Kolreuter inferred that they are really varieties; and no one, as far as I can discover, seems to have doubted that such is the case. He also crossed reciprocally these five varieties with N. glutinosa, and they yielded very sterile hybrids; but those raised from the var. perennis, whether used as the father or mother plant, were not so sterile as the hybrids from the four other varieties. (16/26. Kolreuter was so much struck with this fact that he suspected that a little pollen of N. glutinosa in one of his experiments might have accidentally got mingled with that of var. perennis, and thus aided its fertilising power. But we now know conclusively from Gartner ('Bastarderz.' s. 34, 43) that the pollen of two species never acts CONJOINTLY on a third species; still less will the pollen of a distinct species, mingled with a plant's own pollen, if the latter be present in sufficient quantity, have any effect. The sole effect of mingling two kinds of pollen is to produce in the same capsule seeds which yield plants, some taking after the one and some after the other parent.) So that the sexual capacity of this one variety has certainly been in some degree modified, so as to approach in nature that of N. glutinosa. (16/27. Mr. Scott has made some observations on the absolute sterility of a purple and white primrose (Primula vulgaris) when fertilised by pollen from the common primrose ('Journal of Proc. of Linn. Soc.' volume 8 1864 page 98); but these observations require confirmation. I raised a number of purple-flowered long- styled seedlings from seed kindly sent me by Mr. Scott, and, though they were all in some degree sterile, they were much more fertile with pollen taken from the common primrose than with their own pollen. Mr. Scott has likewise described a red equal-styled cowslip (P. veris ibid page 106), which was found by him to be highly sterile when crossed with the common cowslip; but this was not the case with several equal-styled red seedlings raised by me from his plant. This variety of the cowslip presents the remarkable peculiarity of combining male organs in every respect like those of the short-styled form, with female organs resembling in function and partly in structure those of the long-styled form; so that we have the singular anomaly of the two forms combined in the same flower. Hence it is not surprising that these flowers should be spontaneously self-fertile in a high degree.)

These facts with respect to plants show that in some few cases certain varieties have had their sexual powers so far modified, that they cross together less readily and yield less seed than other varieties of the same species. We shall presently see that the sexual functions of most animals and plants are eminently liable to be affected by the conditions of life to which they are exposed; and hereafter we shall briefly discuss the conjoint bearing of this fact, and others, on the difference in fertility between crossed varieties and crossed species.

## DOMESTICATION ELIMINATES THE TENDENCY TO STERILITY WHICH IS GENERAL WITH SPECIES WHEN CROSSED.

This hypothesis was first propounded by Pallas (16/28. 'Act. Acad. St. Petersburg' 1780 part 2 pages 84, 100.), and has been adopted by several authors. I can find hardly any direct facts in its support; but unfortunately no one has compared, in the case of either animals or plants, the fertility of anciently domesticated varieties, when crossed with a distinct species, with that of the wild parent-species when similarly crossed. No one has compared, for instance, the fertility of Gallus bankiva and of the domesticated fowl, when crossed with a distinct species of Gallus or Phasianus; and the experiment would in all cases be surrounded by many difficulties. Dureau de la Malle, who has so closely studied classical literature, states (16/29. 'Annales des Sc. Nat.' tome 21 1st series page 61.) that in the time of the Romans the common mule was produced with more difficulty than at the present day; but whether this statement may be trusted I know not. A much more important, though somewhat different, case is given by M. Groenland (16/30. 'Bull. Bot. Soc. de France' December 27, 1861 tome 8 page 612.), namely, that plants, known from their intermediate character and sterility to be hybrids between Aegilops and wheat, have perpetuated themselves under culture since 1857, WITH A RAPID BUT VARYING INCREASE OF FERTILITY IN EACH GENERATION. In the fourth generation the plants, still retaining their intermediate character, had become as fertile as common cultivated wheat.

The indirect evidence in favour of the Pallasian doctrine appears to me to be extremely strong. In the earlier chapters I have shown that our various breeds of the dog are descended from several wild species; and this probably is the case with sheep. There can be no doubt that the Zebu or humped Indian ox belongs to a distinct species from European cattle: the latter, moreover, are descended from two forms, which may be called either species or races. We have good evidence that our domesticated pigs belong to at least two specific types, S. scrofa and indicus. Now a widely extended analogy leads to the belief that if these several allied species, when first reclaimed, had been crossed, they would have exhibited, both in their first unions and in their hybrid offspring, some degree of sterility. Nevertheless, the several domesticated races descended from them are now all, as far as can be ascertained, perfectly fertile together. If this reasoning be trustworthy, and it is apparently sound, we must admit the Pallasian doctrine that long-continued domestication tends to eliminate that sterility which is natural to species when crossed in their aboriginal state.

## ON INCREASED FERTILITY FROM DOMESTICATION AND CULTIVATION.

Increased fertility from domestication, without any reference to crossing, may be here briefly considered. This subject bears indirectly on two or three points connected with the modification of organic beings. As Buffon long ago remarked (16/31. Quoted by Isid. Geoffroy St. Hilaire 'Hist. Naturelle Generale' tome 3 page 476. Since this MS. has been sent to press a full discussion on the present subject has appeared in Mr. Herbert Spencer's 'Principles of Biology' volume 2 1867 page 457 et seq.), domestic animals breed oftener in the year and produce more young at a birth than wild animals of the same species; they, also, sometimes breed at an earlier age. The case would hardly have deserved further notice, had not some authors lately attempted to show that fertility increases and decreases in an inverse ratio with the amount of food. This strange doctrine has apparently arisen from individual animals when supplied with an inordinate quantity of food, and from plants of many kinds when grown on excessively rich soil, as on a dunghill, becoming sterile: but to this latter point I shall have occasion presently to return. With hardly an exception, our domesticated animals, which have been long habituated to a regular and copious supply of food, without the labour of searching for it, are more fertile than the corresponding wild animals. It is notorious how frequently cats and dogs breed, and how many young they produce at a birth. The wild rabbit is said generally to breed four times yearly, and to produce each time at most six young; the tame rabbit breeds six or seven times yearly, producing each time from four to eleven young; and Mr. Harrison Weir tells me of a case of eighteen young having been produced at a birth, all of which survived. The ferret, though generally so closely confined, is more prolific than its supposed wild prototype. The wild sow is remarkably prolific; she often breeds twice in the year, and bears from four to eight and sometimes even twelve young; but the domestic sow regularly breeds twice a year, and would breed oftener if permitted; and a sow that produces less than eight at a birth "is worth little, and the sooner she is fattened for the butcher the better." The amount of food affects the fertility of the same individual: thus sheep, which on mountains never produce more than one lamb at a birth, when brought down to lowland pastures frequently bear twins. This difference apparently is not due to the cold of the higher land, for sheep and other domestic animals are said to be extremely prolific in Lapland. Hard living, also, retards the period at which animals conceive; for it has been found disadvantageous in the northern islands of Scotland to allow cows to bear calves before they are four years old. (16/32. For cats and dogs etc. see Bellingeri in 'Annal. des Sc. Nat.' 2nd series, Zoolog. tome 12 page 155. For ferrets Bechstein 'Naturgeschichte Deutschlands' b. 1 1801 s. 786, 795. For rabbits ditto s. 1123, 1131; and Bronn 'Geschichte der Natur.' b. 2 s. 99. For mountain sheep ditto s. 102. For the fertility of the wild sow, see Bechstein 'Naturgesch. Deutschlands' b. 1 1801 s. 534; for the domestic pig Sidney's edition of 'Youatt on the Pig' 1860 page 62. With respect to Lapland see Acerbi 'Travels to the North Cape' English translation volume 2 page 222. About the Highland cows see 'Hogg on Sheep' page 263.)

[Birds offer still better evidence of increased fertility from domestication: the hen of the wild Gallus bankiva lays from six to ten eggs, a number which would be thought nothing of with the domestic hen. The wild duck lays from five to ten eggs; the tame one in the course of the year from eighty to one hundred. The wild grey-lag goose lays from five to eight eggs; the tame from thirteen to eighteen, and she lays a second time; as Mr. Dixon has remarked, "high-feeding, care, and moderate warmth induce a habit of prolificacy which becomes in some measure hereditary." Whether the semi-domesticated dovecote pigeon is more fertile than the wild rock-pigeon, C. livia, I know not; but the more thoroughly domesticated breeds are nearly twice as fertile as dovecotes: the latter, however, when caged and highly fed, become equally fertile with house pigeons. I hear from Judge Caton that the wild turkey in the United States does not breed when a year old, as the domesticated turkeys there invariably do. The peahen alone of domesticated birds is rather more fertile, according to some accounts, when wild in its native Indian home, than in Europe when exposed to our much colder climate. (16/33. For the eggs of Gallus bankiva see Blyth in 'Annals and Mag. of Nat. Hist.' 2nd series volume 1 1848 page 456. For wild and tame ducks Macgillivray 'British Birds' volume 5 page 37; and 'Die Enten' s. 87. For wild geese L. Lloyd 'Scandinavian Adventures' volume 2 1854 page 413; and for tame geese 'Ornamental Poultry' by Rev. E.S. Dixon page 139. On the breeding of Pigeons Pistor 'Das Ganze der Taubenzucht' 1831 s. 48; and Boitard and Corbie 'Les Pigeons' page 158. With respect to peacocks, according to Temminck 'Hist. Nat. Gen. des Pigeons' etc. 1813 tome 2 page 41, the hen lays in India even as many as twenty eggs; but according to Jerdon and another writer quoted in Tegetmeier 'Poultry Book' 1866 pages 280, 282, she there lays only from four to nine or ten eggs: in England she is said, in the 'Poultry Book' to lay five or six, but another writer says from eight to twelve eggs.)

With respect to plants, no one would expect wheat to tiller more, and each ear to produce more grain, in poor than in rich soil; or to get in poor soil a heavy crop of peas or beans. Seeds vary so much in number that it is difficult to estimate them; but on comparing beds of carrots in a nursery garden with wild plants, the former seemed to produce about twice as much seed. Cultivated cabbages yielded thrice as many pods by measure as wild cabbages from the rocks of South Wales. The excess of berries produced by the cultivated asparagus in comparison with the wild plant is enormous. No doubt many highly cultivated plants, such as pears, pineapples, bananas, sugar-cane, etc., are nearly or quite sterile; and I am inclined to attribute this sterility to excess of food and to other unnatural conditions; but to this subject I shall recur.]

In some cases, as with the pig, rabbit, etc., and with those plants which are valued for their seed, the direct selection of the more fertile individuals has probably much increased their fertility; and in all cases this may have occurred indirectly, from the better chance of some of the numerous offspring from the more fertile individuals having been preserved. But with cats, ferrets, and dogs, and with plants like carrots, cabbages, and asparagus, which are not valued for their prolificacy, selection can have played only a subordinate part; and their increased fertility must be attributed to the more favourable conditions of life under which they have long existed.

# CHAPTER 2.XVII.

ON THE GOOD EFFECTS OF CROSSING, AND ON THE EVIL EFFECTS OF CLOSE INTERBREEDING.

DEFINITION OF CLOSE INTERBREEDING. AUGMENTATION OF MORBID TENDENCIES. GENERAL EVIDENCE OF THE GOOD EFFECTS DERIVED FROM CROSSING, AND ON THE EVIL EFFECTS FROM CLOSE INTERBREEDING. CATTLE, CLOSELY INTERBRED; HALF-WILD CATTLE LONG KEPT IN THE SAME PARKS. SHEEP. FALLOW-DEER. DOGS, RABBITS, PIGS. MAN, ORIGIN OF HIS ABHORRENCE OF INCESTUOUS MARRIAGES. FOWLS. PIGEONS. HIVE-BEES. PLANTS, GENERAL CONSIDERATIONS ON THE BENEFITS DERIVED FROM CROSSING. MELONS, FRUIT-TREES, PEAS, CABBAGES, WHEAT, AND FOREST-TREES. ON THE INCREASED SIZE OF HYBRID PLANTS, NOT EXCLUSIVELY DUE TO THEIR STERILITY. ON CERTAIN PLANTS WHICH EITHER NORMALLY OR ABNORMALLY ARE SELF-IMPOTENT, BUT ARE FERTILE, BOTH ON THE MALE AND FEMALE SIDE, WHEN CROSSED WITH DISTINCT INDIVIDUALS EITHER OF THE SAME OR ANOTHER SPECIES. CONCLUSION.

The gain in constitutional vigour, derived from an occasional cross between individuals of the same variety, but belonging to distinct families, or between distinct varieties, has not been so largely or so frequently discussed, as have the evil effects of too close interbreeding. But the former point is the more important of the two, inasmuch as the evidence is more decisive. The evil results from close interbreeding are difficult to detect, for they accumulate slowly, and differ much in degree with different species; whilst the good effects which almost invariably follow a cross are from the first manifest. It should, however, be clearly understood that the advantage of close interbreeding, as far as the retention of character is concerned, is indisputable, and often outweighs the evil of a slight loss of constitutional vigour. In relation to the subject of domestication, the whole question is of some importance, as too close interbreeding interferes with the improvement of old races. It is important as indirectly bearing on Hybridism; and possibly on the extinction of species, when any form has become so rare that only a few individuals remain within a confined area. It bears in an important manner on the influence of free intercrossing, in obliterating individual differences, and thus giving uniformity of character to the individuals of the same race or species; for if additional vigour and fertility be thus gained, the crossed offspring will multiply and prevail, and the ultimate result will be far greater than otherwise would have occurred. Lastly, the question is of high interest, as bearing on mankind. I shall therefore discuss this subject at full length. As the facts which prove the evil effects of close interbreeding are more copious, though less decisive, than those on the good effects of crossing, I shall, under each group of beings, begin with the former.

There is no difficulty in defining what is meant by a cross; but this is by no means easy in regard to "breeding in and in" or "too close interbreeding," because, as we shall see, different species of animals are differently affected by the same degree of interbreeding. The pairing of a father and daughter, or mother and son, or brothers and sisters, if carried on during several generations, is the closest possible form of interbreeding. But some good judges, for instance Sir J. Sebright, believe that the pairing of a brother and sister is much closer than that of parents and children; for when the father is matched with his daughter he crosses, as is said, with only half his own blood. The consequences of close interbreeding carried on for too long a time, are, as is generally believed, loss of size, constitutional vigour, and fertility, sometimes accompanied by a tendency to malformation. Manifest evil does not usually follow from pairing the nearest relations for two, three, or even four generations; but several causes interfere with our detecting the evil— such as the deterioration being very gradual, and the difficulty of distinguishing between such direct evil and the inevitable augmentation of any morbid tendencies which may be latent or apparent in the related parents. On the other hand, the benefit from a cross, even when there has not been any very close interbreeding, is almost invariably at once conspicuous. There is good reason to believe, and this was the opinion of that most experienced observer Sir J. Sebright (17/1. 'The Art of Improving the Breed, etc.' 1809 page 16.), that the evil effects of close interbreeding may be checked or quite prevented by the related individuals being separated for a few generations and exposed to different conditions of life. This conclusion is now held by many breeders; for instance Mr. Carr (17/2. 'The History of the Rise and Progress of the Killerby, etc. Herds' page 41.) remarks, it is a well-known "fact that a change of soil and climate effects perhaps almost as great a change in the constitution as would result from an infusion of fresh blood." I hope to show in a future work that consanguinity by itself counts for nothing, but acts solely from related organisms generally having a similar constitution, and having been exposed in most cases to similar conditions.

That any evil directly follows from the closest interbreeding has been denied by many persons; but rarely by any practical breeder; and never, as far as I know, by one who has largely bred animals which propagate their kind quickly. Many physiologists attribute the evil exclusively to the combination and consequent increase of morbid tendencies common to both parents; and that this is an active source of mischief there can be no doubt. It is unfortunately too notorious that men and various domestic animals endowed with a wretched constitution, and with a strong hereditary disposition to disease, if not actually ill, are fully capable of procreating their kind. Close interbreeding, on the other hand, often induces sterility; and this indicates something quite distinct from the augmentation of morbid tendencies common to both parents. The evidence immediately to be given convinces me that it is a great law of nature, that all organic beings profit from an occasional cross with individuals not closely related to them in blood; and that, on the other hand, long-continued close interbreeding is injurious.

Various general considerations have had much influence in leading me to this conclusion; but the reader will probably rely more on special facts and opinions. The authority of experienced observers, even when they do not advance the grounds of their belief, is of some little value. Now almost all men who have bred many kinds of animals and have written on the subject, such as Sir J. Sebright, Andrew Knight, etc. (17/3. For Andrew Knight see A. Walker on 'Intermarriage' 1838 page 227. Sir J. Sebright 'Treatise' has just been quoted.), have expressed the strongest conviction on the impossibility of long-continued close interbreeding. Those who have compiled works on agriculture, and have associated much with breeders, such as the sagacious Youatt, Low, etc., have strongly declared their opinion to the same effect. Prosper Lucas, trusting largely to French authorities, has come to a similar conclusion. The distinguished German agriculturist Hermann von Nathusius, who has written the most able treatise on this subject which I have met with, concurs; and as I shall have to quote from this treatise, I may state that Nathusius is not only intimately acquainted with works on agriculture in all languages, and knows the pedigrees of our British breeds better than most Englishmen, but has imported many of our improved animals, and is himself an experienced breeder.

Evidence of the evil effects of close interbreeding can most readily be acquired in the case of animals, such as fowls, pigeons, etc., which propagate quickly, and, from being kept in the same place, are exposed to the same conditions. Now I have inquired of very many breeders of these birds, and I have hitherto not met with a single man who was not thoroughly convinced that an occasional cross with another strain of the same sub-variety was absolutely necessary. Most breeders of highly improved or fancy birds value their own strain, and are most unwilling, at the risk, in their opinion, of deterioration, to make a cross. The purchase of a first-rate bird of another strain is expensive, and exchanges are troublesome; yet all breeders, as far as I can hear, excepting those who keep large stocks at different places for the sake of crossing, are driven after a time to take this step.

Another general consideration which has had great influence on my mind is, that with all hermaphrodite animals and plants, which it might have been thought would have perpetually fertilised themselves and been thus subjected for long ages to the closest interbreeding, there is not a single species, as far as I can discover, in which the structure ensures self-fertilisation. On the contrary, there are in a multitude of cases, as briefly stated in the fifteenth chapter, manifest adaptations which favour or inevitably lead to an occasional cross between one hermaphrodite and another of the same species; and these adaptive structures are utterly purposeless, as far as we can see, for any other end.

[With CATTLE there can be no doubt that extremely close interbreeding may be long carried on advantageously with respect to external characters, and with no manifest evil as far as constitution is concerned. The case of Bakewell's Longhorns, which were closely interbred for a long period, has often been quoted; yet Youatt says (17/4. 'Cattle' page 199.) the breed "had acquired a delicacy of constitution inconsistent with common management," and "the propagation of the species was not always certain." But the Shorthorns offer the most striking case of close interbreeding; for instance, the famous bull Favourite (who was himself the offspring of a half-brother and sister from Foljambe) was matched with his own daughter, granddaughter, and great- granddaughter; so that the produce of this last union, or the great-great- granddaughter, had 15-16ths, or 93.75 per cent of the blood of Favourite in her veins. This cow was matched with the bull Wellington, having 62.5 per cent of Favourite blood in his veins, and produced Clarissa; Clarissa was matched with the bull Lancaster, having 68.75 of the same blood, and she yielded valuable offspring. (17/5. I give this on the authority of Nathusius 'Ueber Shorthorn Rindvieh' 1857 s. 71, see also 'Gardeners Chronicle' 1860 page 270. But Mr. J. Storer, a large breeder of cattle, informs me that the parentage of Clarissa is not well authenticated. In the first volume of the 'Herd Book' she was entered as having six descents from Favourite, "which was a palpable mistake," and in all subsequent editions she was spoken of as having only four descents. Mr. Storer doubts even about the four, as no names of the dams are given. Moreover, Clarissa bore "only two bulls and one heifer, and in the next generation her progeny became extinct." Analogous cases of close interbreeding are given in a pamphlet published by Mr. C. Macknight and Dr. H. Madden 'On the True Principles of Breeding' Melbourne Australia 1865.) Nevertheless Collings, who reared these animals, and was a strong advocate for close breeding, once crossed his stock with a Galloway, and the cows from this cross realised the highest prices. Bates's herd was esteemed the most celebrated in the world. For thirteen years he bred most closely in and in; but during the next seventeen years, though he had the most exalted notion of the value of his own stock, he thrice infused fresh blood into his herd: it is said that he did this, not to improve the form of his animals, but on account of their lessened fertility. Mr. Bates's own view, as given by a celebrated breeder (17/6. Mr.

Willoughby Wood in 'Gardener's Chronicle' 1855 page 411; and 1860 page 270. See the very clear tables and pedigrees given in Nathusius 'Rindvieh' s. 72-77.), was, that "to breed in-and-in from a bad stock was ruin and devastation; yet that the practice may be safely followed within certain limits when the parents so related are descended from first-rate animals." We thus see that there has been much close interbreeding with Shorthorns; but Nathusius, after the most careful study of their pedigrees, says that he can find no instance of a breeder who has strictly followed this practice during his whole life. From this study and his own experience, he concludes that close interbreeding is necessary to ennoble the stock; but that in effecting this the greatest care is necessary, on account of the tendency to infertility and weakness. It may be added, that another high authority (17/7. Mr. Wright 'Journal of Royal Agricult. Soc.' volume 7 1846 page 204. Mr. J. Downing (a successful breeder of Shorthorns in Ireland) informs me that the raisers of the great families of Shorthorns carefully conceal their sterility and want of constitution. He adds that Mr. Bates, after he had bred his herd in-and-in for some years, "lost in one season twenty-eight calves solely from want of constitution.") asserts that many more calves are born cripples from Shorthorns than from other and less closely interbred races of cattle.

Although by carefully selecting the best animals (as Nature effectually does by the law of battle) close interbreeding may be long carried on with cattle, yet the good effects of a cross between almost any two breeds is at once shown by the greater size and vigour of the offspring; as Mr. Spooner writes to me, "crossing distinct breeds certainly improves cattle for the butcher." Such crossed animals are of course of no value to the breeder; but they have been raised during many years in several parts of England to be slaughtered (17/8. 'Youatt on Cattle' page 202.); and their merit is now so fully recognised, that at fat-cattle shows a separate class has been formed for their reception. The best fat ox at the great show at Islington in 1862 was a crossed animal.

The half-wild cattle, which have been kept in British parks probably for 400 or 500 years, or even for a longer period, have been advanced by Culley and others as a case of long-continued interbreeding within the limits of the same herd without any consequent injury. With respect to the cattle at Chillingham, the late Lord Tankerville owned that they were bad breeders. (17/9. 'Report British Assoc. Zoolog. Sect.' 1838.) The agent, Mr. Hardy, estimates (in a letter to me, dated May, 1861) that in the herd of about fifty the average number annually slaughtered, killed by fighting, and dying, is about ten, or one in five. As the herd is kept up to nearly the same average number, the annual rate of increase must be likewise about one in five. The bulls, I may add, engage in furious battles, of which battles the present Lord Tankerville has given me a graphic description, so that there will always be rigorous selection of the most vigorous males. I procured in 1855 from Mr. D. Gardner, agent to the Duke of Hamilton, the following account of the wild cattle kept in the Duke's park in Lanarkshire, which is about 200 acres in extent. The number of cattle varies from sixty-five to eighty; and the number annually killed (I presume by all causes) is from eight to ten; so that the annual rate of increase can hardly be more than one in six. Now in South America, where the herds are half-wild, and therefore offer a nearly fair standard of comparison, according to Azara the natural increase of the cattle on an estancia is from one-third to one-fourth of the total number, or one in between three and four and this, no doubt, applies exclusively to adult animals fit for consumption. Hence the half-wild British cattle which have long interbred within the limits of the same herd are relatively far less fertile. Although in an unenclosed country like Paraguay there must be some crossing between the different herds, yet even there the inhabitants believe that the occasional introduction of animals from distant localities is necessary to prevent "degeneration in size and diminution of fertility." (17/10. Azara 'Quadrupedes du Paraguay' tome 2 pages 354, 368.) The decrease in size from ancient times in the Chillingham and Hamilton cattle must have been prodigious, for Professor Rutimeyer has shown that they are almost certainly the descendants of the gigantic Bos primigenius. No doubt this decrease in size may be largely attributed to less favourable conditions of life; yet animals roaming over large parks, and fed during severe winters, can hardly be considered as placed under very unfavourable conditions.

With SHEEP there has often been long-continued interbreeding within the limits of the same flock; but whether the nearest relations have been matched so frequently as in the case of Shorthorn cattle, I do not know. The Messrs. Brown during fifty years have never infused fresh blood into their excellent flock of Leicesters. Since 1810 Mr. Barford has acted on the same principle with the Foscote flock. He asserts that half a century of experience has convinced him that when two nearly related animals are quite sound in constitution, in-and-in breeding does not induce degeneracy; but he adds that he "does not pride himself on breeding from the nearest affinities." In France the Naz flock has been bred for sixty years without the introduction of a single strange ram. (17/11. For the case of the Messrs. Brown see 'Gardener's Chronicle' 1855 page 26. For the Foscote flock 'Gardener's Chronicle' 1860 page 416. For the Naz flock 'Bull. de la Soc. d'Acclimat.' 1860 page 477.) Nevertheless, most great breeders of sheep have protested against close interbreeding prolonged for too great a length of time. (17/12. Nathusius 'Rindvieh' s. 65; 'Youatt on Sheep' page 495.) The most celebrated of recent breeders, Jonas Webb, kept five separate families to work on, thus "retaining the requisite distance of relationship between the sexes" (17/13. 'Gardener's Chronicle' 1861 page 631.); and what is probably of greater importance, the separate flocks will have been exposed to somewhat different conditions.

Although by the aid of careful selection the near interbreeding of sheep may be long continued without any manifest evil, yet it has often been the practice with farmers to cross distinct breeds to obtain animals for the butcher, which plainly shows that good of some kind is derived from this practice. We have excellent evidence on this head from Mr. S. Druce (17/14. 'Journal R. Agricult. Soc.' volume 14 1853 page 212.), who gives in detail the comparative numbers of four pure breeds and of a cross-breed which can be supported on the same ground, and he gives their produce in fleece and carcase. A high authority, Mr. Pusey, sums up the result in money value during an equal length of time, namely (neglecting shillings), for Cotswolds 248 pounds, for Leicesters 223 pounds, for Southdowns 204 pounds, for Hampshire Downs 264 pounds, and for the crossbred 293 pounds. A former celebrated breeder, Lord Somerville, states that his half-breeds from Ryelands and Spanish sheep were larger animals than either the pure Ryelands or pure Spanish sheep. Mr. Spooner concludes his excellent Essay on Crossing by asserting that there is a pecuniary advantage in judicious cross-breeding, especially when the male is larger than the female. (17/15. Lord Somerville 'Facts on Sheep and Husbandry' page 6. Mr. Spooner in 'Journal of Royal Agricult. Soc. of England' volume 20 part 2. See also an excellent paper on the same subject in 'Gardener's Chronicle' 1860 page 321 by Mr. Charles Howard.)

As some of our British parks are ancient, it occurred to me that there must have been long-continued close interbreeding with the fallow-deer (Cervus dama) kept in them; but on inquiry I find that it is a common practice to infuse new blood by procuring bucks from other parks. Mr. Shirley (17/16. 'Some Account of English Deer Parks' by Evelyn P. Shirley 1867.), who has carefully studied the management of deer, admits that in some parks there has been no admixture of foreign blood from a time beyond the memory of man. But he concludes "that in the end the constant breeding in-and-in is sure to tell to the disadvantage of the whole herd, though it may take a very long time to prove it; moreover, when we find, as is very constantly the case, that the introduction of fresh blood has been of the very greatest use to deer, both by improving their size and appearance, and particularly by being of service in removing the taint of 'rickback,' if not of other diseases, to which deer are sometimes subject when the blood has not been changed, there can, I think, be no doubt but that a judicious cross with a good stock is of the greatest consequence, and is indeed essential, sooner or later, to the prosperity of every well-ordered park."

Mr. Meynell's famous foxhounds have been adduced, as showing that no ill effects follow from close interbreeding; and Sir J. Sebright ascertained from him that he frequently bred from father and daughter, mother and son, and sometimes even from brothers and sisters. With greyhounds also there has been much close interbreeding, but the best breeders agree that it may be carried too far. (17/17. Stonehenge 'The Dog' 1867 pages 175-188.) But Sir J. Sebright declares (17/18. 'The Art of Improving the Breed' etc. page 13. With respect to Scotch deerhounds see Scrope 'Art of Deer Stalking' pages 350-353.), that by breeding in-and-in, by which he means matching brothers and sisters, he has actually seen the offspring of strong spaniels degenerate into weak and diminutive lapdogs. The Rev. W.D. Fox has communicated to me the case of a small lot of bloodhounds, long kept in the same family, which had become very bad breeders, and nearly all had a bony enlargement in the tail. A single cross with a distinct strain of bloodhounds restored their fertility, and drove away the tendency to malformation in the tail. I have heard the particulars of another case with bloodhounds, in which the female had to be held to the male. Considering how rapid is the natural increase of the dog, it is difficult to understand the large price of all highly improved breeds, which almost implies long-continued close interbreeding, except on the belief that this process lessens fertility and increases liability to distemper and other diseases. A high authority, Mr. Scrope, attributes the rarity and deterioration in size of the Scotch deerhound (the few individuals formerly existing throughout the country being all related) in large part to close interbreeding.

With all highly-bred animals there is more or less difficulty in getting them to procreate quickly, and all suffer much from delicacy of constitution. A great judge of rabbits (17/19. 'Cottage Gardener' 1861 page 327.) says, "the long-eared does are often too highly bred or forced in their youth to be of much value as breeders, often turning out barren or bad mothers." They often desert their young, so that it is necessary to have nurse-rabbits, but I do not pretend to attribute all these evil results to close interbreeding. (17/20. Mr. Huth gives ('The Marriage of Near Kin' 1875 page 302) from the 'Bulletin de l'Acad. R. de Med. de Belgique' (volume 9 1866 pages 287, 305), several statements made by a M. Legrain with respect to crossing brother and sister rabbits for five or six successive generations with no consequent evil results. I was so much surprised at this account, and at M. Legrain's invariable success in his experiments, that I wrote to a distinguished naturalist in Belgium to inquire whether M. Legrain was a trustworthy observer. In answer, I have heard that, as doubts were expressed about the authenticity of these experiments, a commission of inquiry was appointed, and that at a succeeding meeting of the Society ('Bull. de l'Acad. R. de Med. de Belgique' 1867 3rd series tome 1 no. 1 to 5), Dr. Crocq reported "qu'il etait materiellement impossible que M. Legrain ait fait les experiences qu'il annonce." To this public accusation no satisfactory answer was made.)

With respect to PIGS there is more unanimity amongst breeders on the evil effects of close interbreeding than, perhaps, with any other large animal. Mr. Druce, a great and successful breeder of the Improved Oxfordshires (a crossed race), writes, "without a change of boars of a different tribe, but of the same breed, constitution cannot be preserved." Mr. Fisher Hobbs, the raiser of the celebrated Improved Essex breed, divided his stock into three separate families, by which means he maintained the breed for more than twenty years, "by judicious selection from the THREE DISTINCT FAMILIES." (17/21. Sidney's edition of 'Youatt on the Pig' 1860 page 30; page 33 quotation from Mr. Druce; page 29 on Lord Western's case.) Lord Western was the first importer of a Neapolitan boar and sow. "From this pair he bred in-and-in, until the breed was in danger of becoming extinct, a sure result (as Mr. Sidney remarks) of in-and-in breeding." Lord Western then crossed his Neapolitan pigs with the old Essex, and made the first great step towards the Improved Essex breed. Here is a more interesting case. Mr. J. Wright, well known as a breeder, crossed (17/22. 'Journal of Royal Agricult. Soc. of England' 1846 volume 7 page 205.) the same boar with the daughter, granddaughter, and great-granddaughter, and so on for seven generations. The result was, that in many instances the offspring failed to breed; in others they produced few that lived; and of the latter many were idiotic, without sense, even to suck, and when attempting to move could not walk straight. Now it deserves especial notice, that the two last sows produced by this long course of interbreeding were sent to other boars, and they bore several litters of healthy pigs. The best sow in external appearance produced during the whole seven generations was one in the last stage of descent; but the litter consisted of this one sow. She would not breed to her sire, yet bred at the first trial to a stranger in blood. So that, in Mr. Wright's case, long-continued and extremely close interbreeding did not affect the external form or merit of the young; but with many of them the general constitution and mental powers, and especially the reproductive functions, were seriously affected.

Nathusius gives (17/23. 'Ueber Rindvieh' etc. s. 78. Col. Le Couteur, who has done so much for the agriculture of Jersey, writes to me that from possessing a fine breed of pigs he bred them very closely, twice pairing brothers and sisters, but nearly all the young had fits and died suddenly.) an analogous and even more striking case: he imported from England a pregnant sow of the large Yorkshire breed, and bred the product closely in-and-in for three generations: the result was unfavourable, as the young were weak in constitution, with impaired fertility. One of the latest sows, which he esteemed a good animal, produced, when paired with her own uncle (who was known to be productive with sows of other breeds), a litter of six, and a second time a litter of only five weak young pigs. He then paired this sow with a boar of a small black breed, which he had likewise imported from England; this boar, when matched with sows of his own breed, produced from seven to nine young. Now, the sow of the large breed, which was so unproductive when paired with her own uncle, yielded to the small black boar, in the first litter twenty-one, and in the second litter eighteen young pigs; so that in one year she produced thirty-nine fine young animals!

As in the case of several other animals already mentioned, even when no injury is perceptible from moderately close interbreeding, yet, to quote the words of Mr. Coate (who five times won the annual gold medal of the Smithfield Club Show for the best pen of pigs), "Crosses answer well for profit to the farmer, as you get more constitution and quicker growth; but for me, who sell a great number of pigs for breeding purposes, I find it will not do, as it requires many years to get anything like purity of blood again." (17/24. Sidney on the 'Pig' page 36. See also note page 34. Also Richardson on the 'Pig' 1847 page 26.)]

Almost all the animals as yet mentioned are gregarious, and the males must frequently pair with their own daughters, for they expel the young males as well as all intruders, until forced by old age and loss of strength to yield to some stronger male. It is therefore not improbable that gregarious animals may have been rendered less susceptible than non-social species to the evil consequences of close interbreeding, so that they may be enabled to live in herds without injury to their offspring. Unfortunately we do not know whether an animal like the cat, which is not gregarious, would suffer from close interbreeding in a greater degree than our other domesticated animals. But the pig is not, as far as I can discover, strictly gregarious, and we have seen that it appears eminently liable to the evil effects of close interbreeding. Mr. Huth, in the case of the pig, attributes (page 285) these effects to their having been "cultivated most for their fat," or to the selected individuals having had a weak constitution; but we must remember that it is great breeders who have brought forward the above cases, and who are far more familiar than ordinary men can be, with the causes which are likely to interfere with the fertility of their animals.

The effects of close interbreeding in the case of man is a difficult subject, on which I will say but little. It has been discussed by various authors under many points of view. (17/25. Dr. Dally has published an excellent article (translated in the 'Anthropolog. Review' May 1864 page 65), criticising all writers who have maintained that evil follows from consanguineous marriages. No doubt on this side of the question many advocates have injured their cause by inaccuracies: thus it has been stated (Devay 'Du Danger des Mariages' etc. 1862 page 141) that the marriages of cousins have been prohibited by the legislature of Ohio; but I have been assured, in answer to inquiries made in the United States, that this statement is a mere fable.) Mr. Tylor (17/26. See his interesting work on the 'Early History of Man' 1865 chapter 10.) has shown that with widely different races in the most distant quarters of the world, marriages between relations—even between distant relations—have been strictly prohibited. There are, however, many exceptions to the rule, which are fully given by Mr. Huth (17/27. 'The Marriage of Near Kin' 1875. The evidence given by Mr. Huth would, I think, have been even more valuable than it is on this and some other points, if he had referred solely to the works of men who had long resided in each country referred to, and who showed that they possessed judgment and caution. See also Mr. W. Adam 'On Consanguinity in Marriage' in the 'Fortnightly Review' 1865 page 710. Also Hofacker 'Ueber die Eigenschaften' etc. 1828.) It is a curious problem how these prohibitions arose during early and barbarous times. Mr. Tylor is inclined to attribute them to the evil effects of consanguineous marriages having been observed; and he ingeniously attempts to explain some apparent anomalies in the prohibition not extending equally to the relations on the male and female side. He admits, however, that other causes, such as the extension of friendly alliances, may have come into play. Mr. W. Adam, on the other hand, concludes that related marriages are prohibited and viewed with repugnance, from the confusion which would thus arise in the descent of property, and from other still more recondite reasons. But I cannot accept these views, seeing that incest is held in abhorrence by savages such as those of Australia and South America (17/28. Sir G. Grey 'Journal of Expeditions into Australia' volume 2 page 243; and Dobrizhoffer 'On the Abipones of South America.'), who have no property to bequeath, or fine moral feelings to confuse, and who are not likely to reflect on distant evils to their progeny. According to Mr. Huth the feeling is the indirect result of exogamy, inasmuch as when this practice ceased in any tribe and it became endogamous, so that marriages were strictly confined to the same tribe, it is not unlikely that a vestige of the former practice would still be retained, so that closely-related marriages would be prohibited. With respect to exogamy itself Mr. MacLennan believes that it arose from a scarcity of women, owing to female infanticide, aided perhaps by other causes.

It has been clearly shown by Mr. Huth that there is no instinctive feeling in man against incest any more than in gregarious animals. We know also how readily any prejudice or feeling may rise to abhorrence, as shown by Hindus in regard to objects causing defilement. Although there seems to be no strong inherited feeling in mankind against incest, it seems possible that men during primeval times may have been more excited by strange females than by those with whom they habitually lived; in the same manner as according to Mr. Cupples (17/29. 'Descent of Man' 2nd. edit page 524.), male deerhounds are inclined towards strange females, while the females prefer dogs with whom they have associated. If any such feeling formerly existed in man, this would have led to a preference for marriages beyond the nearest kin, and might have been strengthened by the offspring of such marriages surviving in greater numbers, as analogy would lead us to believe would have occurred.

Whether consanguineous marriages, such as are permitted in civilised nations, and which would not be considered as close interbreeding in the case of our domesticated animals, cause any injury will never be known with certainty until a census is taken with this object in view. My son, George Darwin, has done what is possible at present by a statistical investigation (17/30. 'Journal of Statistical Soc.' June 1875 page 153; and 'Fortnightly Review' June 1875.), and he has come to the conclusion, from his own researches and those of Dr. Mitchell, that the evidence as to any evil thus caused is conflicting, but on the whole points to the evil being very small.

## [BIRDS.

In the case of the FOWL a whole array of authorities could be given against too close interbreeding. Sir J. Sebright positively asserts that he made many trials, and that his fowls, when thus treated, became long in the legs, small in the body, and bad breeders. (17/31. 'The Art of Improving the Breed' page 13.) He produced the famous Sebright Bantams by complicated crosses, and by breeding in-and-in; and since his time there has been much close interbreeding with these animals; and they are now notoriously bad breeders. I have seen Silver Bantams, directly descended from his stock, which had become almost as barren as hybrids; for not a single chicken had been that year hatched from two full nests of eggs. Mr. Hewitt says that with these Bantams the sterility of the male stands, with rare exceptions, in the closest relation with their loss of certain secondary male characters: he adds, "I have noticed, as a general rule, that even the slightest deviation from feminine character in the tail of the male Sebright—say the elongation by only half an inch of the two principal tail feathers—brings with it improved probability of increased fertility." (17/32. 'The Poultry Book' by W.B. Tegetmeier 1866 page 245.)

Mr. Wright states (17/33. 'Journal Royal Agricult. Soc.' 1846 volume 7 page 205; see also Ferguson on the Fowl pages 83, 317; see also 'The Poultry Book' by Tegetmeier 1866 page 135 with respect to the extent to which cock-fighters found that they could venture to breed in-and-in, viz., occasionally a hen with her own son; "but they were cautious not to repeat the in-and-in breeding.") that Mr. Clark, "whose fighting-cocks were so notorious, continued to breed from his own kind till they lost their disposition to fight, but stood to be cut up without making any resistance, and were so reduced in size as to be under those weights required for the best prizes; but on obtaining a cross from Mr. Leighton, they again resumed their former courage and weight." It should be borne in mind that game-cocks before they fought were always weighed, so that nothing was left to the imagination about any reduction or increase of weight. Mr. Clark does not seem to have bred from brothers and sisters, which is the most injurious kind of union; and he found, after repeated trials, that there was a greater reduction in weight in the young from a father paired with his daughter, than from a mother with her son. I may add that Mr. Eyton of Eyton, the well-known ornithologist, who is a large breeder of Grey Dorkings, informs me that they certainly diminish in size, and become less prolific, unless a cross with another strain is occasionally obtained. So it is with Malays, according to Mr. Hewitt, as far as size is concerned. (17/34. 'The Poultry Book' by W.B. Tegetmeier 1866 page 79.)

An experienced writer (17/35. 'The Poultry Chronicle' 1854 volume 1 page 43.) remarks that the same amateur, as is well known, seldom long maintains the superiority of his birds; and this, he adds, undoubtedly is due to all his stock "being of the same blood;" hence it is indispensable that he should occasionally procure a bird of another strain. But this is not necessary with those who keep a stock of fowls at different stations. Thus, Mr. Ballance, who has bred Malays for thirty years, and has won more prizes with these birds than any other fancier in England, says that breeding in-and-in does not necessarily cause deterioration; "but all depends upon how this is managed. My plan has been to keep about five or six distinct runs, and to rear about two hundred or three hundred chickens each year, and select the best birds from each run for crossing. I thus secure sufficient crossing to prevent deterioration." (17/36. 'The Poultry Book' by W.B. Tegetmeier 1866 page 79.)

We thus see that there is almost complete unanimity with poultry-breeders that, when fowls are kept at the same place, evil quickly follows from interbreeding carried on to an extent which would be disregarded in the case of most quadrupeds. Moreover, it is a generally received opinion that cross-bred chickens are the hardiest and most easily reared. (17/37. 'The Poultry Chronicle' volume 1 page 89.) Mr. Tegetmeier, who has carefully attended to poultry of all breeds, says (17/38. 'The Poultry Book' 1866 page 210.) that Dorking hens, allowed to run with Houdan or Creve-coeur cocks, "produce in the early spring chickens that for size, hardihood, early maturity, and fitness for the market, surpass those of any pure breed that we have ever raised." Mr. Hewitt gives it as a general rule with fowls, that crossing the breed increases their size. He makes this remark after stating that hybrids from the pheasant and fowl are considerably larger than either progenitor: so again, hybrids from the male golden pheasant and female common pheasant "are of far larger size than either parent-bird." (17/39. Ibid 1866 page 167; and 'Poultry Chronicle' volume 3 1855 page 15.) To this subject of the increased size of hybrids I shall presently return.

With PIGEONS, breeders are unanimous, as previously stated, that it is absolutely indispensable, notwithstanding the trouble and expense thus caused, occasionally to cross their much-prized birds with individuals of another strain, but belonging, of course, to the same variety. It deserves notice that, when size is one of the desired characters, as with pouters (17/40. 'A Treatise on Fancy Pigeons' by J.M. Eaton page 56.) the evil effects of close interbreeding are much sooner perceived than when small birds, such as short-faced tumblers, are valued. The extreme delicacy of the high fancy breeds, such as these tumblers and improved English carriers, is remarkable; they are liable to many diseases, and often die in the egg or during the first moult; and their eggs have generally to be hatched under foster-mothers. Although these highly-prized birds have invariably been subjected to much close interbreeding, yet their extreme delicacy of constitution cannot perhaps be thus fully explained. Mr. Yarrell informed me that Sir J. Sebright continued closely interbreeding some owl-pigeons, until from their extreme sterility he as nearly as possible lost the whole family. Mr. Brent (17/41. 'The Pigeon Book' page 46.) tried to raise a breed of trumpeters, by crossing a common pigeon, and recrossing the daughter, granddaughter, great-granddaughter, and great-great-granddaughter, with the same male trumpeter, until he obtained a bird with 15/16 of trumpeter's blood; but then the experiment failed, for "breeding so close stopped reproduction." The experienced Neumeister (17/42. 'Das Ganze der Taubenzucht' 1837 s. 18.) also asserts that the offspring from dovecotes and various other breeds are "generally very fertile and hardy birds:" so again MM. Boitard and Corbie (17/43. 'Les Pigeons' 1824 page 35.), after forty-five years' experience, recommend persons to cross their breeds for amusement; for, if they fail to make interesting birds, they will succeed under an economical point of view, "as it is found that mongrels are more fertile than pigeons of pure race."

I will refer only to one other animal, namely, the Hive-bee, because a distinguished entomologist has advanced this as a case of inevitable close interbreeding. As the hive is tenanted by a single female, it might have been thought that her male and female offspring would always have bred together, more especially as bees of different hives are hostile to each other; a strange worker being almost always attacked when trying to enter another hive. But Mr. Tegetmeier has shown (17/44. 'Proc. Entomolog. Soc.' August 6, 1860 page 126.) that this instinct does not apply to drones, which are permitted to enter any hive; so that there is no a priori improbability of a queen receiving a foreign drone. The fact of the union invariably and necessarily taking place on the wing, during the queen's nuptial flight, seems to be a special provision against continued interbreeding. However this may be, experience has shown, since the introduction of the yellow-banded Ligurian race into Germany and England, that bees freely cross: Mr. Woodbury, who introduced Ligurian bees into Devonshire, found during a single season that three stocks, at distances of from one to two miles from his hives, were crossed by his drones. In one case the Ligurian drones must have flown over the city of Exeter, and over several intermediate hives. On another occasion several common black queens were crossed by Ligurian drones at a distance of from one to three and a half miles. (17/45. 'Journal of Horticulture' 1861 pages 39, 77, 158; and 1864 page 206.)

## PLANTS.

When a single plant of a new species is introduced into any country, if propagated by seed, many individuals will soon be raised, so that if the proper insects be present there will be crossing. With newly-introduced trees or other plants not propagated by seed we are not here concerned. With old-established plants it is an almost universal practice occasionally to make exchanges of seed, by which means individuals which have been exposed to different conditions of life,—and this, as we have seen with animals, diminishes the evil from close interbreeding,—will occasionally be introduced into each district.

With respect to individuals belonging to the same sub-variety, Gartner, whose accuracy and experience exceeded that of all other observers, states (17/46. 'Beitrage zur Kenntniss der Befruchtung' 1844 s. 366.) that he has many times observed good effects from this step, especially with exotic genera, of which the fertility is somewhat impaired, such as Passiflora, Lobelia, Fuchsia. Herbert also says (17/47. 'Amaryllidaceae' page 371.), "I am inclined to think that I have derived advantage from impregnating the flower from which I wished to obtain seed with pollen from another individual of the same variety, or at least from another flower, rather than with its own." Again, Professor Lecoq ascertained that crossed offspring are more vigorous and robust than their parents. (17/48. 'De la Fecondation' 2nd edition 1862 page 79.)

General statements of this kind, however, can seldom be fully trusted: I therefore began a long series of experiments, continued for about ten years, which will I think conclusively show the good effects of crossing two distinct plants of the same variety, and the evil effects of long-continued self-fertilisation. A clear light will thus be thrown on such questions, as why flowers are almost invariably constructed so as to permit, or favour, or necessitate the union of two individuals. We shall clearly understand why monoecious and dioecious,—why dichogamous, dimorphic and trimorphic plants exist, and many other such cases. I intend soon to publish an account of these experiments, and I can here give only a few cases in illustration. The plan which I followed was to grow plants in the same pot, or in pots of the same size, or close together in the open ground; carefully to exclude insects; and then to fertilise some of the flowers with pollen from the same flower, and others on the same plant with pollen from a distinct but adjoining plant. In many of these experiments, the crossed plants yielded much more seed than the self-fertilised plants; and I have never seen the reversed case. The self- fertilised and crossed seeds thus obtained were allowed to germinate in the same glass vessel on damp sand; and as the seeds germinated, they were planted in pairs on opposite sides of the same pot, with a superficial partition between them, and were placed so as to be equally exposed to the light. In other cases the self-fertilised and crossed seeds were simply sown on opposite sides of the same small pot. I have, in short, followed different plans, but in every case have taken all the precautions which I could think of, so that the two lots should be equally favoured. The growth of the plants raised from the crossed and self-fertilised seed, were carefully observed from their germination to maturity, in species belonging to fifty-two genera; and the difference in their growth, and in withstanding unfavourable conditions, was in most cases manifest and strongly marked. It is of importance that the two lots of seed should be sown or planted on opposite sides of the same pot, so that the seedlings may struggle against each other; for if sown separately in ample and good soil, there is often but little difference in their growth.

I will briefly describe two of the first cases observed by me. Six crossed and six self-fertilised seeds of Ipomoea purpurea, from plants treated in the manner above described, were planted as soon as they had germinated, in pairs on opposite sides of two pots, and rods of equal thickness were given them to twine up. Five of the crossed plants grew from the first more quickly than the opposed self-fertilised plants; the sixth, however, was weakly and was for a time beaten, but at last its sounder constitution prevailed and it shot ahead of its antagonist. As soon as each crossed plant reached the top of its seven- foot rod its fellow was measured, and the result was that, when the crossed plants were seven feet high the self-fertilised had attained the average height of only five feet four and a half inches. The crossed plants flowered a little before, and more profusely than the self-fertilised plants. On opposite sides of another SMALL pot a large number of crossed and self-fertilised seeds were sown, so that they had to struggle for bare existence; a single rod was given to each lot: here again the crossed plants showed from the first their advantage; they never quite reached the summit of the seven-foot rod, but relatively to the self-fertilised plants their average height was as seven feet to five feet two inches. The experiment was repeated during several succeeding generations, treated in exactly the same manner, and with nearly the same result. In the second generation, the crossed plants, which were again crossed, produced 121 seed-capsules, whilst the self-fertilised, again self-fertilised, produced only 84 capsules.

Some flowers of the Mimulus luteus were fertilised with their own pollen, and others were crossed with pollen from distinct plants growing in the same pot. The seeds were thickly sown on opposite sides of a pot. The seedlings were at first equal in height; but when the young crossed plants were half an inch, the self-fertilised plants were only a quarter of an inch high. But this degree of inequality did not last, for, when the crossed plants were four and a half inches high, the self-fertilised were three inches, and they retained the same relative difference till their growth was complete. The crossed plants looked far more vigorous than the uncrossed, and flowered before them; they produced also a far greater number of capsules. As in the former case, the experiment was repeated during several succeeding generations. Had I not watched these plants of Mimulus and Ipomoea during their whole growth, I could not have believed it possible, that a difference apparently so slight as that of the pollen being taken from the same flower, or from a distinct plant growing in the same pot, could have made so wonderful a difference in the growth and vigour of the plants thus produced. This, under a physiological point of view, is a most remarkable phenomenon.

With respect to the benefit derived from crossing distinct varieties, plenty of evidence has been published. Sageret (17/49. 'Memoire sur les Cucurbitacees' pages 36, 28, 30.) repeatedly speaks in strong terms of the vigour of melons raised by crossing different varieties, and adds that they are more easily fertilised than common melons, and produce numerous good seed. Here follows the evidence of an English gardener (17/50. Loudon's 'Gardener's Mag.' volume 8 1832 page 52.): "I have this summer met with better success in my cultivation of melons, in an unprotected state, from the seeds of hybrids (i.e. mongrels) obtained by cross impregnation, than with old varieties. The offspring of three different hybridisations (one more especially, of which the parents were the two most dissimilar varieties I could select) each yielded more ample and finer produce than any one of between twenty and thirty established varieties."

Andrew Knight (17/51. 'Transact. Hort. Soc.' volume 1 page 25.) believed that his seedlings from crossed varieties of the apple exhibited increased vigour and luxuriance; and M. Chevreul (17/52. 'Annal. des Sc. Nat.' 3rd series, Bot. tome 6 page 189.) alludes to the extreme vigour of some of the crossed fruit-trees raised by Sageret.

By crossing reciprocally the tallest and shortest peas, Knight (17/53. 'Philosophical Transactions' 1799 page 200.) says: "I had in this experiment a striking instance of the stimulative effects of crossing the breeds; for the smallest variety, whose height rarely exceeded two feet, was increased to six feet: whilst the height of the large and luxuriant kind was very little diminished." Mr. Laxton gave me seed-peas produced from crosses between four distinct kinds; and the plants thus raised were extraordinarily vigorous, being in each case from one to two or three feet taller than the parent-forms growing close alongside them.

Wiegmann (17/54. 'Ueber die Bastarderzeugung' 1828 s. 32, 33. For Mr. Chaundy's case see Loudon's 'Gardener's Mag.' volume 7 1831 page 696.) made many crosses between several varieties of cabbage; and he speaks with astonishment of the vigour and height of the mongrels, which excited the amazement of all the gardeners who beheld them. Mr. Chaundy raised a great number of mongrels by planting together six distinct varieties of cabbage. These mongrels displayed an infinite diversity of character; "But the most remarkable circumstance was, that, while all the other cabbages and borecoles in the nursery were destroyed by a severe winter, these hybrids were little injured, and supplied the kitchen when there was no other cabbage to be had."

Mr. Maund exhibited before the Royal Agricultural Society (17/55. 'Gardener's Chronicle' 1846 page 601.) specimens of crossed wheat, together with their parent varieties; and the editor states that they were intermediate in character, "united with that greater vigour of growth, which it appears, in the vegetable as in the animal world, is the result of a first cross." Knight also crossed several varieties of wheat (17/56. 'Philosoph. Transact.' 1799 page 201.), and he says "that in the years 1795 and 1796, when almost the whole crop of corn in the island was blighted, the varieties thus obtained, and these only, escaped in this neighbourhood, though sown in several different soils and situations."

Here is a remarkable case: M. Clotzsch (17/57. Quoted in 'Bull. Bot. Soc. France' volume 2 1855 page 327.) crossed Pinus sylvestris and nigricans, Quercus robur and pedunculata, Alnus glutinosa and incana, Ulmus campestris and effusa; and the cross-fertilised seeds, as well as seeds of the pure parent-trees, were all sown at the same time and in the same place. The result was, that after an interval of eight years, the hybrids were one-third taller than the pure trees!

The facts above given refer to undoubted varieties, excepting the trees crossed by Clotzsch, which are ranked by various botanists as strongly-marked races, sub-species, or species. That true hybrids raised from entirely distinct species, though they lose in fertility, often gain in size and constitutional vigour, is certain. It would be superfluous to quote any facts; for all experimenters, Kolreuter, Gartner, Herbert, Sageret, Lecoq, and Naudin, have been struck with the wonderful vigour, height, size, tenacity of life, precocity, and hardiness of their hybrid productions. Gartner (17/58. Gartner 'Bastarderzeugung' s. 259, 518, 526 et seq.) sums up his conviction on this head in the strongest terms. Kolreuter (17/59. 'Fortsetzung' 1763 s. 29; 'Dritte Fortsetzung' s. 44, 96; 'Act. Acad. St. Petersburg' 1782 part 2 page 251; 'Nova Acta' 1793 pages 391, 394; 'Nova Acta' 1795 pages 316, 323.) gives numerous precise measurements of the weight and height of his hybrids in his comparison with measurements of both parent-forms; and speaks with astonishment of their "statura portentosa," their "ambitus vastissimus ac altitudo valde conspicua." Some exceptions to the rule in the case of very sterile hybrids have, however, been noticed by Gartner and Herbert; but the most striking exceptions are given by Max Wichura (17/60. 'Die

Bastardbefruchtung' etc. 1865 s. 31, 41, 42.) who found that hybrid willows were generally tender in constitution, dwarf, and short-lived.

Kolreuter explains the vast increase in the size of the roots, stems, etc., of his hybrids, as the result of a sort of compensation due to their sterility, in the same way as many emasculated animals are larger than the perfect males. This view seems at first sight extremely probable, and has been accepted by various authors (17/61. Max Wichura fully accepts this view ('Bastardbefruchtung' s. 43), as does the Rev. M.J. Berkeley in 'Journal of Hort. Soc.' January 1866 page 70.); but Gartner (17/62. 'Bastarderzeugung' s. 394, 526, 528.) has well remarked that there is much difficulty in fully admitting it; for with many hybrids there is no parallelism between the degree of their sterility and their increased size and vigour. The most striking instances of luxuriant growth have been observed with hybrids which were not sterile in any extreme degree. In the genus Mirabilis, certain hybrids are unusually fertile, and their extraordinary luxuriance of growth, together with their enormous roots (17/63. Kolreuter 'Nova Acta' 1795 page 316.) have been transmitted to their progeny. The result in all cases is probably in part due to the saving of nutriment and vital force through the sexual organs acting imperfectly or not at all, but more especially to the general law of good being derived from a cross. For it deserves especial attention that mongrel animals and plants, which are so far from being sterile that their fertility is often actually augmented, have, as previously shown, their size, hardiness, and constitutional vigour generally increased. It is not a little remarkable that an accession of vigour and size should thus arise under the opposite contingencies of increased and diminished fertility.

It is a perfectly well ascertained fact (17/64. Gartner 'Bastarderzeugung' s. 430.) that hybrids invariably breed with either pure parent, and not rarely with a distinct species, more readily than with one another. Herbert is inclined to explain even this fact by the advantage derived from a cross; but Gartner more justly accounts for it by the pollen of the hybrid, and probably its ovules, being in some degree vitiated, whereas the pollen and ovules of both pure parents and of any third species are sound. Nevertheless, there are some well-ascertained and remarkable facts, which, as we shall presently see, show that a cross by itself undoubtedly tends to increase or re-establish the fertility of hybrids.

The same law, namely, that the crossed offspring both of varieties and species are larger than the parent-forms, holds good in the most striking manner with hybrid animals as well as with mongrels. Mr. Bartlett, who has had such large experience says, "Among all hybrids of vertebrated animals there is a marked increase of size." He then enumerates many cases with mammals, including monkeys, and with various families of birds. (17/65. Quoted by Dr. Murie in 'Proc. Zoolog. Soc.' 1870 page 40.)]

# ON CERTAIN HERMAPHRODITE PLANTS WHICH, EITHER NORMALLY OR ABNORMALLY, REQUIRE TO BE FERTILISED BY POLLEN FROM A DISTINCT INDIVIDUAL OR SPECIES.

The facts now to be given differ from the foregoing, as self-sterility is not here the result of long-continued close interbreeding. These facts are, however, connected with our present subject, because a cross with a distinct individual is shown to be either necessary or advantageous. Dimorphic and trimorphic plants, though they are hermaphrodites, must be reciprocally crossed, one set of forms by the other, in order to be fully fertile, and in some cases to be fertile in any degree. But I should not have noticed these plants, had it not been for the following cases given by Dr. Hildebrand (17/66. 'Botanische Zeitung' January 1864 s. 3.):—

[Primula sinensis is a reciprocally dimorphic species: Dr. Hildebrand fertilised twenty-eight flowers of both forms, each by pollen of the other form, and obtained the full number of capsules containing on an average 42.7 seed per capsule; here we have complete and normal fertility. He then fertilised forty-two flowers of both forms with pollen of the same form, but taken from a distinct plant, and all produced capsules containing on an average only 19.6 seed. Lastly, and here we come to our more immediate point, he fertilised forty-eight flowers of both forms with pollen of the same form and taken from the same flower, and now he obtained only thirty-two capsules, and these contained on an average 18.6 seed, or one less per capsule than in the former case. So that, with these illegitimate unions, the act of impregnation is less assured, and the fertility slightly less, when the pollen and ovules belong to the same flower, than when belonging to two distinct individuals of the same form. Dr. Hildebrand has recently made analogous experiments on the long-styled form of Oxalis rosea, with the same result. (17/67. 'Monatsbericht Akad. Wissen.' Berlin 1866 s. 372.)]

It has recently been discovered that certain plants, whilst growing in their native country under natural conditions, cannot be fertilised with pollen from the same plant. They are sometimes so utterly self-impotent, that, though they can readily be fertilised by the pollen of a distinct species or even distinct genus, yet, wonderful as is the fact, they never produce a single seed by their own pollen. In some cases, moreover, the plant's own pollen and stigma mutually act on each other in a deleterious manner. Most of the facts to be given relate to orchids, but I will commence with a plant belonging to a widely different family.

[Sixty-three flowers of Corydalis cava, borne on distinct plants, were fertilised by Dr. Hildebrand (17/68. International Hort. Congress, London 1866.) with pollen from other plants of the same species; and fifty-eight capsules were obtained, including on an average 4.5 seed in each. He then fertilised sixteen flowers produced by the same raceme, one with another, but obtained only three capsules, one of which alone contained any good seeds, namely, two in number. Lastly, he fertilised twenty-seven flowers, each with its own pollen; he left also fifty-seven flowers to be spontaneously fertilised, and this would certainly have ensued if it had been possible, for the anthers not only touch the stigma, but the pollen-tubes were seen by Dr. Hildebrand to penetrate it; nevertheless these eighty-four flowers did not produce a single seed-capsule! This whole case is highly instructive, as it shows how widely different the action of the same pollen is, according as it is placed on the stigma of the same flower, or on that of another flower on the same raceme, or on that of a distinct plant.

With exotic Orchids several analogous cases have been observed, chiefly by Mr. John Scott. (17/69. 'Proc. Bot. Soc. of Edinburgh' May 1863: these observations are given in abstract, and others are added, in the 'Journal of Proc. of Linn. Soc.' volume 8 Bot. 1864 page 162.) Oncidium sphacelatum has effective pollen, for Mr. Scott fertilised two distinct species with it; the ovules are likewise capable of impregnation, for they were readily fertilised by the pollen of O. divaricatum; nevertheless, between one and two hundred flowers fertilised by their own pollen did not produce a single capsule, though the stigmas were penetrated by the pollen-tubes. Mr. Robertson Munro, of the Royal Botanic Gardens of Edinburgh, also informs me (1864) that a hundred and twenty flowers of this same species were fertilised by him with their own pollen, and did not produce a capsule, but eight flowers, fertilised by the pollen of O. divaricatum, produced four fine capsules: again, between two and three hundred flowers of O. divaricatum, fertilised by their own pollen, did not set a capsule, but twelve flowers fertilised by O. flexuosum produced eight fine capsules: so that here we have three utterly self-impotent species, with their male and female organs perfect, as shown by their mutual fertilisation. In these cases fertilisation was effected only by the aid of a distinct species. But, as we shall presently see, distinct plants, raised from seed, of Oncidium flexuosum, and probably of the other species, would have been perfectly capable of fertilising each other, for this is the natural process. Again, Mr. Scott found that the pollen of a plant of O. microchilum was effective, for with it he fertilised two distinct species; he found its ovules good, for they could be fertilised by the pollen of one of these species, and by the pollen of a distinct plant of O. microchilum; but they could not be fertilised by pollen of the same plant, though the pollen-tubes penetrated the stigma. An analogous case has been recorded by M. Riviere (17/70. Prof. Lecoq 'De la Fecondation' 2nd edition 1862 page 76.) with two plants of O. cavendishianum, which were both self-sterile, but reciprocally fertilised each other. All these cases refer to the genus Oncidium, but Mr. Scott found that Maxillaria atro-rubens was "totally insusceptible of fertilisation with its own pollen," but fertilised, and was fertilised by, a widely distinct species, viz. M. squalens.

As these orchids had been grown under unnatural conditions in hot-houses, I concluded that their self-sterility was due to this cause. But Fritz Muller informs me that at Desterro, in Brazil, he fertilised above one hundred flowers of the above-mentioned Oncidium flexuosum, which is there endemic, with its own pollen, and with that taken from distinct plants: all the former were sterile, whilst those fertilised by pollen from any OTHER PLANT of the same species were fertile. During the first three days there was no difference in the action of the two kinds of pollen: that placed on stigma of the same plant separated in the usual manner into grains, and emitted tubes which penetrated the column, and the stigmatic chamber shut itself; but only those flowers which had been fertilised by pollen taken from a distinct plant produced seed-capsules. On a subsequent occasion these experiments were repeated on a large scale with the same result. Fritz Muller found that four other endemic species of Oncidium were in like manner utterly sterile with their own pollen, but fertile with that from any other plant: some of them likewise produced seed-capsules when impregnated with pollen of widely distinct genera, such as Cyrtopodium, and Rodriguezia. Oncidium crispum, however, differs from the foregoing species in varying much in its self- sterility; some plants producing fine pods with their own pollen, others failing to do so in two or three instances, Fritz Muller observed that the pods produced by pollen taken from a distinct flower on the same plant, were larger than those produced by the flower's own pollen. In Epidendrum cinnabarinum, an orchid belonging to another division of the family, fine pods were produced by the plant's own pollen, but they contained by weight only about half as much seed as the capsules which had been fertilised by pollen from a distinct plant, and in one instance from a distinct species; moreover, a very large proportion, and in some cases nearly all the seeds produced by the plant's own pollen, were destitute of an embryo. Some self-fertilised capsules of a Maxillaria were in a similar state.

Another observation made by Fritz Muller is highly remarkable, namely, that with various orchids the plant's own pollen not only fails to impregnate the flower, but acts on the stigma, and is acted on, in an injurious or poisonous manner. This is shown by the surface of the stigma in contact with the pollen, and by the pollen itself becoming in from three to five days dark brown, and then decaying. The discoloration and decay are not caused by parasitic cryptograms, which were observed by Fritz Muller in only a single instance. These changes are well shown by placing on the same stigma, at the same time, the plant's own pollen and that from a distinct plant of the same species, or of another species, or even of another and widely remote genus. Thus, on the stigma of Oncidium flexuosum, the plant's own pollen and that from a distinct plant were placed side by side, and in five days' time the latter was perfectly fresh, whilst the plant's own pollen was brown. On the other hand, when the pollen of a distinct plant of the Oncidium flexuosum and of the Epidendrum zebra (nov. spec.?) were placed together on the same stigma, they behaved in exactly the same manner, the grains separating, emitting tubes, and penetrating the stigma, so that the two pollen-masses, after an interval of eleven days, could not be distinguished except by the difference of their caudicles, which, of course, undergo no change. Fritz Muller has, moreover, made a large number of crosses between orchids belonging to distinct species and genera, and he finds that in all cases when the flowers are not fertilised their footstalks first begin to wither; and the withering slowly spreads upwards until the germens fall off, after an interval of one or two weeks, and in one instance of between six and seven weeks; but even in this latter case, and in most other cases, the pollen and stigma remained in appearance fresh. Occasionally, however, the pollen becomes brownish, generally on the external surface, and not in contact with the stigma, as is invariably the case when the plant's own pollen is applied.

Fritz Muller observed the poisonous action of the plant's own pollen in the above-mentioned Oncidium flexuosum, O. unicorne, pubes (?), and in two other unnamed species. Also in two species of Rodriguezia, in two of Notylia, in one of Burlingtonia, and of a fourth genus in the same group. In all these cases, except the last, it was proved that the flowers were, as might have been expected, fertile with pollen from a distinct plant of the same species. Numerous flowers of one species of Notylia were fertilised with pollen from the same raceme; in two days' time they all withered, the germens began to shrink, the pollen-masses became dark brown, and not one pollen-grain emitted a tube. So that in this orchid the injurious action of the plant's own pollen is more rapid than with Oncidium flexuosum. Eight other flowers on the same raceme were fertilised with pollen from a distinct plant of the same species: two of these were dissected, and their stigmas were found to be penetrated by numberless pollen-tubes; and the germens of the other six flowers became well developed. On a subsequent occasion many other flowers were fertilised with their own pollen, and all fell off dead in a few days; whilst some flowers on the same raceme which had been left simply unfertilised adhered and long remained fresh. We have seen that in cross-unions between extremely distinct orchids the pollen long remains undecayed; but Notylia behaved in this respect differently; for when its pollen was placed on the stigma of Oncidium flexuosum, both the stigma and pollen quickly became dark brown, in the same manner as if the plant's own pollen had been applied.

Fritz Muller suggests that, as in all these cases the plant's own pollen is not only impotent (thus effectually preventing self-fertilisation), but likewise prevents, as was ascertained in the case of the Notylia and Oncidium flexuosum, the action of subsequently applied pollen from a distinct individual, it would be an advantage to the plant to have its own pollen rendered more and more deleterious; for the germens would thus quickly be killed, and dropping off, there would be no further waste in nourishing a part which ultimately could be of no avail.

The same naturalist found in Brazil three plants of a Bignonia growing near together. He fertilised twenty-nine flowerets on one of them with their own pollen, and they did not set a single capsule. Thirty flowers were then fertilised with pollen from a distinct plant, one of the three, and they yielded only two capsules. Lastly, five flowers were fertilised with pollen from a fourth plant growing at a distance, and all five produced capsules. Fritz Muller thinks that the three plants which grew near one another were probably seedlings from the same parent, and that from being closely related, they acted very feebly on one another. This view is extremely probable, for he has since shown in a remarkable paper (17/71. 'Jenaische Zeitschrift fur Naturwiss.' b. 7 page 22 1872 and page 441 1873. A large part of this paper has been translated in the 'American Naturalist' 1874 page 223.), that in the case of some Brazilian species of Abutilon, which are self-sterile, and between which he raised some complex hybrids, that these, if near relatives, were much less fertile inter se, than when not closely related.]

We now come to cases closely analogous with those just given, but different in so far that only certain individuals of the species are self-sterile. This self-impotence does not depend on the pollen or ovules being in an unfit state for fertilisation, for both have been found effective in union with other plants of the same or of a distinct species. The fact of plants having acquired so peculiar a constitution, that they can be fertilised more readily by the pollen of a distinct species than by their own, is exactly the reverse of what occurs with all ordinary species. For in the latter the two sexual elements of the same individual plant are of course capable of freely acting on each other; but are so constituted that they are more or less impotent when brought into union with the sexual elements of a distinct species, and produce more or less sterile hybrids.

[Gartner experimented on two plants of Lobelia fulgens, brought from separate places, and found (17/72. 'Bastarderzeugung' s. 64, 357.) that their pollen was good, for he fertilised with it L. cardinalis and syphilitica; their ovules were likewise good, for they were fertilised by the pollen of these same two species; but these two plants of L. fulgens could not be fertilised by their own pollen, as can generally be effected with perfect ease with this species. Again, the pollen of a plant of Verbascum nigrum grown in a pot was found by Gartner (17/73. Ibid s. 357.) capable of fertilising V. lychnitis and V. austriacum; the ovules could be fertilised by the pollen of V. thapsus; but the flowers could not be fertilised by their own pollen. Kolreuter, also (17/74. 'Zweite Fortsetzung' s. 10; 'Dritte Forts.' s. 40. Mr. Scott likewise fertilised fifty-four flowers of Verbascum phoeniceum, including two varieties, with their own pollen, and not a single capsule was produced. Many of the pollen-grains emitted their tubes, but only a few of them penetrated the stigmas; some slight effect however was produced, as many of the ovaries became somewhat developed: 'Journal Asiatic Soc. Bengal' 1867 page 150.), gives the case of three garden plants of Verbascum phoeniceum, which bore during two years many flowers; these he fertilised successfully with the pollen of no less than four distinct species, but they produced not a seed with their own apparently good pollen; subsequently these same plants, and others raised from seed, assumed a strangely fluctuating condition, being temporarily sterile on the male or female side, or on both sides, and sometimes fertile on both sides; but two of the plants were perfectly fertile throughout the summer.

With Reseda odorata I have found certain individuals quite sterile with their own pollen, and so it is with the indigenous Reseda lutea. The self-sterile plants of both species were perfectly fertile when crossed with pollen from any other individual of the same species. These observations will hereafter be published in another work, in which I shall also show that seeds sent to me by Fritz Muller produced by plants of Eschscholtzia californica which were quite self-sterile in Brazil, yielded in this country plants which were only slightly self-sterile.

It appears (17/75. Duvernoy quoted by Gartner 'Bastarderzeugung' s. 334) that certain flowers on certain plants of Lilium candidum can be fertilised more freely by pollen from a distinct individual than by their own. So, again, with the varieties of the potato. Tinzmann (17/76. 'Gardener's Chronicle' 1846 page 183.), who made many trials with this plant, says that pollen from another variety sometimes "exerts a powerful influence, and I have found sorts of potatoes which would not bear seed from impregnation with the pollen of their own flowers would bear it when impregnated with other pollen." It does not, however, appear to have been proved that the pollen which failed to act on the flower's own stigma was in itself good.

In the genus Passiflora it has long been known that several species do not produce fruit, unless fertilised by pollen taken from distinct species: thus, Mr. Mowbray (17/77. 'Transact. Hort. Soc.' volume 7 1830 page 95.) found that he could not get fruit from P. alata and racemosa except by reciprocally fertilising them with each other's pollen; and similar facts have been observed in Germany and France. (17/78. Prof. Lecoq 'De la Fecondation' 1845 page 70; Gartner 'Bastarderzeugung' s. 64.) I have received two accounts of P. quadrangularis never producing fruit from its own pollen, but doing so freely when fertilised in one case with the pollen of P. coerulea, and in another case with that of P. edulis. But in three other cases this species fruited freely when fertilised with its own pollen; and the writer in one case attributed the favourable result to the temperature of the house having been raised from 5 deg to 10 deg Fahr. above the former temperature, after the flowers were fertilised. (17/79. 'Gardener's Chronicle' 1868 page 1341.) With respect to P. laurifolia, a cultivator of much experience has recently remarked (17/80. 'Gardener's Chronicle' 1866 page 1068.) that the flowers "must be fertilised with the pollen of P. coerulea, or of some other common kind, as their own pollen will not fertilise them." But the fullest details on this subject have been given by Messrs. Scott and Robertson Munro (17/81. 'Journal of Proc. of Linn. Soc.' volume 8 1864 page 1168. Mr. Robertson Munro in 'Trans. Bot. Soc.' of Edinburgh volume 9 page 399.): plants of Passiflora racemosa, coerulea, and alata flowered profusely during many years in the Botanic Gardens of Edinburgh, and, though repeatedly fertilised with their own pollen, never produced any seed; yet this occurred at once with all three species when they were crossed together in various ways. In the case of P. coerulea three plants, two of which grew in the Botanic Gardens, were all rendered fertile, merely by impregnating each with pollen of one of the others. The same result was attained in the same manner with P. alata, but with only one plant out of three. As so many self-sterile species of Passiflora have been mentioned, it should be stated that the flowers of the annual P. gracilis are nearly as fertile with their own pollen as with that from a distinct plant; thus sixteen flowers spontaneously self-fertilised produced fruit, each containing on an average 21.3 seed, whilst fruit from fourteen crossed flowers contained 24.1 seed.

Returning to P. alata, I have received (1866) some interesting details from Mr. Robertson Munro. Three plants, including one in England, have already been mentioned which were inveterately self-sterile, and Mr. Munro informs me of several others which, after repeated trials during many years, have been found in the same predicament. At some other places, however, this species fruits readily when fertilised with its own pollen. At Taymouth Castle there is a plant which was formerly grafted by Mr. Donaldson on a distinct species, name unknown, and ever since the operation it has produced fruit in abundance by its own pollen; so that this small and unnatural change in the state of this plant has restored its self-fertility! Some of the seedlings from the Taymouth Castle plant were found to be not only sterile with their own pollen, but with each other's pollen, and with the pollen of distinct species. Pollen from the Taymouth plant failed to fertilise certain plants of the same species, but was successful on one plant in the Edinburgh Botanic Gardens. Seedlings were raised from this latter union, and some of their flowers were fertilised by Mr. Munro with their own pollen; but they were found to be as self-impotent as the mother-plant had always proved, except when fertilised by the grafted Taymouth plant, and except, as we shall see, when fertilised by her own seedlings. For Mr. Munro fertilised eighteen flowers on the self-impotent mother-plant with pollen from these her own self-impotent seedlings, and obtained, remarkable as the fact is, eighteen fine capsules full of excellent seed! I have met with no case in regard to plants which shows so well as this of P. alata, on what small and mysterious causes complete fertility or complete sterility depends.]

The facts hitherto given relate to the much-lessened or completely destroyed fertility of pure species when impregnated with their own pollen, in comparison with their fertility when impregnated by distinct individuals or distinct species; but closely analogous facts have been observed with hybrids.

[Herbert states (17/82. 'Amaryllidaceae' 1837 page 371; 'Journal of Hort. Soc.' volume 2 1847 page 19.) that having in flower at the same time nine hybrid Hippeastrums, of complicated origin, descended from several species, he found that "almost every flower touched with pollen from another cross produced seed abundantly, and those which were touched with their own pollen either failed entirely, or formed slowly a pod of inferior size, with fewer seeds." In the 'Horticultural Journal' he adds that "the admission of the pollen of another cross-bred Hippeastrum (however complicated the cross) to any one flower of the number, is almost sure to check the fructification of the others." In a letter written to me in 1839, Dr. Herbert says that he had already tried these experiments during five consecutive years, and he subsequently repeated them, with the same invariable result. He was thus led to make an analogous trial on a pure species, namely, on the Hippeastrum aulicum, which he had lately imported from Brazil: this bulb produced four flowers, three of which were fertilised by their own pollen, and the fourth by the pollen of a triple cross between H. bulbulosum, reginae, and vittatum; the result was, that "the ovaries of the three first flowers soon ceased to grow, and after a few days perished entirely: whereas the pod impregnated by the hybrid made vigorous and rapid progress to maturity, and bore good seed, which vegetated freely." This is, indeed, as Herbert remarks, "a strange truth," but not so strange as it then appeared.

As a confirmation of these statements, I may add that Mr. M. Mayes (17/83. Loudon's 'Gardener's Magazine' volume 11 1835 page 260.) after much experience in crossing the species of Amaryllis (Hippeastrum), says, "neither the species nor the hybrids will, we are well aware, produce seed so abundantly from their own pollen as from that of others." So, again, Mr. Bidwell, in New South Wales (17/84. 'Gardener's Chronicle' 1850 page 470.) asserts that Amaryllis belladonna bears many more seeds when fertilised by the pollen of Brunswigia (Amaryllis of some authors) josephinae or of B. multiflora, than when fertilised by its own pollen. Mr. Beaton dusted four flowers of a Cyrtanthus with their own pollen, and four with the pollen of Vallota (Amaryllis) purpurea; on the seventh day "those which received their own pollen slackened their growth, and ultimately perished; those which were crossed with the Vallota held on." (17/85. 'Journal Hort. Soc.' volume 5 page 135. The seedlings thus raised were given to the Hort. Soc.; but I find, on inquiry, that they unfortunately died the following winter.) These latter cases, however, relate to uncrossed species, like those before given with respect to Passiflora, Orchids, etc., and are here referred to only because the plants belong to the same group of Amaryllidaceae.

In the experiments on the hybrid Hippeastrums, if Herbert had found that the pollen of two or three kinds alone had been more efficient on certain kinds than their own pollen, it might have been argued that these, from their mixed parentage, had a closer mutual affinity than the others; but this explanation is inadmissible, for the trials were made reciprocally backwards and forwards on nine different hybrids; and a cross, whichever way taken, always proved highly beneficial. I can add a striking and analogous case from experiments made by the Rev. A. Rawson, of Bromley Common, with some complex hybrids of Gladiolus. This skilful horticulturist possessed a number of French varieties, differing from each other only in the colour and size of the flowers, all descended from Gandavensis, a well-known old hybrid, said to be descended from G. natalensis by the pollen of G. oppositiflorus. (17/86. Mr. D. Beaton in 'Journal of Hort.' 1861 page 453. Lecoq however ('De la Fecond.' 1862 page 369) states that this hybrid is descended from G. psittacinus and cardinalis; but this is opposed to Herbert's experience, who found that the former species could not be crossed.) Mr. Rawson, after repeated trials, found that none of the varieties would set seed with their own pollen, although taken from distinct plants of the same variety (which had, of course, been propagated by bulbs), but that they all seeded freely with pollen from any other variety. To give two examples: Ophir did not produce a capsule with its own pollen, but when fertilised with that of Janire, Brenchleyensis, Vulcain and Linne, it produced ten fine capsules; but the pollen of Ophir was good, for when Linne was fertilised by it seven capsules were produced. This latter variety, on the other hand, was utterly barren with its own pollen, which we have seen was perfectly efficient on Ophir. Altogether, Mr. Rawson, in the year 1861 fertilised twenty-six flowers borne by four varieties with pollen taken from other varieties, and every single flower produced a fine seed-capsule; whereas fifty-two flowers on the same plants, fertilised at the same time with their own pollen, did not yield a single seed-capsule. Mr. Rawson fertilised, in some cases, the alternate flowers, and in other cases all those down one side of the spike, with pollen of other varieties, and the remaining flowers with their own pollen. I saw these plants when the capsules were nearly mature, and their curious arrangement at once brought full conviction to the mind that an immense advantage had been derived from crossing these hybrids.

Lastly, I have heard from Dr. E. Bornet, of Antibes, who has made numerous experiments in crossing the species of Cistus, but has not yet published the results, that, when any of these hybrids are fertile, they may be said to be, in regard to function, dioecious; "for the flowers are always sterile when the pistil is fertilised by pollen taken from the same flower or from flowers on the same plant. But they are often fertile if pollen be employed from a distinct individual of the same hybrid nature, or from a hybrid made by a reciprocal cross."]

## CONCLUSION.

That plants should be self-sterile, although both sexual elements are in a fit state for reproduction, appears at first sight opposed to all analogy. With respect to the species, all the individuals of which are in this state, although living under their natural conditions, we may conclude that their self-sterility has been acquired for the sake of effectually preventing self- fertilisation. The case is closely analogous with that of dimorphic and trimorphic or heterostyled plants, which can be fully fertilised only by plants belonging to a different form, and not, as in the foregoing cases, indifferently by any other individual of the species. Some of these hetero- styled plants are completely sterile with pollen taken from the same plant or from the same form. With respect to species living under their natural conditions, of which only certain individuals are self-sterile (as with Reseda lutea), it is probable that these have been rendered self-sterile to ensure occasional cross-fertilisation, whilst other individuals have remained self- fertile to ensure the propagation of the species. The case seems to be parallel with that of plants which produce, as Hermann Muller has discovered, two forms—one bearing more conspicuous flowers with their structure adapted for cross-fertilisation by insects, and the other form with less conspicuous flowers adapted for self-fertilisation. The self-sterility, however, of some of the foregoing plants is incidental on the conditions to which they have been subjected, as with the Eschscholtzia, the Verbascum phoeniceum (the sterility of which varied according to the season), and with the Passiflora alata, which recovered its self-fertility when grafted on a different stock.

It is interesting to observe in the above several cases the graduated series from plants which, when fertilised by their own pollen, yield the full number of seeds, but with the seedlings a little dwarfed in stature—to plants which when self-fertilised yield few seeds—to those which yield none, but have their ovaria somewhat developed—and, lastly, to those in which the plant's own pollen and stigma mutually act on one another like poison. It is also interesting to observe on how slight a difference in the nature of the pollen or of the ovules complete self-sterility or complete self-fertility must depend in some of the above cases. Every individual of the self-sterile species appears to be capable of producing the full complement of seed when fertilised by the pollen of any other individual (though judging from the facts given with respect to Abutilon the nearest kin must be excepted); but not one individual can be fertilised by its own pollen. As every organism differs in some slight degree from every other individual of the same species, so no doubt it is with their pollen and ovules; and in the above cases we must believe that complete self-sterility and complete self-fertility depend on such slight differences in the ovules and pollen, and not their having been differentiated in some special manner in relation to one another; for it is impossible that the sexual elements of many thousand individuals should have been specialised in relation to every other individual. In some, however, of the above cases, as with certain Passifloras, an amount of differentiation between the pollen and ovules sufficient for fertilisation is gained only by employing pollen from a distinct species; but this is probably the result of such plants having been rendered somewhat sterile from the unnatural conditions to which they have been exposed.

Exotic animals confined in menageries are sometimes in nearly the same state as the above-described self-impotent plants; for, as we shall see in the following chapter, certain monkeys, the larger carnivora, several finches, geese, and pheasants, cross together, quite as freely as, or even more freely than the individuals of the same species breed together. Cases will, also, be given of sexual incompatibility between certain, male and female domesticated animals, which, nevertheless, are fertile when matched with any other individual of the same kind.

In the early part of this chapter it was shown that the crossing of individuals belonging to distinct families of the same race, or to different races or species, gives increased size and constitutional vigour to the offspring, and, except in the case of crossed species, increased fertility. The evidence rests on the universal testimony of breeders (for it should be observed that I am not here speaking of the evil results of close interbreeding), and is practically exemplified in the higher value of cross- bred animals for immediate consumption. The good results of crossing have also been demonstrated with some animals and with numerous plants, by actual weight and measurement. Although animals of pure blood will obviously be deteriorated by crossing, as far as their characteristic qualities are concerned, there seems to be no exception to the rule that advantages of the kind just mentioned are thus gained, even when there has not been any previous close interbreeding; and the rule applies to such animals as cattle and sheep, which can long resist breeding in-and-in between the nearest blood-relations.

In the case of crossed species, although size, vigour, precocity, and hardiness are, with rare exceptions, gained, fertility, in a greater or less degree, is lost; but the gain in the above respects can hardly be attributed to the principle of compensation; for there is no close parallelism between the increased size and vigour of hybrid offspring and their sterility. Moreover, it has been clearly proved that mongrels which are perfectly fertile gain these same advantages as well as sterile hybrids.

With the higher animals no special adaptations for ensuring occasional crosses between distinct families seem to exist. The eagerness of the males, leading to severe competition between them, is sufficient; for even with gregarious animals, the old and dominant males will be dispossessed after a time and it would be a mere chance if a closely related member of the same family were to be the victorious successor. The structure of many of the lower animals, when they are hermaphrodites, is such as to prevent the ovules being fertilised by the male element of the same individual; so that the concourse of two individuals is necessary. In other cases the access of the male element of a distinct individual is at least possible. With plants, which are affixed to the ground and cannot wander from place to place like animals, the numerous adaptations for cross-fertilisation are wonderfully perfect, as has been admitted by every one who has studied the subject.

The evil consequences of long-continued close interbreeding are not so easily recognised as the good effects from crossing, for the deterioration is gradual. Nevertheless, it is the general opinion of those who have had most experience, especially with animals which propagate quickly, that evil does inevitably follow sooner or later, but at different rates with different animals. No doubt a false belief may, like a superstition, prevail widely; yet it is difficult to suppose that so many acute observers have all been deceived at the expense of much cost and trouble. A male animal may sometimes be paired with his daughter, granddaughter, and so on, even for seven generations, without any manifest bad result: but the experiment has never been tried of matching brothers and sisters, which is considered the closest form of interbreeding, for an equal number of generations. There is good reason to believe that by keeping the members of the same family in distinct bodies, especially if exposed to somewhat different conditions of life, and by occasionally crossing these families, the evil results of interbreeding may be much diminished or quite eliminated. These results are loss of constitutional vigour, size, and fertility; but there is no necessary deterioration in the general form of the body, or in other good qualities. We have seen that with pigs first-rate animals have been produced after long-continued close interbreeding, though they had become extremely infertile when paired with their near relations. The loss of fertility, when it occurs, seems never to be absolute, but only relative to animals of the same blood; so that this sterility is to a certain extent analogous with that of self-impotent plants which cannot be fertilised by their own pollen, but are perfectly fertile with pollen of any other individual of the same species. The fact of infertility of this peculiar nature being one of the results of long-continued interbreeding, shows that interbreeding does not act merely by combining and augmenting various morbid tendencies common to both parents; for animals with such tendencies, if not at the time actually ill, can generally propagate their kind. Although offspring descended from the nearest blood-relations are not necessarily deteriorated in structure, yet some authors believe that they are eminently liable to malformations; and this is not improbable, as everything which lessens the vital powers acts in this manner. Instances of this kind have been recorded in the case of pigs, bloodhounds, and some other animals.

Finally, when we consider the various facts now given which plainly show that good follows from crossing, and less plainly that evil follows from close interbreeding, and when we bear in mind that with very many organisms elaborate provisions have been made for the occasional union of distinct individuals, the existence of a great law of nature is almost proved; namely, that the crossing of animals and plants which are not closely related to each other is highly beneficial or even necessary, and that interbreeding prolonged during many generations is injurious.

# CHAPTER 2.XVIII.

## ON THE ADVANTAGES AND DISADVANTAGES OF CHANGED CONDITIONS OF LIFE: STERILITY FROM VARIOUS CAUSES.

ON THE GOOD DERIVED FROM SLIGHT CHANGES IN THE CONDITIONS OF LIFE. STERILITY FROM CHANGED CONDITIONS, IN ANIMALS, IN THEIR NATIVE COUNTRY AND IN MENAGERIES. MAMMALS, BIRDS, AND INSECTS. LOSS OF SECONDARY SEXUAL CHARACTERS AND OF INSTINCTS. CAUSES OF STERILITY. STERILITY OF DOMESTICATED ANIMALS FROM CHANGED CONDITIONS. SEXUAL INCOMPATIBILITY OF INDIVIDUAL ANIMALS. STERILITY OF PLANTS FROM CHANGED CONDITIONS OF LIFE. CONTABESCENCE OF THE ANTHERS. MONSTROSITIES AS A CAUSE OF STERILITY. DOUBLE FLOWERS. SEEDLESS FRUIT. STERILITY FROM THE EXCESSIVE DEVELOPMENT OF THE ORGANS OF VEGETATION. FROM LONG-CONTINUED PROPAGATION BY BUDS. INCIPIENT STERILITY THE PRIMARY CAUSE OF DOUBLE FLOWERS AND SEEDLESS FRUIT.

## ON THE GOOD DERIVED FROM SLIGHT CHANGES IN THE CONDITIONS OF LIFE.

In considering whether any facts were known which might throw light on the conclusion arrived at in the last chapter, namely, that benefits ensue from crossing, and that it is a law of nature that all organic beings should occasionally cross, it appeared to me probable that the good derived from slight changes in the conditions of life, from being an analogous phenomenon, might serve this purpose. No two individuals, and still less no two varieties, are absolutely alike in constitution and structure; and when the germ of one is fertilised by the male element of another, we may believe that it is acted on in a somewhat similar manner as an individual when exposed to slightly changed conditions. Now, every one must have observed the remarkable influence on convalescents of a change of residence, and no medical man doubts the truth of this fact. Small farmers who hold but little land are convinced that their cattle derive great benefit from a change of pasture. In the case of plants, the evidence is strong that a great advantage is derived from exchanging seeds, tubers, bulbs, and cuttings from one soil or place to another as different as possible.

[The belief that plants are thus benefited, whether or not well founded, has been firmly maintained from the time of Columella, who wrote shortly after the Christian era, to the present day; and it now prevails in England, France, and Germany. (18/1. For England see below. For Germany see Metzger 'Getreidearten' 1841 s. 63. For France Loiseleur-Deslongchamps ('Consid. sur les Cereales' 1843 page 200) gives numerous references on this subject. For Southern France see Godron 'Florula Juvenalis' 1854 page 28.) A sagacious observer, Bradley, writing in 1724 (18/2. 'A General Treatise of Husbandry' volume 3 page 58.), says, "When we once become Masters of a good Sort of Seed, we should at least put it into Two or Three Hands, where the Soils and Situations are as different as possible; and every Year the Parties should change with one another; by which Means, I find the Goodness of the Seed will be maintained for several Years. For Want of this Use many Farmers have failed in their Crops and been great Losers." He then gives his own practical experience on this head. A modern writer (18/3. 'Gardener's Chronicle and Agricult. Gazette' 1858 page 247; and for the second statement, Ibid 1850 page 702. On this same subject see also Rev. D. Walker 'Prize Essay of Highland Agricult. Soc.' volume 2 page 200. Also Marshall 'Minutes of Agriculture' November 1775.) asserts, "Nothing can be more clearly established in agriculture than that the continual growth of any one variety in the same district makes it liable to deterioration either in quality or quantity." Another writer states that he sowed close together in the same field two lots of wheat-seed, the product of the same original stock, one of which had been grown on the same land and the other at a distance, and the difference in favour of the crop from the latter seed was remarkable. A gentleman in Surrey who has long made it his business to raise wheat to sell for seed, and who has constantly realised in the market higher prices than others, assures me that he finds it indispensable continually to change his seed; and that for this purpose he keeps two farms differing much in soil and elevation.

With respect to the tubers of the potato, I find that at the present day the practice of exchanging sets is almost everywhere followed. The great growers of potatoes in Lancashire formerly used to get tubers from Scotland, but they found that "a change from the moss-lands, and vice versa, was generally sufficient." In former times in France the crop of potatoes in the Vosges had become reduced in the course of fifty or sixty years in the proportion from 120-150 to 30-40 bushels; and the famous Oberlin attributed the surprising good which he effected in large part to changing the sets. (18/4. Oberlin 'Memoirs' English translation page 73. For Lancashire see Marshall 'Review of Reports' 1808 page 295.)

A well-known practical gardener, Mr. Robson (18/5. 'Cottage Gardener' 1856 page 186. For Mr. Robson's subsequent statements see 'Journal of Horticulture' February 18, 1866 page 121. For Mr. Abbey's remarks on grafting etc. Ibid July 18, 1865 page 44.) positively states that he has himself witnessed decided advantage from obtaining bulbs of the onion, tubers of the potato, and various seeds, all of the same kind, from different soils and distant parts of England. He further states that with plants propagated by cuttings, as with the Pelargonium, and especially the Dahlia, manifest advantage is derived from getting plants of the same variety, which have been cultivated in another place; or, "where the extent of the place allows, to take cuttings from one description of soil to plant on another, so as to afford the change that seems so necessary to the well-being of the plants." He maintains that after a time an exchange of this nature is "forced on the grower, whether he be prepared for it or not." Similar remarks have been made by another excellent gardener, Mr. Fish, namely, that cuttings of the same variety of Calceolaria, which he obtained from a neighbour, "showed much greater vigour than some of his own that were "treated in exactly the same manner," and he attributed this solely to his own plants having become "to a certain extent worn out or tired of their quarters." Something of this kind apparently occurs in grafting and budding fruit-trees; for, according to Mr. Abbey, grafts or buds generally take with greater facility on a distinct variety or even species, or on a stock previously grafted, than on stocks raised from seeds of the variety which is to be grafted; and he believes this cannot be altogether explained by the stocks in question being better adapted to the soil and climate of the place. It should, however, be added, that varieties grafted or budded on very distinct kinds, though they may take more readily and grow at first more vigorously than when grafted on closely allied stocks, afterwards often become unhealthy.

I have studied M. Tessier's careful and elaborate experiments (18/6. 'Mem. de l'Acad. des Sciences' 1790 page 209.) made to disprove the common belief that good is derived from a change of seed; and he certainly shows that the same seed may with care be cultivated on the same farm (it is not stated whether on exactly the same soil) for ten consecutive years without loss. Another excellent observer, Colonel Le Couteur (18/7. 'On the Varieties of Wheat' page 52.) has come to the same conclusion; but then he expressly adds, if the same seed be used, "that which is grown on land manured from the mixen one year becomes seed for land prepared with lime, and that again becomes seed for land dressed with ashes, then for land dressed with mixed manure, and so on." But this in effect is a systematic exchange of seed, within the limits of the same farm.]

On the whole the belief, which has long been held by many cultivators, that good follows from exchanging seed, tubers, etc., seems to be fairly well founded. It seems hardly credible that the advantage thus derived can be due to the seeds, especially if very small ones, obtaining in one soil some chemical element deficient in the other and in sufficient quantity to influence the whole after-growth of the plant. As plants after once germinating are fixed to the same spot, it might have been anticipated that they would show the good effects of a change more plainly than do animals which continually wander about; and this apparently is the case. Life depending on, or consisting in, an incessant play of the most complex forces, it would appear that their action is in some way stimulated by slight changes in the circumstances to which each organism is exposed. All forces throughout nature, as Mr. Herbert Spencer (18/8. Mr. Spencer has fully and ably discussed this whole subject in his 'Principles of Biology' 1864 volume 2 chapter 10. In the first edition of my 'Origin of Species' 1859 page 267, I spoke of the good effects from slight changes in the conditions of life and from cross-breeding, and of the evil effects from great changes in the conditions and from crossing widely distinct forms, as a series of facts "connected together by some common but unknown bond, which is essentially related to the principle of life.) remarks, tend towards an equilibrium, and for the life of each organism it is necessary that this tendency should be checked. These views and the foregoing facts probably throw light, on the one hand, on the good effects of crossing the breed, for the germ will be thus slightly modified or acted on by new forces; and on the other hand, on the evil effects of close interbreeding prolonged during many generations, during which the germ will be acted on by a male having almost identically the same constitution.

## STERILITY FROM CHANGED CONDITIONS OF LIFE.

I will now attempt to show that animals and plants, when removed from their natural conditions, are often rendered in some degree infertile or completely barren; and this occurs even when the conditions have not been greatly changed. This conclusion is not necessarily opposed to that at which we have just arrived, namely, that lesser changes of other kinds are advantageous to organic beings. Our present subject is of some importance, from having an intimate connection with the causes of variability. Indirectly it perhaps bears on the sterility of species when crossed: for as, on the one hand, slight changes in the conditions of life are favourable to plants and animals, and the crossing of varieties adds to the size, vigour, and fertility of their offspring; so, on the other hand, certain other changes in the conditions of life cause sterility; and as this likewise ensues from crossing much-modified forms or species, we have a parallel and double series of facts, which apparently stand in close relation to each other.

It is notorious that many animals, though perfectly tamed, refuse to breed in captivity. Isidore Geoffroy St.-Hilaire (18/9. 'Essais de Zoologie Generale' 1841 page 256.) consequently has drawn a broad distinction between tamed animals which will not breed under captivity, and truly domesticated animals which breed freely—generally more freely, as shown in the sixteenth chapter, than in a state of nature. It is possible and generally easy to tame most animals; but experience has shown that it is difficult to get them to breed regularly, or even at all. I shall discuss this subject in detail; but will give only those cases which seem most illustrative. My materials are derived from notices scattered through various works, and especially from a Report, kindly drawn up for me by the officers of the Zoological Society of London, which has especial value, as it records all the cases, during nine years from 1838-46, in which the animals were seen to couple but produced no offspring, as well as the cases in which they never, as far as known, coupled. This MS. Report I have corrected by the annual Reports subsequently published up to the year 1865. (18/10. Since the appearance of the first edition of this work, Mr. Sclater has published ('Proc. Zoolog. Soc.' 1868 page 623) a list of the species of mammals which have bred in the gardens from 1848 to 1867 inclusive. Of the Artiodactyla 85 species have been kept, and of these 1 species in 1.9 have bred at least once during the 20 years; of 28 Marsupialia, 1 in 2.5 have bred; of 74 Carnivora, 1 in 3.0 have bred; of 52 Rodentia, 1 in 4.7 have bred; and of Quadrumana 75 species have been kept, and 1 in 6.2 have bred.) Many facts are given on the breeding of the animals in that magnificent work, 'Gleanings from the Menageries of Knowsley Hall' by Dr. Gray. I made, also, particular inquiries from the experienced keeper of the birds in the old Surrey Zoological Gardens. I should premise that a slight change in the treatment of animals sometimes makes a great difference in their fertility; and it is probable that the results observed in different menageries would differ. Indeed, some animals in our Zoological Gardens have become more productive since the year 1846. It is, also, manifest from F. Cuvier's account of the Jardin des Plantes (18/11. Du Rut 'Annales du Museum' 1807 tome 9 page 120.) that the animals formerly bred much less freely there than with us; for instance, in the Duck tribe, which is highly prolific, only one species had at that period produced young.

[The most remarkable cases, however, are afforded by animals kept in their native country, which, though perfectly tamed, quite healthy, and allowed some freedom, are absolutely incapable of breeding. Rengger (18/12. 'Saugethiere von Paraguay' 1830 s. 49, 106, 118, 124, 201, 208, 249, 265, 327.), who in Paraguay particularly attended to this subject, specifies six quadrupeds in this condition; and he mentions two or three others which most rarely breed. Mr. Bates, in his admirable work on the Amazons, strongly insists on similar cases (18/13. 'The Naturalist on the Amazons' 1863 volume 1 pages 99, 193; volume 2 page 113.); and he remarks, that the fact of thoroughly tamed native mammals and birds not breeding when kept by the Indians, cannot be wholly accounted for by their negligence or indifference, for the turkey and fowl are kept and bred by various remote tribes. In almost every part of the world—for instance, in the interior of Africa, and in several of the Polynesian islands —the natives are extremely fond of taming the indigenous quadrupeds and birds; but they rarely or never succeed in getting them to breed.

The most notorious case of an animal not breeding in captivity is that of the elephant. Elephants are kept in large numbers in their native Indian home, live to old age, and are vigorous enough for the severest labour; yet, with a very few exceptions, they have never been known even to couple, though both males and females have their proper periodical seasons. If, however, we proceed a little eastward to Ava, we hear from Mr. Crawfurd (18/14. 'Embassy to the Court of Ava' volume 1 page 534.) that their "breeding in the domestic state, or at least in the half-domestic state in which the female elephants are generally kept, is of everyday occurrence;" and Mr. Crawfurd informs me that he believes that the difference must be attributed solely to the females being allowed to roam the forest with some degree of freedom. The captive rhinoceros, on the other hand, seems from Bishop Heber's account (18/15. 'Journal' volume 1 page 213.) to breed in India far more readily than the elephant. Four wild species of the horse genus have bred in Europe, though here exposed to a great change in their natural habits of life; but the species have generally been crossed one with another. Most of the members of the pig family breed readily in our menageries; even the Red River hog (Potamochoerus penicillatus), from the sweltering plains of West Africa, has bred twice in the Zoological Gardens. Here also the Peccary (Dicotyles torquatus) has bred several times; but another species, the D. labiatus, though rendered so tame as to be half-domesticated, is said to breed so rarely in its native country of Paraguay, that according to Rengger (18/16. 'Saugethiere' s. 327.) the fact requires confirmation. Mr. Bates remarks that the tapir, though often kept tame in Amazonia by the Indians, never breeds.

Ruminants generally breed quite freely in England, though brought from widely different climates, as may be seen in the Annual Reports of the Zoological Gardens, and in the Gleanings from Lord Derby's menagerie.

The Carnivora, with the exception of the Plantigrade division, breed (though with capricious exceptions) about half as freely as ruminants. Many species of Felidae have bred in various menageries, although imported from diverse climates and closely confined. Mr. Bartlett, the present superintendent of the Zoological Gardens (18/17. On the Breeding of the Larger Felidae 'Proc. Zoolog. Soc.' 1861 page 140.) remarks that the lion appears to breed more frequently and to bring forth more young at a birth than any other species of the family. He adds that the tiger has rarely bred; "but there are several well-authenticated instances of the female tiger breeding with the lion." Strange as the fact may appear, many animals under confinement unite with distinct species and produce hybrids quite as freely as, or even more freely than, with their own species. On inquiring from Dr. Falconer and others, it appears that the tiger when confined in India does not breed, though it has been known to couple. The chetah (Felis jubata) has never been known by Mr. Bartlett to breed in England, but it has bred at Frankfort; nor does it breed in India, where it is kept in large numbers for hunting; but no pains would be taken to make them breed, as only those animals which have hunted for themselves in a state of nature are serviceable and worth training. (18/18. Sleeman's 'Rambles in India' volume 2 page 10.) According to Rengger, two species of wild cats in Paraguay, though thoroughly tamed, have never bred. Although so many of the Felidae breed readily in the Zoological Gardens, yet conception by no means always follows union: in the nine-year Report, various species are specified which were observed to couple seventy-three times, and no doubt this must have passed many times unnoticed; yet from the seventy- three unions only fifteen births ensued. The Carnivora in the Zoological Gardens were formerly less freely exposed to the air and cold than at present, and this change of treatment, as I was assured by the former superintendent, Mr. Miller, greatly increased their fertility. Mr. Bartlett, and there cannot be a more capable judge, says, "it is remarkable that lions breed more freely in travelling collections than in the Zoological Gardens; probably the constant excitement and irritation produced by moving from place to place, or change of air, may have considerable influence in the matter."

Many members of the Dog family breed readily when confined. The Dhole is one of the most untamable animals in India, yet a pair kept there by Dr. Falconer produced young. Foxes, on the other hand, rarely breed, and I have never heard of such an occurrence with the European fox: the silver fox of North America (Canis argentatus), however, has bred several times in the Zoological Gardens. Even the otter has bred there. Every one knows how readily the semi- domesticated ferret breeds, though shut up in miserably small cages; but other species of Viverra and Paradoxurus absolutely refuse to breed in the Zoological Gardens. The Genetta has bred both here and in the Jardin des Plantes, and produced hybrids. The Herpestes fasciatus has likewise bred; but I was formerly assured that the H. griseus, though many were kept in the Gardens, never bred.

The Plantigrade Carnivora breed under confinement much less freely than other Carnivora, although no reason can be assigned for this fact. In the nine-year Report it is stated that the bears had been seen in the Zoological Gardens to couple freely, but previously to 1848 had most rarely conceived. In the Reports published since this date three species have produced young (hybrids in one case), and, wonderful to relate, the white Polar bear has produced young. The badger (Meles taxus) has bred several times in the Gardens; but I have not heard of this occurring elsewhere in England, and the event must be very rare, for an instance in Germany has been thought worth recording. (18/19. Wiegmann 'Archiv. fur Naturgesch.' 1837 s. 162.) In Paraguay the native Nasua, though kept in pairs during many years and perfectly tamed, has never been known, according to Rengger, to breed or show any sexual passion; nor, as I hear from Mr. Bates, does this animal, or the Cercoleptes, breed in Amazonia. Two other plantigrade genera, Procyon and Gulo, though often kept tame in Paraguay, never breed there. In the Zoological Gardens species of Nasua and Procyon have been seen to couple; but they did not produce young.

As domesticated rabbits, guinea-pigs, and white mice breed so abundantly when closely confined under various climates, it might have been thought that most other members of the Rodent order would have bred in captivity, but this is not the case. It deserves notice, as showing how the capacity to breed sometimes goes by affinity, that the one native rodent of Paraguay, which there breeds FREELY and has yielded successive generations, is the Cavia aperea; and this animal is so closely allied to the guinea-pig, that it has been erroneously thought to be the parent form. (18/20. Rengger 'Saugethiere' etc. s. 276. On the parentage of the guinea-pig, see also Isid. Geoffroy St.- Hilaire 'Hist. Nat. Gen.' I sent to Mr. H. Denny of Leeds the lice which I collected from the wild aperea in La Plata, and he informs me that they belong to a genus distinct from those found on the guinea-pig. This is important evidence that the aperea is not the parent of the guinea-pig; and is worth giving, as some authors erroneously suppose that the guinea-pig since being domesticated has become sterile when crossed with the aperea.) In the Zoological Gardens, some rodents have coupled, but have never produced young; some have neither coupled nor bred; but a few have bred, as the porcupine more than once, the Barbary mouse, lemming, chinchilla, and agouti (Dasyprocta aguti) several times. This latter animal has also produced young in Paraguay, though they were born dead and ill-formed; but in Amazonia, according to Mr. Bates, it never breeds, though often kept tame about the houses. Nor does the paca (Coelogenys paca) breed there. The common hare when confined has, I believe, never bred in Europe; though, according to a recent statement, it has crossed with the rabbit. (18/21. Although the existence of the Leporides, as described by Dr. Broca ('Journal de Phys.' tome 2 page 370), has been positively denied, yet Dr. Pigeaux ('Annals and Mag. of Nat. Hist.' volume 20 1867 page 75) affirms that the hare and rabbit have produced hybrids.) I have never heard of the dormouse breeding in confinement. But squirrels offer a more curious case: with one exception, no species has bred in the Zoological Gardens, yet as many as fourteen individuals of S. palmarum were kept together during several years. The S. cinera has been seen to couple, but it did not produce young; nor has this species, when rendered extremely tame in its native country, North America, been ever known to breed. (18/22. 'Quadrupeds of North America' by Audubon and Bachman 1846 page 268.) At Lord Derby's menagerie squirrels of many kinds were kept in numbers, but Mr. Thompson, the superintendent, told me that none had ever bred there, or elsewhere as far as he knew. I have never heard of the English squirrel breeding in confinement. But the species which has bred more than once in the Zoological Gardens is the one which perhaps might have been least expected, namely, the flying squirrel (Sciuropterus volucella): it has, also, bred several times near Birmingham; but the female never produced more than two young at a birth, whereas in its native American home she bears from three to six young. (18/23. Loudon's 'Mag. of Nat. Hist.' volume 9 1836 page 571; Audubon and Bachman 'Quadrupeds of North America' page 221.)

Monkeys, in the nine-year Report from the Zoological Gardens, are stated to unite most freely, but during this period, though many individuals were kept, there were only seven births. I have heard of only one American monkey, the Ouistiti, breeding in Europe. (18/24. Flourens 'De l'Instinct' etc. 1845 page 88.) A Macacus, according to Flourens, bred in Paris; and more than one species of this genus has produced young in London, especially the Macacus rhesus, which everywhere shows a special capacity to breed under confinement. Hybrids have been produced both in Paris and London from this same genus. The Arabian baboon, or Cynocephalus hamadryas (18/25. See 'Annual Reports Zoolog. Soc.' 1855, 1858, 1863, 1864; 'Times' newspaper August 10, 1847; Flourens 'De l'Instinct' page 85.), and a Cercopithecus have bred in the Zoological Gardens, and the latter species at the Duke of Northumberland's. Several members of the family of Lemurs have produced hybrids in the Zoological Gardens. It is much more remarkable that monkeys very rarely breed when confined in their native country; thus the Cay (Cebus azara) is frequently and completely tamed in Paraguay, but Rengger (18/26. 'Saugethiere' etc. s. 34, 49.) says that it breeds so rarely, that he never saw more than two females which had produced young. A similar observation has been made with respect to the monkeys which are frequently tamed by the aborigines in Brazil. (18/27. Art. Brazil 'Penny Cyclop.' page 363.) In Amazonia, these animals are so often kept in a tame state, that Mr. Bates in walking through the streets of Para counted thirteen species; but, as he asserts, they have never been known to breed in captivity. (18/28. 'The Naturalist on the Amazons' volume 1 page 99.)

## BIRDS.

Birds offer in some respects better evidence than quadrupeds, from their breeding more rapidly and being kept in greater numbers. (18/29. A list of the species of birds which have bred in the Zoological Gardens from 1848 to 1867 inclusive has been published by Mr. Sclater in 'Proc. Zoolog. Soc.' 1869 page 626, since the first edition of this work appeared. Of Columbae 51 species have been kept, and of Anseres 80 species, and in both these families 1 species in 2.6 have bred at least once in the 20 years. Of Gallinae 83 species have been kept and 1 in 27 have bred; of 57 Grallae 1 in 9 have bred; of 110 Prehensores 1 in 22 have bred; of 178 Passeres 1 in 25.4 have bred; of 94 Accipitres 1 in 47 have bred; of 25 Picariae and of 35 Herodiones not one species in either group has bred.) We have seen that carnivorous animals are more fertile under confinement than most other mammals. The reverse holds good with carnivorous birds. It is said (18/30. 'Encyclop. of Rural Sports' page 691.) that as many as eighteen species have been used in Europe for hawking, and several others in Persia and India (18/31. According to Sir A. Burnes 'Cabool' etc. page 51, eight species are used for hawking in Sinde.); they have been kept in their native country in the finest condition, and have been flown during six, eight, or nine years (18/32. Loudon's 'Mag. of Nat. Hist.' volume 6 1833 page 110.); yet there is no record of their having ever produced young. As these birds were formerly caught whilst young, at great expense, being imported from Iceland, Norway, and Sweden, there can be little doubt that, if possible, they would have been propagated. In the Jardin des Plantes, no bird of prey has been known to couple. (18/33. F. Cuvier 'Annal. du Museum' tome 9 page 128.) No hawk, vulture, or owl has ever produced fertile eggs in the Zoological Gardens, or in the old Surrey Gardens, with the exception, in the former place on one occasion, of a condor and a kite (Milvus niger). Yet several species, namely, the Aquila fusca, Haliaetus leucocephalus, Falco tinnunculus, F. subbuteo, and Buteo vulgaris, have been seen to couple in the Zoological Gardens. Mr. Morris (18/34. 'The Zoologist' volume 7-8 1849-50 page 2648.) mentions as a unique fact that a kestrel (Falco tinnunculus) bred in an aviary. The one kind of owl which has been known to couple in the Zoological Gardens was the Eagle Owl (Bubo maximus); and this species shows a special inclination to breed in captivity; for a pair at Arundel Castle, kept more nearly in a state of nature "than ever fell to the lot of an animal deprived of its liberty" (18/35. Knox 'Ornithological Rambles in Sussex' page 91.), actually reared their young. Mr. Gurney has given another instance of this same owl breeding in confinement; and he records the case of a second species of owl, the Strix passerina, breeding in captivity. (18/36. 'The Zoologist' volume 7-8 1849-50 page 2566; volume 9-10 1851-2 page 3207.)

Of the smaller graminivorous birds, many kinds have been kept tame in their native countries, and have lived long; yet, as the highest authority on cage-birds (18/37. Bechstein 'Naturgesch. der Stubenvogel' 1840 s. 20.) remarks, their propagation is "uncommonly difficult." The canary-bird shows that there is no inherent difficulty in these birds breeding freely in confinement; and Audubon says (18/38. 'Ornithological Biography' volume 5 page 517.) that the Fringilla (Spiza) ciris of North America breeds as perfectly as the canary. The difficulty with the many finches which have been kept in confinement is all the more remarkable as more than a dozen species could be named which have yielded hybrids with the canary; but hardly any of these, with the exception of the siskin (Fringilla spinus), have reproduced their own kind. Even the bullfinch (Loxia pyrrhula) has bred as frequently with the canary, though belonging to a distinct genus, as with its own species. (18/39. A case is recorded in 'The Zoologist' volume 1-2 1843-45 page 453. For the siskin breeding, volume 3-4 1845-46 page 1075. Bechstein 'Stubenvogel' s. 139 speaks of bullfinches making nests, but rarely producing young.) With respect to the skylark (Alauda arvensis), I have heard of birds living for seven years in an aviary, which never produced young; and a great London bird-fancier assured me that he had never known an instance of their breeding; nevertheless one case has been recorded. (18/40. Yarrell 'Hist. British Birds' 1839 volume 1 page 412.) In the nine-year Report from the Zoological Society, twenty-four insessorial species are enumerated which had not bred, and of these only four were known to have coupled.

Parrots are singularly long-lived birds; and Humboldt mentions the curious fact of a parrot in South America, which spoke the language of an extinct Indian tribe, so that this bird preserved the sole relic of a lost language. Even in this country there is reason to believe (18/41. Loudon's 'Mag. of Nat. History' volume 19 1836 page 347.) that parrots have lived to the age of nearly one hundred years; yet they breed so rarely, though many have been kept in Europe, that the event has been thought worth recording in the gravest publications. (18/42. 'Memoires du Museum d'Hist. Nat.' tome 10 page 314: five cases of parrots breeding in France are here recorded. See also 'Report Brit. Assoc. Zoolog.' 1843.) Nevertheless, when Mr. Buxton turned out a large number of parrots in Norfolk, three pairs bred and reared ten young birds in the course of two seasons; and this success may be attributed to their free life. (18/43. 'Annals and Mag. of Nat. Hist.' November 1868 page 311.) According to Bechstein (18/44. 'Stubenvogel' s. 105, 83.) the African Psittacus erithacus breeds oftener than any other species in Germany: the P. macoa occasionally lays fertile eggs, but rarely succeeds in hatching them; this bird, however, has the instinct of incubation sometimes so strongly developed, that it will hatch the eggs of fowls or pigeons. In the Zoological Gardens and in the old Surrey Gardens some few species have coupled, but, with the exception of three species of parakeets, none have bred. It is a much more remarkable fact that in Guiana parrots of two kinds, as I am informed by Sir R. Schomburgk, are often taken from the nests by the Indians and reared in large numbers; they are so tame that they fly freely about the houses, and come when called to be fed, like pigeons; yet he has never heard of a single instance of their breeding. (18/45. Dr. Hancock remarks ('Charlesworth's Mag. of Nat. Hist.' volume 2 1838 page 492) "it is singular that, amongst the numerous useful birds that are indigenous to Guiana, none are found to propagate among the Indians; yet the common fowl is reared in abundance throughout the country.") In Jamaica, a resident naturalist, Mr. R. Hill (18/46. 'A Week at Pert Royal' 1855 page 7.), says, "no birds more readily submit to human dependence than the parrot-tribe, but no instance of a parrot breeding in this tame life has been known yet." Mr. Hill specifies a number of other native birds kept tame in the West Indies, which never breed in this state.

The great pigeon family offers a striking contrast with the parrots: in the nine-year Report thirteen species are recorded as having bred, and, what is more noticeable, only two were seen to couple without any result. Since the above date every annual Report gives many cases of various pigeons breeding. The two magnificent crowned pigeons (Goura coronata and victoriae) produced hybrids; nevertheless, of the former species more than a dozen birds were kept, as I am informed by Mr. Crawfurd, in a park at Penang, under a perfectly well-adapted climate, but never once bred. The Columba migratoria in its native country, North America, invariably lays two eggs, but in Lord Derby's menagerie never more than one. The same fact has been observed with the C. leucocephala. (18/47. Audubon 'American Ornithology' volume 5 pages 552, 557.)

Gallinaceous birds of many genera likewise show an eminent capacity for breeding under captivity. This is particularly the case with pheasants, yet our English species seldom lays more than ten eggs in confinement; whilst from eighteen to twenty is the usual number in the wild state. (18/48. Mowbray on 'Poultry' 7th edition page 133.) With the Gallinaceae, as with all other orders, there are marked and inexplicable exceptions in regard to the fertility of certain species and genera under confinement. Although many trials have been made with the common partridge, it has rarely bred, even when reared in large aviaries; and the hen will never hatch her own eggs. (18/49. Temminck 'Hist. Nat. Gen. des Pigeons' etc. 1813 tome 3 pages 288, 382; 'Annals and Mag. of Nat. Hist.' volume 12 1843 page 453. Other species of partridge have occasionally bred; as the red-legged (P. rubra), when kept in a large court in France (see Journal de Physique' tome 25 page 294), and in the Zoological Gardens in 1856.) The American tribe of Guans or Cracidae are tamed with remarkable ease, but are very shy breeders in this country (18/50. Rev. E.S. Dixon 'The Dovecote' 1851 pages 243-252.); but with care various species were formerly made to breed rather freely in Holland. (18/51. Temminck 'Hist. Nat. Gen. des Pigeons' etc. tome 2 pages 456, 458; tome 3 pages 2, 13, 47.) Birds of this tribe are often kept in a perfectly tamed condition in their native country by the Indians, but they never breed. (18/52. Bates 'The Naturalist on the Amazons' volume 1 page 193; volume 2 page 112.) It might have been expected that grouse from their habits of life would not have bred in captivity, more especially as they are said soon to languish and die. (18/53. Temminck 'Hist. Nat. Gen.' etc. tome 2 page 125. For Tetrao urogallus see L. Lloyd 'Field Sports of North of Europe' volume 1 pages 287, 314; and Bull. de la Soc. d'Acclimat.' tome 7 1860 page 600. For T. scoticus Thompson 'Nat. Hist. of Ireland' volume 2 1850 page 49. For T. cupido 'Boston Journal of Nat. Hist.' volume 3 page 199.) But many cases are recorded of their breeding: the capercailzie (Tetrao urogallus) has bred in the Zoological Gardens; it breeds without much difficulty when confined in Norway, and in Russia five successive generations have been reared: Tetrao tetrix has likewise bred in Norway; T. scoticus in Ireland; T. umbellus at Lord Derby's; and T. cupido in North America.

It is scarcely possible to imagine a greater change in habits than that which the members of the ostrich family must suffer, when cooped up in small enclosures under a temperate climate, after freely roaming over desert and tropical plains or entangled forests; yet almost all the kinds have frequently produced young in the various European menageries, even the mooruk (Casuarius bennetii) from New Ireland. The African ostrich, though perfectly healthy and living long in the South of France, never lays more than from twelve to fifteen eggs, though in its native country it lays from twenty-five to thirty. (18/54. Marcel de Serres 'Annales des Sc. Nat.' 2nd series Zoolog. tome 13 page 175.) Here we have another instance of fertility impaired, but not lost, under confinement, as with the flying squirrel, the hen-pheasant, and two species of American pigeons.

Most Waders can be tamed, as the Rev. E.S. Dixon informs me, with remarkable facility; but several of them are short-lived under confinement, so that their sterility in this state is not surprising. The cranes breed more readily than other genera: Grus montigresia has bred several times in Paris and in the Zoological Gardens, as has G. cinerea at the latter place, and G. antigone at Calcutta. Of other members of this great order, Tetrapteryx paradisea has bred at Knowsley, a Porphyrio in Sicily, and the Gallinula chloropus in the Zoological Gardens. On the other hand, several birds belonging to this order will not breed in their native country, Jamaica; and the Psophia, though often kept by the Indians of Guiana about their houses, "is seldom or never known to breed." (18/55. Dr. Hancock in 'Charlesworth's Mag. of Nat. Hist.' volume 2 1838 page 491; R. Hill 'A Week at Port Royal' page 8; 'Guide to the Zoological Gardens' by P.L. Sclater 1859 pages 11, 12; 'The Knowsley Menagerie' by D. Gray 1846 pl. 14; E. Blyth 'Report Asiatic Soc. of Bengal' May 1855.)

The members of the great Duck family breed as readily in confinement as do the Columbae and Gallinae and this, considering their aquatic and wandering habits, and the nature of their food, could not have been anticipated. Even some time ago above two dozen species had bred in the Zoological Gardens; and M. Selys-Longchamps has recorded the production of hybrids from forty-four different members of the family; and to these Professor Newton has added a few more cases. (18/56. Prof. Newton in 'Proc. Zoolog. Soc.' 1860 page 336.) "There is not," says Mr. Dixon (18/57. 'The Dovecote and Aviary' page 428.), "in the wide world, a goose which is not in the strict sense of the word domesticable;" that is, capable of breeding under confinement;

but this statement is probably too bold. The capacity to breed sometimes varies in individuals of the same species; thus Audubon (18/58. 'Ornithological Biography' volume 3 page 9.) kept for more than eight years some wild geese (Anser canadensis), but they would not mate; whilst other individuals of the same species produced young during the second year. I know of but one instance in the whole family of a species which absolutely refuses to breed in captivity, namely, the Dendrocygna viduata, although, according to Sir R. Schomburgk (18/59. 'Geograph. Journal' volume 13 1844 page 32.), it is easily tamed, and is frequently kept by the Indians of Guiana. Lastly, with respect to Gulls, though many have been kept in the Zoological Gardens and in the old Surrey Gardens, no instance was known before the year 1848 of their coupling or breeding; but since that period the herring gull (Larus argentatus) has bred many times in the Zoological Gardens and at Knowsley.

There is reason to believe that insects are affected by confinement like the higher animals. It is well known that the Sphingidae rarely breed when thus treated. An entomologist (18/60. Loudon's 'Mag. of Nat. Hist.' volume 5 1832 page 153.) in Paris kept twenty-five specimens of Saturnia pyri, but did not succeed in getting a single fertile egg. A number of females of Orthosia munda and of Mamestra suasa reared in confinement were unattractive to the males. (18/61. 'Zoologist' volumes 5-6 1847-48 page 1660.) Mr. Newport kept nearly a hundred individuals of two species of Vanessa, but not one paired; this, however, might have been due to their habit of coupling on the wing. (18/62. 'Transact. Entomolog. Soc.' volume 4 1845 page 60.) Mr. Atkinson could never succeed in India in making the Tarroo silk-moth breed in confinement. (18/63. 'Transact. Linn. Soc.' volume 7 page 40.) It appears that a number of moths, especially the Sphingidae, when hatched in the autumn out of their proper season, are completely barren; but this latter case is still involved in some obscurity. (18/64. See an interesting paper by Mr. Newman in the 'Zoologist' 1857 page 5764; and Dr. Wallace in 'Proc. Entomolog. Soc.' June 4, 1860 page 119.)]

Independently of the fact of many animals under confinement not coupling, or, if they couple, not producing young, there is evidence of another kind that their sexual functions are disturbed. For many cases have been recorded of the loss by male birds when confined of their characteristic plumage. Thus the common linnet (Linota cannabina) when caged does not acquire the fine crimson colour on its breast, and one of the buntings (Emberiza passerina) loses the black on its head. A Pyrrhula and an Oriolus have been observed to assume the quiet plumage of the hen-bird; and the Falco albidus returned to the dress of an earlier age. (18/65. Yarrell 'British Birds' volume 1 page 506; Bechstein 'Stubenvogel' s. 185; 'Philosoph. Transact.' 1772 page 271. Bronn 'Geschichte der Natur' b. 2 s. 96 has collected a number of cases. For the case of the deer see 'Penny Cyclop.' volume 8 page 350.) Mr. Thompson, the superintendent of the Knowsley menagerie, informed me that he had often observed analogous facts. The horns of a male deer (Cervus canadensis) during the voyage from America were badly developed; but subsequently in Paris perfect horns were produced.

When conception takes place under confinement, the young are often born dead, or die soon, or are ill-formed. This frequently occurs in the Zoological Gardens, and, according to Rengger, with native animals confined in Paraguay. The mother's milk often fails. We may also attribute to the disturbance of the sexual functions the frequent occurrence of that monstrous instinct which leads the mother to devour her own offspring,—a mysterious case of perversion, as it at first appears.

Sufficient evidence has now been advanced to prove that animals when first confined are eminently liable to suffer in their reproductive systems. We feel at first naturally inclined to attribute the result to loss of health, or at least to loss of vigour; but this view can hardly be admitted when we reflect how healthy, long-lived, and vigorous many animals are under captivity, such as parrots, and hawks when used for hawking, cheetahs when used for hunting, and elephants. The reproductive organs themselves are not diseased; and the diseases, from which animals in menageries usually perish, are not those which in any way affect their fertility. No domestic animal is more subject to disease than the sheep, yet it is remarkably prolific. The failure of animals to breed under confinement has been sometimes attributed exclusively to a failure in their sexual instincts: this may occasionally come into play, but there is no obvious reason why this instinct should be especially liable to be affected with perfectly tamed animals, except, indeed, indirectly through the reproductive system itself being disturbed. Moreover, numerous cases have been given of various animals which couple freely under confinement, but never conceive; or, if they conceive and produce young, these are fewer in number than is natural to the species. In the vegetable kingdom instinct of course can play no part; and we shall presently see that plants when removed from their natural conditions are affected in nearly the same manner as animals. Change of climate cannot be the cause of the loss of fertility, for, whilst many animals imported into Europe from extremely different climates breed freely, many others when confined in their native land are completely sterile. Change of food cannot be the chief cause; for ostriches, ducks, and many other animals, which must have undergone a great change in this respect, breed freely. Carnivorous birds when confined are extremely sterile, whilst most carnivorous mammals, except plantigrades, are moderately fertile. Nor can the amount of food be the cause; for a sufficient supply will certainly be given to valuable animals; and there is no reason to suppose that much more food would be given to them than to our choice domestic productions which retain their full fertility. Lastly, we may infer from the case of the elephant, cheetah, various hawks, and of many animals which are allowed to lead an almost free life in their native land, that want of exercise is not the sole cause.

It would appear that any change in the habits of life, whatever these habits may be, if great enough, tends to affect in an inexplicable manner the powers of reproduction. The result depends more on the constitution of the species than on the nature of the change; for certain whole groups are affected more than others; but exceptions always occur, for some species in the most fertile groups refuse to breed, and some in the most sterile groups breed freely. Those animals which usually breed freely under confinement, rarely breed, as I was assured, in the Zoological Gardens, within a year or two after their first importation. When an animal which is generally sterile under confinement happens to breed, the young apparently do not inherit this power: for had this been the case, various quadrupeds and birds, which are valuable for exhibition, would have become common. Dr. Broca even affirms (18/66. 'Journal de Physiologie' tome 2 page 347.) that many animals in the Jardin des Plantes, after having produced young for three or four successive generations, become sterile; but this may be the result of too close interbreeding. It is a remarkable circumstance that many mammals and birds have produced hybrids under confinement quite as readily as, or even more readily than, they have procreated their own kind. Of this fact many instances have been given (18/67. For additional evidence on this subject see F. Cuvier in 'Annales du Museum' tome 12 page 119.); and we are thus reminded of those plants which when cultivated refuse to be fertilised by their own pollen, but can easily be fertilised by that of a distinct species. Finally, we must conclude, limited as the conclusion is, that changed conditions of life have an especial power of acting injuriously on the reproductive system. The whole case is quite peculiar, for these organs, though not diseased, are thus rendered incapable of performing their proper functions, or perform them imperfectly.

## [STERILITY OF DOMESTICATED ANIMALS FROM CHANGED CONDITIONS.

With respect to domesticated animals, as their domestication mainly depends on the accident of their breeding freely under captivity, we ought not to expect that their reproductive system would be affected by any moderate degree of change. Those orders of quadrupeds and birds, of which the wild species breed most readily in our menageries, have afforded us the greatest number of domesticated productions. Savages in most parts of the world are fond of taming animals (18/68. Numerous instances could be given. Thus Livingstone ('Travels' page 217) states that the King of the Barotse, an inland tribe which never had any communication with white men, was extremely fond of taming animals, and every young antelope was brought to him. Mr. Galton informs me that the Damaras are likewise fond of keeping pets. The Indians of South America follow the same habit. Capt. Wilkes states that the Polynesians of the Samoan Islands tamed pigeons; and the New Zealanders, as Mr. Mantell informs me, kept various kinds of birds.); and if any of these regularly produced young, and were at the same time useful, they would be at once domesticated. If, when their masters migrated into other countries, they were in addition found capable of withstanding various climates, they would be still more valuable; and it appears that the animals which breed readily in captivity can generally withstand different climates. Some few domesticated animals, such as the reindeer and camel, offer an exception to this rule. Many of our domesticated animals can bear with undiminished fertility the most unnatural conditions; for instance, rabbits, guinea-pigs, and ferrets breed in miserably confined hutches. Few European dogs of any kind withstand the climate of India without degenerating, but as long as they survive, they retain, as I hear from Dr. Falconer, their fertility; so it is, according to Dr. Daniell, with English dogs taken to Sierra Leone. The fowl, a native of the hot jungles of India, becomes more fertile than its parent-stock in every quarter of the world, until we advance as far north as Greenland and Northern Siberia, where this bird will not breed. Both fowls and pigeons, which I received during the autumn direct from Sierra Leone, were at once ready to couple. (18/69. For analogous cases with the fowl see Reaumur 'L'Art de faire Eclore' etc. 1749 page 243; and Col. Sykes in 'Proc. Zoolog. Soc.' 1832 etc. With respect to the fowl not breeding in northern regions see Latham 'Hist. of Birds' volume 8 1823 page 169.) I have, also, seen pigeons breeding as freely as the common kinds within a year after their importation from the upper Nile. The guinea-fowl, an aboriginal of the hot and dry deserts of Africa, whilst living under our damp and cool climate, produces a large supply of eggs.

Nevertheless, our domesticated animals under new conditions occasionally show signs of lessened fertility. Roulin asserts that in the hot valleys of the equatorial Cordillera sheep are not fully fecund (18/70. "Mem. par divers Savans" 'Acad. des Sciences' tome 6 1835 page 347.); and according to Lord Somerville (18/71. 'Youatt on Sheep' page 181.) the merino-sheep which he imported from Spain were not at first perfectly fertile, it is said (18/72. J. Mills 'Treatise on Cattle' 1776 page 72.) that mares brought up on dry food in the stable, and turned out to grass, do not at first breed. The peahen, as we have seen, is said not to lay so many eggs in England as in India. It was long before the canary-bird was fully fertile, and even now first-rate breeding birds are not common. (18/73. Bechstein 'Stubenvogel' s. 242.) In the hot and dry province of Delhi, as I hear from Dr. Falconer, the eggs of the turkey, though placed under a hen, are extremely liable to fail. According to Roulin, geese taken to the lofty plateau of Bogota, at first laid seldom, and then only a few eggs; of these scarcely a fourth were hatched, and half the young birds died; in the second generation they were more fertile; and when Roulin wrote they were becoming as fertile as our geese in Europe. With respect to the valley of Quito, Mr. Orton says (18/74. 'The Andes and the Amazon' 1870 page 107.) "the only geese in the valley are a few imported from Europe, and these refuse to propagate." In the Philippine Archipelago the goose, it is asserted, will not breed or even lay eggs. (18/75. Crawford 'Descriptive Dict. of the Indian Islands' 1856 page 145.) A more curious case is that of the fowl, which, according to Roulin, when first introduced would not breed at Cusco in Bolivia, but subsequently became quite fertile; and the English Game fowl, lately introduced, had not as yet arrived at its full fertility, for to raise two or three chickens from a nest of eggs was thought fortunate. In Europe close confinement has a marked effect on the fertility of the fowl: it has been found in France that with fowls allowed considerable freedom only twenty per cent of the eggs failed; when allowed less freedom forty per cent failed; and in close confinement sixty out of the hundred were not hatched. (18/76. 'Bull. de la Soc. d'Acclimat.' tome 9 1862 pages 380, 384.) So we see that unnatural and changed conditions of life produce some effect on the fertility of our most thoroughly domesticated animals, in the same manner, though in a far less degree, as with captive wild animals.

It is by no means rare to find certain males and females which will not breed together, though both are known to be perfectly fertile with other males and females. We have no reason to suppose that this is caused by these animals having been subjected to any change in their habits of life; therefore such cases are hardly related to our present subject. The cause apparently lies in an innate sexual incompatibility of the pair which are matched. Several instances have been communicated to me by Mr. W.C. Spooner (well known for his essay on Cross-breeding), by Mr. Eyton of Eyton, by Mr. Wicksted and other breeders, and especially by Mr. Waring of Chelsfield, in relation to horses, cattle, pigs, foxhounds, other dogs, and pigeons. (18/77. For pigeons see Dr. Chapuis 'Le Pigeon Voyageur Belge' 1865 page 66.) In these cases, females, which either previously or subsequently were proved to be fertile, failed to breed with certain males, with whom it was particularly desired to match them. A change in the constitution of the female may sometimes have occurred before she was put to the second male; but in other cases this explanation is hardly tenable, for a female, known not to be barren, has been unsuccessfully paired seven or eight times with the same male likewise known to be perfectly fertile. With cart-mares, which sometimes will not breed with stallions of pure blood, but subsequently have bred with cart-stallions, Mr. Spooner is inclined to attribute the failure to the lesser sexual power of the racehorse. But I have heard from the greatest breeder of racehorses at the present day, through Mr. Waring, that "it frequently occurs with a mare to be put several times during one or two seasons to a particular stallion of acknowledged power, and yet prove barren; the mare afterwards breeding at once with some other horse." These facts are worth recording, as they show, like so many previous facts, on what slight constitutional differences the fertility of an animal often depends.]

## STERILITY OF PLANTS FROM CHANGED CONDITIONS OF LIFE, AND FROM OTHER CAUSES.

In the vegetable kingdom cases of sterility frequently occur, analogous with those previously given in the animal kingdom. But the subject is obscured by several circumstances, presently to be discussed, namely, the contabescence of the anthers, as Gartner has named a certain affection—monstrosities—doubleness of the flower—much-enlarged fruit—and long-continued or excessive propagation by buds.

[It is notorious that many plants in our gardens and hot-houses, though preserved in the most perfect health, rarely or never produce seed. I do not allude to plants which run to leaves, from being kept too damp, or too warm, or too much manured; for these do not flower, and the case may be wholly different. Nor do I allude to fruit not ripening from want of heat or rotting from too much moisture. But many exotic plants, with their ovules and pollen appearing perfectly sound, will not set any seed. The sterility in many cases, as I know from my own observation, is simply due to the absence of the proper insects for carrying the pollen to the stigma. But after excluding the several cases just specified, there are many plants in which the reproductive system has been seriously affected by the altered conditions of life to which they have been subjected.

It would be tedious to enter on many details. Linnaeus long ago observed (18/78. 'Swedish Acts' volume 1 1739 page 3. Pallas makes the same remark in his 'Travels' English translation volume 1 page 292.) that Alpine plants, although naturally loaded with seed, produce either few or none when cultivated in gardens. But exceptions often occur: the Draba sylvestris, one of our most thoroughly Alpine plants, multiplies itself by seed in Mr. H.C. Watson's garden, near London; and Kerner, who has particularly attended to the cultivation of Alpine plants, found that various kinds, when cultivated, spontaneously sowed themselves. (18/79. A. Kerner 'Die Cultur der Alpenpflanzen' 1864 s. 139; Watson 'Cybele Britannica' volume 1 page 131; Mr. D. Cameron also has written on the culture of Alpine plants in 'Gard. Chronicle' 1848 pages 253, 268, and mentions a few which seed.) Many plants which naturally grow in peat-earth are entirely sterile in our gardens. I have noticed the same fact with several liliaceous plants, which nevertheless grew vigorously.

Too much manure renders some kinds utterly sterile, as I have myself observed. The tendency to sterility from this cause runs in families; thus, according to Gartner (18/80. 'Beitrage zur Kenntniss der Befruchtung' 1844 s. 333.), it is hardly possible to give too much manure to most Gramineae, Cruciferae, and Leguminosae, whilst succulent and bulbous-rooted plants are easily affected. Extreme poverty of soil is less apt to induce sterility; but dwarfed plants of Trifolium minus and repens, growing on a lawn often mown and never manured, were found by me not to produce any seed. The temperature of the soil, and the season at which plants are watered, often have a marked effect on their fertility, as was observed by Kolreuter in the case of Mirabilis. (18/81. 'Nova Acta Petrop.' 1793 page 391.) Mr. Scott, in the Botanic Gardens of Edinburgh, observed that Oncidium divaricatum would not set seed when grown in a basket in which it throve, but was capable of fertilisation in a pot where it was a little damper. Pelargonium fulgidum, for many years after its introduction, seeded freely; it then became sterile; now it is fertile (18/82. 'Cottage Gardener' 1856 pages 44, 109.) if kept in a dry stove during the winter. Other varieties of pelargonium are sterile and others fertile without our being able to assign any cause. Very slight changes in the position of a plant, whether planted on a bank or at its base, sometimes make all the difference in its producing seed. Temperature apparently has a much more powerful influence on the fertility of plants than on that of animals. Nevertheless it is wonderful what changes some few plants will withstand with undiminished fertility: thus the Zephyranthes candida, a native of the moderately warm banks of the Plata, sows itself in the hot dry country near Lima, and in Yorkshire resists the severest frosts, and I have seen seeds gathered from pods which had been covered with snow during three weeks. (18/83. Dr. Herbert 'Amaryllidaceae' page 176.) Berberis wallichii, from the hot Khasia range in India, is uninjured by our sharpest frosts, and ripens its fruit under our cool summers. Nevertheless, I presume we must attribute to change of climate the sterility of many foreign plants; thus, the Persian and Chinese lilacs (Syringa persica and chinensis), though perfectly hardy here, never produce a seed; the common lilac (S. vulgaris) seeds with us moderately well, but in parts of Germany the capsules never contain seed. (18/84. Gartner 'Beitrage zur Kenntniss' etc. s. 560, 564.) Some few of the cases, given in the last chapter, of self-impotent plants, might have been here introduced, as their state seems due to the conditions to which they have been subjected.

The liability of plants to be affected in their fertility by slightly changed conditions is the more remarkable, as the pollen when once in process of formation is not easily injured; a plant may be transplanted, or a branch with flower-buds be cut off and placed in water, and the pollen will be matured. Pollen, also, when once mature, may be kept for weeks or even months. (18/85. 'Gardener's Chronicle' 1844 page 215; 1850 page 470. Faivre gives a good resume on this subject in his 'La Variabilite des Especes' 1868 page 155.) The female organs are more sensitive, for Gartner (18/86. 'Beitrage zur

Kenntniss' etc. s. 252, 338.) found that dicotyledonous plants, when carefully removed so that they did not in the least flag, could seldom be fertilised; this occurred even with potted plants if the roots had grown out of the hole at the bottom. In some few cases, however, as with Digitalis, transplantation did not prevent fertilisation; and according to the testimony of Mawz, Brassica rapa, when pulled up by its roots and placed in water, ripened its seed. Flower-stems of several monocotyledonous plants when cut off and placed in water likewise produce seed. But in these cases I presume that the flowers had been already fertilised, for Herbert (18/87. 'Journal of Hort. Soc.' volume 2 1847 page 83.) found with the Crocus that the plants might be removed or mutilated after the act of fertilisation, and would still perfect their seeds; but that, if transplanted before being fertilised, the application of pollen was powerless.

Plants which have been long cultivated can generally endure with undiminished fertility various and great changes; but not in most cases so great a change of climate as domesticated animals. It is remarkable that many plants under these circumstances are so much affected that the proportion and the nature of their chemical ingredients are modified, yet their fertility is unimpaired. Thus, as Dr. Falconer informs me, there is a great difference in the character of the fibre in hemp, in the quantity of oil in the seed of the Linum, in the proportion of narcotin to morphine in the poppy, in gluten to starch in wheat, when these plants are cultivated on the plains and on the mountains of India; nevertheless, they all remain fully fertile.

## CONTABESCENCE.

Gartner has designated by this term a peculiar condition of the anthers in certain plants, in which they are shrivelled, or become brown and tough, and contain no good pollen. When in this state they exactly resemble the anthers of the most sterile hybrids. Gartner (18/88. 'Beitrage zur Kenntniss' etc. s. 117 et seq.; Kolreuter 'Zweite Fortsetzung' s. 10, 121; 'Dritte Fortsetzung' s. 57. Herbert 'Amaryllidaceae' page 355. Wiegmann 'Ueber die Bastarderzeugung' s. 27.), in his discussion on this subject, has shown that plants of many orders are occasionally thus affected; but the Caryophyllaceae and Liliaceae suffer most, and to these orders, I think, the Ericaceae may be added. Contabescence varies in degree, but on the same plant all the flowers are generally affected to nearly the same extent. The anthers are affected at a very early period in the flower-bud, and remain in the same state (with one recorded exception) during the life of the plant. The affection cannot be cured by any change of treatment, and is propagated by layers, cuttings, etc., and perhaps even by seed. In contabescent plants the female organs are seldom affected, or merely become precocious in their development. The cause of this affection is doubtful, and is different in different cases. Until I read Gartner's discussion I attributed it, as apparently did Herbert, to the unnatural treatment of the plants; but its permanence under changed conditions, and the female organs not being affected, seem incompatible with this view. The fact of several endemic plants becoming contabescent in our gardens seems, at first sight, equally incompatible with this view; but Kolreuter believes that this is the result of their transplantation. The contabescent plants of Dianthus and Verbascum, found wild by Wiegmann, grew on a dry and sterile bank. The fact that exotic plants are eminently liable to this affection also seems to show that it is in some manner caused by their unnatural treatment. In some instances, as with Silene, Gartner's view seems the most probable, namely, that it is caused by an inherent tendency in the species to become dioecious. I can add another cause, namely, the illegitimate unions of heterostyled plants, for I have observed seedlings of three species of Primula and of Lythrum salicaria, which had been raised from plants illegitimately fertilised by their own-form pollen, with some or all their anthers in a contabescent state. There is perhaps an additional cause, namely, self-fertilisation; for many plants of Dianthus and Lobelia, which had been raised from self-fertilised seeds, had their anthers in this state; but these instances are not conclusive, as both genera are liable from other causes to this affection.

Cases of an opposite nature likewise occur, namely, plants with the female organs struck with sterility, whilst the male organs remain perfect. Dianthus japonicus, a Passiflora, and Nicotiana, have been described by Gartner (18/89. 'Bastarderzengung' s. 356.) as being in this unusual condition.

## MONSTROSITIES AS A CAUSE OF STERILITY.

Great deviations of structure, even when the reproductive organs themselves are not seriously affected, sometimes cause plants to become sterile. But in other cases plants may become monstrous to an extreme degree and yet retain their full fertility. Gallesio, who certainly had great experience (18/90. 'Teoria della Riproduzione' 1816 page 84; 'Traite du Citrus' 1811 page 67.), often attributes sterility to this cause; but it may be suspected that in some of his cases sterility was the cause, and not the result, of the monstrous growths. The curious St. Valery apple, although it bears fruit, rarely produces seed. The wonderfully anomalous flowers of Begonia frigida, formerly described, though they appear fit for fructification, are sterile. (18/91. Mr. C.W. Crocker in 'Gardener's Chronicle' 1861 page 1092.) Species of Primula in which the calyx is brightly coloured are said (18/92. Verlot 'Des Varietes' 1865 page 80.) to be often sterile, though I have known them to be fertile. On the other hand, Verlot gives several cases of proliferous flowers which can be propagated by seed. This was the case with a poppy, which had become monopetalous by the union of its petals. (18/93. Verlot ibid page 88.) Another extraordinary poppy, with the stamens replaced by numerous small supplementary capsules, likewise reproduces itself by seed. This has also occurred with a plant of Saxifraga geum, in which a series of adventitious carpels, bearing ovules on their margins, had been developed between the stamens and the normal carpels (18/94. Prof. Allman, Brit. Assoc., quoted in the 'Phytologist' volume 2 page 483. Prof. Harvey, on the authority of Mr. Andrews, who discovered the plant, informed me that this monstrosity could be propagated by seed. With respect to the poppy see Prof. Goeppert as quoted in 'Journal of Horticulture' July 1, 1863 page 171.) Lastly, with respect to peloric flowers, which depart wonderfully from the natural structure,— those of Linaria vulgaris seem generally to be more or less sterile, whilst those before described of Antirrhinum majus, when artificially fertilised with their own pollen, are perfectly fertile, though sterile when left to themselves, for bees are unable to crawl into the narrow tubular flower. The peloric flowers of Corydalis solida, according to Godron (18/95. 'Comptes Rendus' December 19, 1864 page 1039.), are sometimes barren and sometimes fertile; whilst those of Gloxinia are well known to yield plenty of seed. In our greenhouse Pelargoniums, the central flower of the truss is often peloric, and Mr. Masters informs me that he tried in vain during several years to get seed from these flowers. I likewise made many vain attempts, but sometimes succeeded in fertilising them with pollen from a normal flower of another variety; and conversely I several times fertilised ordinary flowers with peloric pollen. Only once I succeeded in raising a plant from a peloric flower fertilised by pollen from a peloric flower borne by another variety; but the plant, it may be added, presented nothing particular in its structure. Hence we may conclude that no general rule can be laid down; but any great deviation from the normal structure, even when the reproductive organs themselves are not seriously affected, certainly often leads to sexual impotence.

**DOUBLE FLOWERS.**

When the stamens are converted into petals, the plant becomes on the male side sterile; when both stamens and pistils are thus changed, the plant becomes completely barren. Symmetrical flowers having numerous stamens and petals are the most liable to become double, as perhaps follows from all multiple organs being the most subject to variability. But flowers furnished with only a few stamens, and others which are asymmetrical in structure, sometimes become double, as we see with the double gorse or Ulex, and Antirrhinum. The Compositae bear what are called double flowers by the abnormal development of the corolla of their central florets. Doubleness is sometimes connected with prolification (18/96. 'Gardener's Chronicle' 1866 page 681.), or the continued growth of the axis of the flower. Doubleness is strongly inherited. No one has produced, as Lindley remarks (18/97. 'Theory of Horticulture' page 333.), double flowers by promoting the perfect health of the plant. On the contrary, unnatural conditions of life favour their production. There is some reason to believe that seeds kept during many years, and seeds believed to be imperfectly fertilised, yield double flowers more freely than fresh and perfectly fertilised seed. (18/98. Mr. Fairweather 'Transact. Hort. Soc.' volume 3 page 406: Bosse quoted by Bronn 'Geschichte der Natur' b. 2 s. 77. On the effects of the removal of the anthers see Mr. Leitner in Silliman's 'North American Journ. of Science' volume 23 page 47; and Verlot 'Des Varietes' 1865 page 84.) Long-continued cultivation in rich soil seems to be the commonest exciting cause. A double narcissus and a double Anthemis nobilis, transplanted into very poor soil, has been observed to become single (18/99. Lindley's 'Theory of Horticulture' page 3?3.); and I have seen a completely double white primrose rendered permanently single by being divided and transplanted whilst in full flower. It has been observed by Professor E. Morren that doubleness of the flowers and variegation of the leaves are antagonistic states; but so many exceptions to the rule have lately been recorded (18/100. 'Gardener's Chronicle' 1865 page 626; 1866 pages 290, 730; and Verlot 'Des Varietes' page 75.), that, though general, it cannot be looked at as invariable. Variegation seems generally to result from a feeble or atrophied condition of the plant, and a large proportion of the seedlings raised from parents, if both are variegated, usually perish at an early age; hence we may perhaps infer that doubleness, which is the antagonistic state, commonly arises from a plethoric condition. On the other hand, extremely poor soil sometimes, though rarely, appears to cause doubleness: I formerly described (18/101. 'Gardener's Chronicle' 1843 page 628. In this article I suggested the theory above given on the doubleness of flowers. This view is adopted by Carriere 'Production et Fix. des Varietes' 1865 page 67.) some completely double, bud-like, flowers produced in large numbers by stunted wild plants of Gentiana amarella growing on a poor chalky bank. I have also noticed a distinct tendency to doubleness in the flowers of a Ranunculus, Horse-chestnut, and Bladder-nut (Ranunculus repens, Aesculus pavia, and Staphylea), growing under very unfavourable conditions. Professor Lehmann (18/102. Quoted by Gartner 'Bastarderzeugung' s. 567.) found several wild plants growing near a hot spring with double flowers. With respect to the cause of doubleness, which arises, as we see, under widely different circumstances, I shall presently attempt to show that the most probable view is that unnatural conditions first give a tendency to sterility, and that then, on the principle of compensation, as the reproductive organs do not perform their proper functions, they either become developed into petals, or additional petals are formed. This view has lately been supported by Mr. Laxton (18/103. 'Gardener's Chronicle' 1866 page 901.) who advances the case of some common peas, which, after long-continued heavy rain, flowered a second time, and produced double flowers.

## SEEDLESS FRUIT.

Many of our most valuable fruits, although consisting in a homological sense of widely different organs, are either quite sterile, or produce extremely few seeds. This is notoriously the case with our best pears, grapes, and figs, with the pine-apple, banana, bread-fruit, pomegranate, azarole, date-palms, and some members of the orange-tribe. Poorer varieties of these same fruits either habitually or occasionally yield seed. (18/104. Lindley 'Theory of Horticulture' pages 175-179; Godron 'De l'Espece' tome 2 page 106; Pickering 'Races of Man;' Gallesio 'Teoria della Riproduzione' 1816 pages 101-110. Meyen 'Reise um Erde' Th. 2 s. 214 states that at Manilla one variety of the banana is full of seeds: and Chamisso (Hooker's 'Bot. Misc.' volume 1 page 310) describes a variety of the bread-fruit in the Mariana Islands with small fruit, containing seeds which are frequently perfect. Burnes in his 'Travels in Bokhara' remarks on the pomegranate seeding in Mazenderan, as a remarkable peculiarity.) Most horticulturists look at the great size and anomalous development of the fruit as the cause, and sterility as the result; but the opposite view, as we shall presently see, is more probable.

## STERILITY FROM THE EXCESSIVE DEVELOPMENT OF THE ORGANS OF GROWTH OR VEGETATION.

Plants which from any cause grow too luxuriantly, and produce leaves, stems, runners, suckers, tubers, bulbs, etc., in excess, sometimes do not flower, or if they flower do not yield seed. To make European vegetables under the hot climate of India yield seed, it is necessary to check their growth; and, when one-third grown, they are taken up, and their stems and tap-roots are cut or mutilated. (18/105. Ingledew in 'Transact. of Agricult. and Hort. Soc. of India' volume 2.) So it is with hybrids; for instance, Prof. Lecoq (18/106. 'De la Fecondation' 1862 page 308.) had three plants of Mirabilis, which, though they grew luxuriantly and flowered, were quite sterile; but after beating one with a stick until a few branches alone were left, these at once yielded good seed. The sugar-cane, which grows vigorously and produces a large supply of succulent stems, never, according to various observers, bears seed in the West Indies, Malaga, India, Cochin China, Mauritius, or the Malay Archipelago. (18/107. Hooker 'Bot. Misc.' volume 1 page 99; Gallesio 'Teoria della Riproduzione' page 110. Dr. J. de Cordemoy in 'Transact. of the R. Soc. of Mauritius' new series volume 6 1873 pages 60-67, gives a large number of cases of plants which never seed, including several species indigenous in Mauritius.) Plants which produce a large number of tubers are apt to be sterile, as occurs, to a certain extent, with the common potato; and Mr. Fortune informs me that the sweet potato (Convolvulus batatas) in China never, as far as he has seen, yields seed. Dr. Royle remarks (18/108. 'Transact. Linn. Soc.' volume 17 page 563.) that in India the Agave vivipara, when grown in rich soil, invariably produces bulbs, but no seeds; whilst a poor soil and dry climate lead to an opposite result. In China, according to Mr. Fortune, an extraordinary number of little bulbs are developed in the axils of the leaves of the yam, and this plant does not bear seed. Whether in these cases, as in those of double flowers and seedless fruit, sexual sterility from changed conditions of life is the primary cause which leads to the excessive development of the organs of vegetation, is doubtful; though some evidence might be advanced in favour of this view. It is perhaps a more probable view that plants which propagate themselves largely by one method, namely by buds, have not sufficient vital power or organised matter for the other method of sexual generation.

Several distinguished botanists and good practical judges believe that long-continued propagation by cuttings, runners, tubers, bulbs, etc., independently of any excessive development of these parts, is the cause of many plants failing to produce flowers, or producing only barren flowers,—it is as if they had lost the habit of sexual generation. (18/109. Godron 'De l'Espece' tome 2 page 106; Herbert on Crocus in 'Journal of Hort. Soc.' volume 1 1846 page 254: Dr. Wight, from what he has seen in India, believes in this view; 'Madras Journal of Lit. and Science' volume 4 1836 page 61.) That many plants when thus propagated are sterile there can be no doubt, but as to whether the long continuance of this form of propagation is the actual cause of their sterility, I will not venture, from the want of sufficient evidence, to express an opinion.

That plants may be propagated for long periods by buds, without the aid of sexual generation, we may safely infer from this being the case with many plants which must have long survived in a state of nature. As I have had occasion before to allude to this subject, I will here give such cases as I have collected. Many alpine plants ascend mountains beyond the height at which they can produce seed. (18/110. Wahlenberg specifies eight species in this state on the Lapland Alps: see Appendix to Linnaeus 'Tour in Lapland' translated by Sir J.E. Smith volume 2 pages 274-280.) Certain species of Poa and Festuca, when growing on mountain-pastures, propagate themselves, as I hear from Mr. Bentham, almost exclusively by bulblets. Kalm gives a more curious instance (18/111. 'Travels in North America' English translation volume 3 page 175.) of several American trees, which grow so plentifully in marshes or in thick woods, that they are certainly well adapted for these stations, yet scarcely ever produce seeds; but when accidentally growing on the outside of the marsh or wood, are loaded with seed. The common ivy is found in Northern Sweden and Russia, but flowers and fruits only in the southern provinces. The Acorus calamus extends over a large portion of the globe, but so rarely perfects fruit that this has been seen only by a few botanists; according to Caspary, all its pollen-grains are in a worthless condition. (18/112. With respect to the ivy and Acorus see Dr. Broomfield in the 'Phytologist' volume 3 page 376. Also Lindley and Vaucher on the Acorus and see Caspary as below.) The Hypericum calycinum, which propagates itself so freely in our shrubberies by rhizomes, and is naturalised in Ireland, blossoms profusely, but rarely sets any seed, and this only during certain years; nor did it set any when fertilised in my garden by pollen from plants growing at a distance. The Lysimachia nummularia, which is furnished with long runners, so seldom produces seed-capsules, that Prof. Decaisne (18/113. 'Annal. des Sc. Nat.' 3rd series Zool. tome 4 page 280. Prof. Decaisne refers also to analogous cases with mosses and lichens near Paris.), who has especially attended to this plant, has never seen it in fruit. The Carex rigida often fails to perfect its seed in Scotland, Lapland, Greenland, Germany, and New Hampshire in the United States. (18/114. Mr. Tuckermann in Silliman's 'American Journal of Science' volume 65 page 1.) The periwinkle (Vinca minor), which spreads largely by runners, is said scarcely ever to produce fruit in England (18/115. Sir J.E. Smith 'English Flora' volume 1 page 339.); but this plant requires insect-aid for its fertilisation, and the proper insects may be absent or rare. The Jussiaea grandiflora has become naturalised in Southern France, and has spread by its rhizomes so extensively as to impede the navigation of the waters, but never produces fertile seed. (18/116. G. Planchon 'Flora de Montpellier' 1864 page 20.) The horse-radish (Cochleria armoracia) spreads pertinaciously and is naturalised in various parts of Europe; though it bears flowers, these rarely produce capsules: Professor Caspary informs me that he has watched this plant since 1851, but has never seen its fruit; 65 per cent of its pollen-grains are bad. The common Ranunculus ficaria rarely bears seed in England, France, or Switzerland; but in 1863 I observed seeds on several plants growing near my house. (18/117. On the non-production of seeds in England see Mr. Crocker in 'Gardener's Weekly Magazine' 1852 page 70; Vaucher 'Hist. Phys. Plantes d'Europe' tome 1 page 33; Lecoq 'Geograph. Bot. d'Europe' tome 4 page 466; Dr. D. Clos in 'Annal. des Sc. Nat.' 3rd series Bot. tome 17 1852 page 129: this latter author refers to other analogous cases. See more especially on this plant and on other allied cases Prof. Caspary "Die Nuphar" 'Abhand. Naturw. Gesellsch. zu Halle' b. 11 1870 page 40, 78.) Other cases analogous with the foregoing could be given; for instance, some kinds of mosses and lichens have never been seen to fructify in France.

Some of these endemic and naturalised plants are probably rendered sterile from excessive multiplication by buds, and their consequent incapacity to produce and nourish seed. But the sterility of others more probably depends on the peculiar conditions under which they live, as in the case of the ivy in the northern part of Europe, and of the trees in the swamps of the United States; yet these plants must be in some respects eminently well adapted for the stations which they occupy, for they hold their places against a host of competitors.]

Finally, the high degree of sterility which often accompanies the doubling of flowers, or an excessive development of fruit, seldom supervenes at once. An incipient tendency is observed, and continued selection completes the result. The view which seems the most probable, and which connects together all the foregoing facts and brings them within our present subject, is, that changed and unnatural conditions of life first give a tendency to sterility; and in consequence of this, the organs of reproduction being no longer able fully to perform their proper functions, a supply of organised matter, not required for the development of the seed, flows either into these organs and renders them foliaceous, or into the fruit, stems, tubers, etc., increasing their size and succulency. But it is probable that there exists, independently of any incipient sterility, an antagonism between the two forms of reproduction, namely, by seed and buds, when either is carried to an extreme degree. That incipient sterility plays an important part in the doubling of flowers, and in the other cases just specified, I infer chiefly from the following facts. When fertility is lost from a wholly different cause, namely, from hybridism, there is a strong tendency, as Gartner (18/118. 'Bastarderzeugung' s. 565. Kolreuter 'Dritte Fortsetzung' s. 73, 87, 119) also shows that when two species, one single and the other double, are crossed, the hybrids are apt to be extremely double.) affirms, for flowers to become double, and this tendency is inherited. Moreover, it is notorious that with hybrids the male organs become sterile before the female organs, and with double flowers the stamens first become foliaceous. This latter fact is well shown by the male flowers of dioecious plants, which, according to Gallesio (18/119. 'Teoria della Riproduzione Veg.' 1816 page 73.) first become double. Again, Gartner (18/120. 'Bastarderzeugung' s. 573.) often insists that the flowers of even utterly sterile hybrids, which do not produce any seed, generally yield perfect capsules or fruit,—a fact which has likewise been repeatedly observed by Naudin with the Cucurbitaceae; so that the production of fruit by plants rendered sterile through any cause is intelligible. Kolreuter has also expressed his unbounded astonishment at the size and development of the tubers in certain hybrids; and all experimentalists (18/121. Ibid s. 527.) have remarked on the strong tendency in hybrids to increase by roots, runners, and suckers. Seeing that hybrid plants, which from their nature are more or less sterile, thus tend to produce double flowers; that they have the parts including the seed, that is the fruit, perfectly developed, even when containing no seed; that they sometimes yield gigantic roots; that they almost invariably tend to increase largely by suckers and other such means;—seeing this, and knowing, from the many facts given in the earlier parts of this chapter, that almost all organic beings when exposed to unnatural conditions tend to become more or less sterile, it seems much the most probable view that with cultivated plants sterility is the exciting cause, and double flowers, rich seedless fruit, and in some cases largely-developed organs of vegetation, etc., are the indirect results—these results having been in most cases largely increased through continued selection by man.

# CHAPTER 2.XIX.

SUMMARY OF THE FOUR LAST CHAPTERS, WITH REMARKS ON HYBRIDISM.

ON THE EFFECTS OF CROSSING. THE INFLUENCE OF DOMESTICATION ON FERTILITY. CLOSE INTERBREEDING. GOOD AND EVIL RESULTS FROM CHANGED CONDITIONS OF LIFE. VARIETIES WHEN CROSSED NOT INVARIABLY FERTILE. ON THE DIFFERENCE IN FERTILITY BETWEEN CROSSED SPECIES AND VARIETIES. CONCLUSIONS WITH RESPECT TO HYBRIDISM. LIGHT THROWN ON HYBRIDISM BY THE ILLEGITIMATE PROGENY OF HETEROSTYLED PLANTS. STERILITY OF CROSSED SPECIES DUE TO DIFFERENCES CONFINED TO THE REPRODUCTIVE SYSTEM. NOT ACCUMULATED THROUGH NATURAL SELECTION. REASONS WHY DOMESTIC VARIETIES ARE NOT MUTUALLY STERILE. TOO MUCH STRESS HAS BEEN LAID ON THE DIFFERENCE IN FERTILITY BETWEEN CROSSED SPECIES AND CROSSED VARIETIES. CONCLUSION.

It was shown in the fifteenth chapter that when individuals of the same variety, or even of a distinct variety, are allowed freely to intercross, uniformity of character is ultimately acquired. Some few characters, however, are incapable of fusion, but these are unimportant, as they are often of a semi-monstrous nature, and have suddenly appeared. Hence, to preserve our domesticated breeds true, or to improve them by methodical selection, it is obviously necessary that they should be kept separate. Nevertheless, a whole body of individuals may be slowly modified, through unconscious selection, as we shall see in a future chapter, without separating them into distinct lots. Domestic races have often been intentionally modified by one or two crosses, made with some allied race, and occasionally even by repeated crosses with very distinct races; but in almost all such cases, long-continued and careful selection has been absolutely necessary, owing to the excessive variability of the crossed offspring, due to the principle of reversion. In a few instances, however, mongrels have retained a uniform character from their first production.

When two varieties are allowed to cross freely, and one is much more numerous than the other, the former will ultimately absorb the latter. Should both varieties exist in nearly equal numbers, it is probable that a considerable period would elapse before the acquirement of a uniform character; and the character ultimately acquired would largely depend on prepotency of transmission and on the conditions of life; for the nature of these conditions would generally favour one variety more than another, so that a kind of natural selection would come into play. Unless the crossed offspring were slaughtered by man without the least discrimination, some degree of unmethodical selection would likewise come into action. From these several considerations we may infer, that when two or more closely allied species first came into the possession of the same tribe, their crossing will not have influenced, in so great a degree as has often been supposed, the character of the offspring in future times; although in some cases it probably has had a considerable effect.

Domestication, as a general rule, increases the prolificness of animals and plants. It eliminates the tendency to sterility which is common to species when first taken from a state of nature and crossed. On this latter head we have no direct evidence; but as our races of dogs, cattle, pigs etc., are almost certainly descended from aboriginally distinct stocks, and as these races are now fully fertile together, or at least incomparably more fertile than most species when crossed, we may with entire confidence accept this conclusion.

Abundant evidence has been given that crossing adds to the size, vigour, and fertility of the offspring. This holds good when there has been no previous close interbreeding. It applies to the individuals of the same variety but belonging to different families, to distinct varieties, sub-species, and even to species. In the latter case, though size is gained, fertility is lost; but the increased size, vigour, and hardiness of many hybrids cannot be accounted for solely on the principle of compensation from the inaction of the reproductive system. Certain plants whilst growing under their natural conditions, others when cultivated, and others of hybrid origin, are completely self-impotent, though perfectly healthy; and such plants can be stimulated to fertility only by being crossed with other individuals of the same or of a distinct species.

On the other hand, long-continued close interbreeding between the nearest relations diminishes the constitutional vigour, size, and fertility of the offspring; and occasionally leads to malformations, but not necessarily to general deterioration of form or structure. This failure of fertility shows that the evil results of interbreeding are independent of the augmentation of morbid tendencies common to both parents, though this augmentation no doubt is often highly injurious. Our belief that evil follows from close interbreeding rests to a certain extent on the experience of practical breeders, especially of those who have reared many animals of quickly propagating kinds; but it likewise rests on several carefully recorded experiments. With some animals close interbreeding may be carried on for a long period with impunity by the selection of the most vigorous and healthy individuals; but sooner or later evil follows. The evil, however, comes on so slowly and gradually that it easily escapes observation, but can be recognised by the almost instantaneous manner in which size, constitutional vigour, and fertility are regained when animals that have long been interbred are crossed with a distinct family.

These two great classes of facts, namely, the good derived from crossing, and the evil from close interbreeding, with the consideration of the innumerable adaptations throughout nature for compelling, or favouring, or at least permitting, the occasional union of distinct individuals, taken together, lead to the conclusion that it is a law of nature that organic beings shall not fertilise themselves for perpetuity. This law was first plainly hinted at in 1799, with respect to plants, by Andrew Knight (19/1. 'Transactions Phil. Soc.' 1799 page 202. For Kolreuter see 'Mem. de l'Acad. de St.-Petersbourg' tome 3 1809 published 1811 page 197. In reading C.K. Sprengel's remarkable work, 'Das entdeckte Geheimniss' etc. 1793, it is curious to observe how often this wonderfully acute observer failed to understand the full meaning of the structure of the flowers which he has so well described, from not always having before his mind the key to the problem, namely, the good derived from the crossing of distinct individual plants.) and, not long afterwards, that sagacious observer Kolreuter, after showing how well the Malvaceae are adapted for crossing, asks, "an id aliquid in recessu habeat, quod hujuscemodi flores nunquam proprio suo pulvere, sed semper eo aliarum su speciei impregnentur, merito quaritur? Certe natura nil facit frustra." Although we may demur to Kolreuter's saying that nature does nothing in vain, seeing how many rudimentary and useless organs there are, yet undoubtedly the argument from the innumerable contrivances, which favour crossing, is of the greatest weight. The most important result of this law is that it leads to uniformity of character in the individuals of the same species. In the case of certain hermaphrodites, which probably intercross only at long intervals of time, and with unisexual animals inhabiting somewhat separated localities, which can only occasionally come into contact and pair, the greater vigour and fertility of the crossed offspring will ultimately tend to give uniformity of character. But when we go beyond the limits of the same species, free intercrossing is barred by the law of sterility.

In searching for facts which might throw light on the cause of the good effects from crossing, and of the evil effects from close interbreeding, we have seen that, on the one hand, it is a widely prevalent and ancient belief, that animals and plants profit from slight changes in their condition of life; and it would appear that the germ, in a somewhat analogous manner, is more effectually stimulated by the male element, when taken from a distinct individual, and therefore slightly modified in nature, than when taken from a male having the same identical constitution. On the other hand, numerous facts have been given, showing that when animals are first subjected to captivity, even in their native land, and although allowed much liberty, their reproductive functions are often greatly impaired or quite annulled. Some groups of animals are more affected than others, but with apparently capricious exceptions in every group. Some animals never or rarely couple under confinement; some couple freely, but never or rarely conceive. The secondary male characters, the maternal functions and instincts, are occasionally affected. With plants, when first subjected to cultivation, analogous facts have been observed. We probably owe our double flowers, rich seedless fruits, and in some cases greatly developed tubers, etc., to incipient sterility of the above nature combined with a copious supply of nutriment. Animals which have long been domesticated, and plants which have long been cultivated, can generally withstand, with unimpaired fertility, great changes in their conditions of life; though both are sometimes slightly affected. With animals the somewhat rare capacity of breeding freely under confinement, together with their utility, mainly determine the kinds which have been domesticated.

We can in no case precisely say what is the cause of the diminished fertility of an animal when first captured, or of a plant when first cultivated; we can only infer that it is caused by a change of some kind in the natural conditions of life. The remarkable susceptibility of the reproductive system to such changes,—a susceptibility not common to any other organ,—apparently has an important bearing on Variability, as we shall see in a future chapter.

It is impossible not to be struck with the double parallelism between the two classes of facts just alluded to. On the one hand, slight changes in the conditions of life, and crosses between slightly modified forms or varieties, are beneficial as far as prolificness and constitutional vigour are concerned. On the other hand, changes in the conditions greater in degree, or of a different nature, and crosses between forms which have been slowly and greatly modified by natural means,—in other words, between species,—are highly injurious, as far as the reproductive system is concerned, and in some few instances as far as constitutional vigour is concerned. Can this parallelism be accidental? Does it not rather indicate some real bond of connection? As a fire goes out unless it be stirred up, so the vital forces are always tending, according to Mr. Herbert Spencer, to a state of equilibrium, unless disturbed and renovated through the action of other forces.

In some few cases varieties tend to keep distinct, by breeding at different seasons, by great difference in size, or by sexual preference. But the crossing of varieties, far from diminishing, generally adds to the fertility of the first union and of the mongrel offspring. Whether all the more widely distinct domestic varieties are invariably quite fertile when crossed, we do not positively know; much time and trouble would be requisite for the necessary experiments, and many difficulties occur, such as the descent of the various races from aboriginally distinct species, and the doubts whether certain forms ought to be ranked as species or varieties. Nevertheless, the wide experience of practical breeders proves that the great majority of varieties, even if some should hereafter prove not to be indefinitely fertile inter se, are far more fertile when crossed, than the vast majority of closely allied natural species. A few remarkable cases have, however, been given on the authority of excellent observers, showing that with plants certain forms, which undoubtedly must be ranked as varieties, yield fewer seeds when crossed than is natural to the parent-species. Other varieties have had their reproductive powers so far modified that they are either more or less fertile than their parents, when crossed with a distinct species.

Nevertheless, the fact remains indisputable that domesticated varieties, of animals and of plants, which differ greatly from one another in structure, but which are certainly descended from the same aboriginal species, such as the races of the fowl, pigeon, many vegetables, and a host of other productions, are extremely fertile when crossed; and this seems to make a broad and impassable barrier between domestic varieties and natural species. But, as I will now attempt to show, the distinction is not so great and overwhelmingly important as it at first appears.

## ON THE DIFFERENCE IN FERTILITY BETWEEN VARIETIES AND SPECIES WHEN CROSSED.

This work is not the proper place for fully treating the subject of hybridism, and I have already given in my 'Origin of Species' a moderately full abstract. I will here merely enumerate the general conclusions which may be relied on, and which bear on our present point.

## FIRSTLY.

The laws governing the production of hybrids are identical, or nearly identical, in the animal and vegetable kingdoms.

## SECONDLY.

The sterility of distinct species when first united, and that of their hybrid offspring, graduate, by an almost infinite number of steps, from zero, when the ovule is never impregnated and a seed-capsule is never formed, up to complete fertility. We can only escape the conclusion that some species are fully fertile when crossed, by determining to designate as varieties all the forms which are quite fertile. This high degree of fertility is, however, rare. Nevertheless, plants, which have been exposed to unnatural conditions, sometimes become modified in so peculiar a manner, that they are much more fertile when crossed with a distinct species than when fertilised by their own pollen. Success in effecting a first union between two species, and the fertility of their hybrids, depend in an eminent degree on the conditions of life being favourable. The innate sterility of hybrids of the same parentage and raised from the same seed-capsule often differs much in degree.

## THIRDLY.

The degree of sterility of a first cross between two species does not always run strictly parallel with that of their hybrid offspring. Many cases are known of species which can be crossed with ease, but yield hybrids excessively sterile; and conversely some which can be crossed with great difficulty, but produce fairly fertile hybrids. This is an inexplicable fact, on the view that species have been specially endowed with mutual sterility in order to keep them distinct.

## FOURTHLY.

The degree of sterility often differs greatly in two species when reciprocally crossed; for the first will readily fertilise the second; but the latter is incapable, after hundreds of trials, of fertilising the former. Hybrids produced from reciprocal crosses between the same two species likewise sometimes differ in their degree of sterility. These cases also are utterly inexplicable on the view of sterility being a special endowment.

## FIFTHLY.

The degree of sterility of first crosses and of hybrids runs, to a certain extent, parallel with the general or systematic affinity of the forms which are united. For species belonging to distinct genera can rarely, and those belonging to distinct families can never, be crossed. The parallelism, however, is far from complete; for a multitude of closely allied species will not unite, or unite with extreme difficulty, whilst other species, widely different from one another, can be crossed with perfect facility. Nor does the difficulty depend on ordinary constitutional differences, for annual and perennial plants, deciduous and evergreen trees, plants flowering at different seasons, inhabiting different stations, and naturally living under the most opposite climates, can often be crossed with ease. The difficulty or facility apparently depends exclusively on the sexual constitution of the species which are crossed; or on their sexual elective affinity, i.e. Wahlverwandtschaft of Gartner. As species rarely or never become modified in one character, without being at the same time modified in many characters, and as systematic affinity includes all visible similarities and dissimilarities, any difference in sexual constitution between two species would naturally stand in more or less close relation with their systematic position.

## SIXTHLY.

The sterility of species when first crossed, and that of hybrids, may possibly depend to a certain extent on distinct causes. With pure species the reproductive organs are in a perfect condition, whilst with hybrids they are often plainly deteriorated. A hybrid embryo which partakes of the constitution of its father and mother is exposed to unnatural conditions, as long as it is nourished within the womb, or egg, or seed of the mother-form; and as we know that unnatural conditions often induce sterility, the reproductive organs of the hybrid might at this early age be permanently affected. But this cause has no bearing on the infertility of first unions. The diminished number of the offspring from first unions may often result, as is certainly sometimes the case, from the premature death of most of the hybrid embryos. But we shall immediately see that a law of an unknown nature apparently exists, which leads to the offspring from unions, which are infertile, being themselves more or less infertile; and this at present is all that can be said.

## SEVENTHLY.

Hybrids and mongrels present, with the one great exception of fertility, the most striking accordance in all other respects; namely, in the laws of their resemblance to their two parents, in their tendency to reversion, in their variability, and in being absorbed through repeated crosses by either parent-form.

After arriving at these conclusions, I was led to investigate a subject which throws considerable light on hybridism, namely, the fertility of heterostyled or dimorphic and trimorphic plants, when illegitimately united. I have had occasion several times to allude to these plants, and I may here give a brief abstract of my observations. Several plants belonging to distinct orders present two forms, which exist in about equal numbers, and which differ in no respect except in their reproductive organs; one form having a long pistil with short stamens, the other a short pistil with long stamens; both with differently sized pollen-grains. With trimorphic plants there are three forms likewise differing in the lengths of their pistils and stamens, in the size and colour of the pollen-grains, and in some other respects; and as in each of the three forms there are two sets of stamens, there are altogether six sets of stamens and three kinds of pistils. These organs are so proportioned in length to one another that, in any two of the forms, half the stamens in each stand on a level with the stigma of the third form. Now I have shown, and the result has been confirmed by other observers, that, in order to obtain full fertility with these plants, it is necessary that the stigma of the one form should be fertilised by pollen taken from the stamens of corresponding height in the other form. So that with dimorphic species two unions, which may be called legitimate, are fully fertile, and two, which may be called illegitimate, are more or less infertile. With trimorphic species six unions are legitimate, or fully fertile, and twelve are illegitimate, or more or less infertile. (19/2. My observations 'On the Character and hybrid-like nature of the offspring from the illegitimate union of Dimorphic and Trimorphic Plants' were published in the 'Journal of the Linnean Soc.' volume 10 page 393. The abstract here given is nearly the same with that which appeared in the 6th edition of my 'Origin of Species.')

The infertility which may be observed in various dimorphic and trimorphic plants, when illegitimately fertilised, that is, by pollen taken from stamens not corresponding in height with the pistil, differs much in degree, up to absolute and utter sterility; just in the same manner as occurs in crossing distinct species. As the degree of sterility in the latter case depends in an eminent degree on the conditions of life being more or less favourable, so I have found it with illegitimate unions. It is well known that if pollen of a distinct species be placed on the stigma of a flower, and its own pollen be afterwards, even after a considerable interval of time, placed on the same stigma, its action is so strongly prepotent that it generally annihilates the effect of the foreign pollen; so it is with the pollen of the several forms of the same species, for legitimate pollen is strongly prepotent over illegitimate pollen, when both are placed on the same stigma. I ascertained this by fertilising several flowers, first illegitimately, and twenty-four hours afterwards legitimately, with pollen taken from a peculiarly coloured variety, and all the seedlings were similarly coloured; this shows that the legitimate pollen, though applied twenty-four hours subsequently, had wholly destroyed or prevented the action of the previously applied illegitimate pollen. Again, as, in making reciprocal crosses between the same two species, there is occasionally a great difference in the result, so the same thing occurs with trimorphic plants; for instance, the mid-styled form of Lythrum salicaria could be illegitimately fertilised with the greatest ease by pollen from the longer stamens of the short-styled form, and yielded many seeds; but the short-styled form did not yield a single seed when fertilised by the longer stamens of the mid-styled form.

In all these respects the forms of the same undoubted species, when illegitimately united, behave in exactly the same manner as do two distinct species when crossed. This led me carefully to observe during four years many seedlings, raised from several illegitimate unions. The chief result is that these illegitimate plants, as they may be called, are not fully fertile. It is possible to raise from dimorphic species, both long-styled and short-styled illegitimate plants, and from trimorphic plants all three illegitimate forms. These can then be properly united in a legitimate manner. When this is done, there is no apparent reason why they should not yield as many seeds as did their parents when legitimately fertilised. But such is not the case; they are all infertile, but in various degrees; some being so utterly and incurably sterile that they did not yield during four seasons a single seed or even seed-capsule. These illegitimate plants, which are so sterile, although united with each other in a legitimate manner, may be strictly compared with hybrids when crossed inter se, and it is well known how sterile these latter generally are. When, on the other hand, a hybrid is crossed with either pure parent- species, the sterility is usually much lessened: and so it is when an illegitimate plant is fertilised by a legitimate plant. In the same manner as the sterility of hybrids does not always run parallel with the difficulty of making the first cross between the two parent-species, so the sterility of certain illegitimate plants was unusually great, whilst the sterility of the union from which they were derived was by no means great. With hybrids raised from the same seed-capsule the degree of sterility is innately variable, so it is in a marked manner with illegitimate plants. Lastly, many hybrids are profuse and persistent flowerers, whilst other and more sterile hybrids produce few flowers, and are weak, miserable dwarfs; exactly similar cases occur with the illegitimate offspring of various dimorphic and trimorphic plants.

Although there is the closest identity in character and behaviour between illegitimate plants and hybrids, it is hardly an exaggeration to maintain that the former are hybrids, but produced within the limits of the same species by the improper union of certain forms, whilst ordinary hybrids are produced from an improper union between so-called distinct species. We have already seen that there is the closest similarity in all respects between first illegitimate unions, and first crosses between distinct species. This will perhaps be made more fully apparent by an illustration:—we may suppose that a botanist found two well-marked varieties (and such occur) of the long-styled form of the trimorphic Lithrum salicaria, and that he determined to try by crossing whether they were specifically distinct. He would find that they yielded only about one-fifth of the proper number of seed, and that they behaved in all the other above-specified respects as if they had been two distinct species. But to make the case sure, he would raise plants from his supposed hybridised seed, and he would find that the seedlings were miserably dwarfed and utterly sterile, and that they behaved in all other respects like ordinary hybrids, he might then maintain that he had actually proved, in accordance with the common view, that his two varieties were as good and as distinct species as any in the world; but he would be completely mistaken.

The facts now given on dimorphic and trimorphic plants are important, because they show us, first, that the physiological test of lessened fertility, both in first crosses and in hybrids, is no criterion of specific distinction; secondly, because we may conclude that there is some unknown bond which connects the infertility of illegitimate unions with that of their illegitimate offspring, and we are led to extend the same view to first crosses and hybrids; thirdly, because we find, and this seems to me of especial importance, that two or three forms of the same species may exist and may differ in no respect whatever, either in structure or in constitution, relatively to external conditions, and yet be sterile when united in certain ways. For we must remember that it is the union of the sexual elements of individuals of the same form, for instance, of two long-styled forms, which results in sterility; whilst it is the union of the sexual element proper to two distinct forms which is fertile. Hence the case appears at first sight exactly the reverse of what occurs in the ordinary unions of the individuals of the same species, and with crosses between distinct species. It is, however, doubtful whether this is really so; but I will not enlarge on this obscure subject.

We may, however, infer as probable from the consideration of dimorphic and trimorphic plants, that the sterility of distinct species when crossed, and of their hybrid progeny, depends exclusively on the nature of their sexual elements, and not on any difference in their structure or general constitution. We are also led to this same conclusion by considering reciprocal crosses, in which the male of one species cannot be united, or only with great difficulty, with the female of a second species, whilst the converse cross can be effected with perfect facility. That excellent observer, Gartner, likewise concluded that species when crossed are sterile owing to differences confined to their reproductive systems.

On the principle which makes it necessary for man, whilst he is selecting and improving his domestic varieties, to keep them separate, it would clearly be advantageous to varieties in a state of nature, that is to incipient species, if they could be kept from blending, either through sexual aversion, or by becoming mutually sterile. Hence it at one time appeared to me probable, as it has to others, that this sterility might have been acquired through natural selection. On this view we must suppose that a shade of lessened fertility first spontaneously appeared, like any other modification, in certain individuals of a species when crossed with other individuals of the same species; and that successive slight degrees of infertility, from being advantageous, were slowly accumulated. This appears all the more probable, if we admit that the structural differences between the forms of dimorphic and trimorphic plants, as the length and curvature of the pistil, etc., have been co-adapted through natural selection; for if this be admitted, we can hardly avoid extending the same conclusion to their mutual infertility. Sterility, moreover, has been acquired through natural selection for other and widely different purposes, as with neuter insects in reference to their social economy. In the case of plants, the flowers on the circumference of the truss in the guelder rose (Viburnum opulus) and those on the summit of the spike in the feather-hyacinth (Muscari comosum) have been rendered conspicuous, and apparently in consequence sterile, in order that insects might easily discover and visit the perfect flowers. But when we endeavour to apply the principle of natural selection to the acquirement by distinct species of mutual sterility, we meet with great difficulties. In the first place, it may be remarked that separate regions are often inhabited by groups of species or by single species, which when brought together and crossed are found to be more or less sterile; now it could clearly have been no advantage to such separated species to have been rendered mutually sterile, and consequently this could not have been effected through natural selection; but it may perhaps be argued, that, if a species were rendered sterile with some one compatriot, sterility with other species would follow as a necessary consequence. In the second place, it is as much opposed to the theory of natural selection, as to the theory of special creation, that in reciprocal crosses the male element of one form should have been rendered utterly impotent on a second form, whilst at the same time the male element of this second form is enabled freely to fertilise the first form; for this peculiar state of the reproductive system could not possibly have been advantageous to either species.

In considering the probability of natural selection having come into action in rendering species mutually sterile, one of the greatest difficulties will be found to lie in the existence of many graduated steps from slightly lessened fertility to absolute sterility. It may be admitted, on the principle above explained, that it would profit an incipient species if it were rendered in some slight degree sterile when crossed with its parent-form or with some other variety; for thus fewer bastardised and deteriorated offspring would be produced to commingle their blood with the new species in process of formation. But he who will take the trouble to reflect on the steps by which this first degree of sterility could be increased through natural selection to that higher degree which is common to so many species, and which is universal with species which have been differentiated to a generic or family rank, will find the subject extraordinarily complex. After mature reflection it seems to me that this could not have been effected through natural selection. Take the case of any two species which, when crossed, produce few and sterile offspring; now, what is there which could favour the survival of those individuals which happened to be endowed in a slightly higher degree with mutual infertility, and which thus approached by one small step towards absolute sterility? Yet an advance of this kind, if the theory of natural selection be brought to bear, must have incessantly occurred with many species, for a multitude are mutually quite barren. With sterile neuter insects we have reason to believe that modifications in their structure and fertility have been slowly accumulated by natural selection, from an advantage having been thus indirectly given to the community to which they belonged over other communities of the same species; but an individual animal not belonging to a social community, if rendered slightly sterile when crossed with some other variety, would not thus itself gain any advantage or indirectly give any advantage to the other individuals of the same variety, thus leading to their preservation.

But it would be superfluous to discuss this question in detail; for with plants we have conclusive evidence that the sterility of crossed species must be due to some principle, quite independent of natural selection. Both Gartner and Kolreuter have proved that in general including numerous species, a series can be formed from species which when crossed yield fewer and fewer seeds, to species which never produce a single seed, but yet are affected by the pollen of certain other species, for the germen swells. It is here manifestly impossible to select the more sterile individuals, which have already ceased to yield seeds; so that this acme of sterility, when the germen alone is affected, cannot have been gained through selection; and from the laws governing the various grades of sterility being so uniform throughout the animal and vegetable kingdoms, we may infer that the cause, whatever it may be, is the same or nearly the same in all cases.

As species have not been rendered mutually infertile through the accumulative action of natural selection, and as we may safely conclude, from the previous as well as from other and more general considerations, that they have not been endowed through an act of creation with this quality, we must infer that it has arisen incidentally during their slow formation in connection with other and unknown changes in their organisation. By a quality arising incidentally, I refer to such cases as different species of animals and plants being differently affected by poisons to which they are not naturally exposed; and this difference in susceptibility is clearly incidental on other and unknown differences in their organisation. So again the capacity in different kinds of trees to be grafted on each other, or on a third species, differs much, and is of no advantage to these trees, but is incidental on structural or functional differences in their woody tissues. We need not feel surprise at sterility incidentally resulting from crosses between distinct species,—the modified descendants of a common progenitor,—when we bear in mind how easily the reproductive system is affected by various causes—often by extremely slight changes in the conditions of life, by too close interbreeding, and by other agencies. It is well to bear in mind such cases as that of the Passiflora alata, which recovered its self-fertility from being grafted on a distinct species—the cases of plants which normally or abnormally are self-impotent, but can readily be fertilised by the pollen of a distinct species—and lastly the cases of individual domesticated animals which evince towards each other sexual incompatibility.

We now at last come to the immediate point under discussion: how is it that, with some few exceptions in the case of plants, domesticated varieties, such as those of the dog, fowl, pigeon, several fruit-trees, and culinary vegetables, which differ from each other in external characters more than many species, are perfectly fertile when crossed, or even fertile in excess, whilst closely allied species are almost invariably in some degree sterile? We can, to a certain extent, give a satisfactory answer to this question. Passing over the fact that the amount of external difference between two species is no sure guide to their degree of mutual sterility, so that similar differences in the case of varieties would be no sure guide, we know that with species the cause lies exclusively in differences in their sexual constitution. Now the conditions to which domesticated animals and cultivated plants have been subjected have had so little tendency towards modifying the reproductive system in a manner leading to mutual sterility, that we have very good grounds for admitting the directly opposite doctrine of Pallas, namely, that such conditions generally eliminate this tendency; so that the domesticated descendants of species, which in their natural state would have been in some degree sterile when crossed, become perfectly fertile together. With plants, so far is cultivation from giving a tendency towards mutual sterility, that in several well-authenticated cases, already often alluded to, certain species have been affected in a very different manner, for they have become self- impotent, whilst still retaining the capacity of fertilising, and being fertilised by, distinct species. If the Pallasian doctrine of the elimination of sterility through long-continued domestication be admitted, and it can hardly be rejected, it becomes in the highest degree improbable that similar circumstances should commonly both induce and eliminate the same tendency; though in certain cases, with species having a peculiar constitution, sterility might occasionally be thus induced. Thus, as I believe, we can understand why with domesticated animals varieties have not been produced which are mutually sterile; and why with plants only a few such cases have been observed, namely, by Gartner, with certain varieties of maize and verbascum, by other experimentalists with varieties of the gourd and melon, and by Kolreuter with one kind of tobacco.

With respect to varieties which have originated in a state of nature, it is almost hopeless to expect to prove by direct evidence that they have been rendered mutually sterile; for if even a trace of sterility could be detected, such varieties would at once be raised by almost every naturalist to the rank of distinct species. If, for instance, Gartner's statement were fully confirmed, that the blue and red flowered forms of the pimpernel (Anagallis arvensis) are sterile when crossed, I presume that all the botanists who now maintain on various grounds that these two forms are merely fleeting varieties, would at once admit that they were specifically distinct.

The real difficulty in our present subject is not, as it appears to me, why domestic varieties have not become mutually infertile when crossed, but why this has so generally occurred with natural varieties as soon as they have been modified in a sufficient and permanent degree to take rank as species. We are far from precisely knowing the cause; but we can see that the species, owing to their struggle for existence with numerous competitors, must have been exposed to more uniform conditions of life during long periods of time than domestic varieties have been, and this may well make a wide difference in the result. For we know how commonly wild animals and plants, when taken from their natural conditions and subjected to captivity, are rendered sterile; and the reproductive functions of organic beings which have always lived and been slowly modified under natural conditions would probably in like manner be eminently sensitive to the influence of an unnatural cross. Domesticated productions, on the other hand, which, as shown by the mere fact of their domestication, were not originally highly sensitive to changes in their conditions of life, and which can now generally resist with undiminished fertility repeated changes of conditions, might be expected to produce varieties, which would be little liable to have their reproductive powers injuriously affected by the act of crossing with other varieties which had originated in a like manner.

Certain naturalists have recently laid too great stress, as it appears to me, on the difference in fertility between varieties and species when crossed. Some allied species of trees cannot be grafted on one another, whilst all varieties can be so grafted. Some allied animals are affected in a very different manner by the same poison, but with varieties no such case until recently was known; whilst now it has been proved that immunity from certain poisons sometimes stands in correlation with the colour of the individuals of the same species. The period of gestation generally differs much in distinct species, but with varieties until lately no such difference had been observed. Here we have various physiological differences, and no doubt others could be added, between one species and another of the same genus, which do not occur, or occur with extreme rarity, in the case of varieties; and these differences are apparently wholly or in chief part incidental on other constitutional differences, just in the same manner as the sterility of crossed species is incidental on differences confined to the sexual system. Why, then, should these latter differences, however serviceable they may indirectly be in keeping the inhabitants of the same country distinct, be thought of such paramount importance, in comparison with other incidental and functional differences? No sufficient answer to this question can be given. Hence the fact that widely distinct domestic varieties are, with rare exceptions, perfectly fertile when crossed, and produce fertile offspring, whilst closely allied species are, with rare exceptions, more or less sterile, is not nearly so formidable an objection as it appears at first to the theory of the common descent of allied species.

# CHAPTER 2.XX.

SELECTION BY MAN.

SELECTION A DIFFICULT ART. METHODICAL, UNCONSCIOUS, AND NATURAL SELECTION. RESULTS OF METHODICAL SELECTION. CARE TAKEN IN SELECTION. SELECTION WITH PLANTS. SELECTION CARRIED ON BY THE ANCIENTS AND BY SEMI-CIVILISED PEOPLE. UNIMPORTANT CHARACTERS OFTEN ATTENDED TO. UNCONSCIOUS SELECTION. AS CIRCUMSTANCES SLOWLY CHANGE, SO HAVE OUR DOMESTICATED ANIMALS CHANGED THROUGH THE ACTION OF UNCONSCIOUS SELECTION. INFLUENCE OF DIFFERENT BREEDERS ON THE SAME SUB-VARIETY. PLANTS AS AFFECTED BY UNCONSCIOUS SELECTION. EFFECTS OF SELECTION AS SHOWN BY THE GREAT AMOUNT OF DIFFERENCE IN THE PARTS MOST VALUED BY MAN.

The power of Selection, whether exercised by man, or brought into play under nature through the struggle for existence and the consequent survival of the fittest, absolutely depends on the variability of organic beings. Without variability nothing can be effected; slight individual differences, however, suffice for the work, and are probably the chief or sole means in the production of new species. Hence our discussion on the causes and laws of variability ought in strict order to have preceded the present subject, as well as inheritance, crossing, etc.; but practically the present arrangement has been found the most convenient. Man does not attempt to cause variability; though he unintentionally effects this by exposing organisms to new conditions of life, and by crossing breeds already formed. But variability being granted, he works wonders. Unless some degree of selection be exercised, the free commingling of the individuals of the same variety soon obliterates, as we have previously seen, the slight differences which arise, and gives uniformity of character to the whole body of individuals. In separated districts, long- continued exposure to different conditions of life may produce new races without the aid of selection; but to this subject of the direct action of the conditions of life I shall recur in a future chapter.

When animals or plants are born with some conspicuous and firmly inherited new character, selection is reduced to the preservation of such individuals, and to the subsequent prevention of crosses; so that nothing more need be said on the subject. But in the great majority of cases a new character, or some superiority in an old character, is at first faintly pronounced, and is not strongly inherited; and then the full difficulty of selection is experienced.

Indomitable patience, the finest powers of discrimination, and sound judgment must be exercised during many years. A clearly predetermined object must be kept steadily in view. Few men are endowed with all these qualities, especially with that of discriminating very slight differences; judgment can be acquired only by long experience; but if any of these qualities be wanting, the labour of a life may be thrown away. I have been astonished when celebrated breeders, whose skill and judgment have been proved by their success at exhibitions, have shown me their animals, which appeared all alike, and have assigned their reasons for matching this and that individual. The importance of the great principle of Selection mainly lies in this power of selecting scarcely appreciable differences, which nevertheless are found to be transmissible, and which can be accumulated until the result is made manifest to the eyes of every beholder.

The principle of selection may be conveniently divided into three kinds. METHODICAL SELECTION is that which guides a man who systematically endeavours to modify a breed according to some predetermined standard. UNCONSCIOUS SELECTION is that which follows from men naturally preserving the most valued and destroying the less valued individuals, without any thought of altering the breed; and undoubtedly this process slowly works great changes. Unconscious selection graduates into methodical, and only extreme cases can be distinctly separated; for he who preserves a useful or perfect animal will generally breed from it with the hope of getting offspring of the same character; but as long as he has not a predetermined purpose to improve the breed, he may be said to be selecting unconsciously. (20/1. The term "unconscious selection" has been objected to as a contradiction; but see some excellent observations on this head by Prof. Huxley ('Nat. Hist. Review' October 1864 page 578) who remarks that when the wind heaps up sand-dunes it sifts and UNCONSCIOUSLY SELECTS from the gravel on the beach grains of sand of equal size.) Lastly, we have NATURAL SELECTION, which implies that the individuals which are best fitted for the complex, and in the course of ages changing conditions to which they are exposed, generally survive and procreate their kind. With domestic productions, natural selection comes to a certain extent into action, independently of, and even in opposition to, the will of man.

## METHODICAL SELECTION.

What man has effected within recent times in England by methodical selection is clearly shown by our exhibitions of improved quadrupeds and fancy birds. With respect to cattle, sheep, and pigs, we owe their great improvement to a long series of well-known names—Bakewell, Coiling, Ellman, Bates, Jonas Webb, Lords Leicester and Western, Fisher Hobbs, and others. Agricultural writers are unanimous on the power of selection: any number of statements to this effect could be quoted; a few will suffice. Youatt, a sagacious and experienced observer, writes (20/2. 'On Sheep' 1838 page 60.) the principle of selection is "that which enables the agriculturist, not only to modify the character of his flock, but to change it altogether." A great breeder of Shorthorns (20/3. Mr. J. Wright on Shorthorn Cattle in 'Journal of Royal Agricult. Soc.' volume 7 pages 208, 209.) says, "In the anatomy of the shoulder modern breeders have made great improvement on the Ketton shorthorns by correcting the defect in the knuckle or shoulder-joint, and by laying the top of the shoulder more snugly in the crop, and thereby filling up the hollow behind it...The eye has its fashion at different periods: at one time the eye high and outstanding from the head, and at another time the sleepy eye sunk into the head; but these extremes have merged into the medium of a full, clear and prominent eye with a placid look."

Again, hear what an excellent judge of pigs (20/4. H.D. Richardson 'On Pigs' 1847 page 44.) says: "The legs should be no longer than just to prevent the animal's belly from trailing on the ground. The leg is the least profitable portion of the hog, and we therefore require no more of it than is absolutely necessary for the support of the rest." Let any one compare the wild-boar with any improved breed, and he will see how effectually the legs have been shortened.

Few persons, except breeders, are aware of the systematic care taken in selecting animals, and of the necessity of having a clear and almost prophetic vision into futurity. Lord Spencer's skill and judgment were well known; and he writes (20/5. 'Journal of Royal Agricult. Soc.' volume 1 page 24.), "It is therefore very desirable, before any man commences to breed either cattle or sheep, that he should make up his mind to the shape and qualities he wishes to obtain, and steadily pursue this object." Lord Somerville, in speaking of the marvellous improvement of the New Leicester sheep, effected by Bakewell and his successors, says, "It would seem as if they had first drawn a perfect form, and then given it life." Youatt (20/6. 'On Sheep' pages 520, 319.) urges the necessity of annually drafting each flock, as many animals will certainly degenerate "from the standard of excellence which the breeder has established in his own mind." Even with a bird of such little importance as the canary, long ago (1780-1790) rules were established, and a standard of perfection was fixed according to which the London fanciers tried to breed the several sub- varieties. (20/7. Loudon's 'Mag. of Nat. Hist.' volume 8 1835 page 618.) A great winner of prizes at the Pigeon-shows (20/8. 'A treatise on the Art of Breeding the Almond Tumbler' 1851 page 9.), in describing the short-faced Almond Tumbler, says, "There are many first-rate fanciers who are particularly partial to what is called the goldfinch-beak, which is very beautiful; others say, take a full-size round cherry then take a barleycorn, and judiciously placing and thrusting it into the cherry, form as it were your beak; and that is not all, for it will form a good head and beak, provided, as I said before, it is judiciously done; others take an oat; but as I think the goldfinch-beak the handsomest, I would advise the inexperienced fancier to get the head of a goldfinch, and keep it by him for his observation." Wonderfully different as are the beaks of the rock pigeon and goldfinch, the end has undoubtedly been nearly gained, as far as external shape and proportions are concerned.

Not only should our animals be examined with the greatest care whilst alive, but, as Anderson remarks (20/9. 'Recreations in Agriculture' volume 2 page 409.) their carcases should be scrutinised, "so as to breed from the descendants of such only as, in the language of the butcher, cut up well." The "grain of the meat" in cattle, and its being well marbled with fat (20/10. 'Youatt on Cattle' pages 191, 227.), and the greater or less accumulation of fat in the abdomen of our sheep, have been attended to with success. So with poultry, a writer (20/11. Ferguson 'Prize Poultry' 1854 page 208.), speaking of Cochin-China fowls, which are said to differ much in the quality of their flesh, says, "the best mode is to purchase two young brother-cocks, kill, dress, and serve up one; if he be indifferent, similarly dispose of the other, and try again; if, however, he be fine and well-flavoured, his brother will not be amiss for breeding purposes for the table."

The great principle of the division of labour has been brought to bear on selection. In certain districts (20/12. Wilson in 'Transact. Highland Agricult. Soc.' quoted in 'Gardener's Chronicle' 1844 page 29.) "the breeding of bulls is confined to a very limited number of persons, who by devoting their whole attention to this department, are able from year to year to furnish a class of bulls which are steadily improving the general breed of the district." The rearing and letting of choice rams has long been, as is well known, a chief source of profit to several eminent breeders. In parts of Germany this principle is carried with merino sheep to an extreme point. (20/13. Simmonds quoted in 'Gardener's Chronicle' 1855 page 637. And for the second quotation see 'Youatt on Sheep' page 171.) So "important is the proper selection of breeding animals considered, that the best flock-masters do not trust to their own judgment or to that of their shepherds, but employ persons called 'sheep-classifiers' who make it their special business to attend to this part of the management of several flocks, and thus to preserve, or if possible to improve, the best qualities of both parents in the lambs." In Saxony, "when the lambs are weaned, each in his turn is placed upon a table that his wool and form may be minutely observed. The finest are selected for breeding and receive a first mark. When they are one year old, and prior to shearing them, another close examination of those previously marked takes place: those in which no defect can be found receive a second mark, and the rest are condemned. A few months afterwards a third and last scrutiny is made; the prime rams and ewes receive a third and final mark, but the slightest blemish is sufficient to cause the rejection of the animal." These sheep are bred and valued almost exclusively for the fineness of their wool; and the result corresponds with the labour bestowed on their selection. Instruments have been invented to measure accurately the thickness of the fibres; and "an Austrian fleece has been produced of which twelve hairs equalled in thickness one from a Leicester sheep."

Throughout the world, wherever silk is produced, the greatest care is bestowed on selecting the cocoons from which the moths for breeding are to be reared. A careful cultivator (20/14. Robinet 'Vers a Soie' 1848 page 271.) likewise examines the moths themselves, and destroys those that are not perfect. But what more immediately concerns us is that certain families in France devote themselves to raising eggs for sale. (20/15. Quatrefages 'Les Maladies du Ver a Soie' 1859 page 101.) In China, near Shanghai, the inhabitants of two small districts have the privilege of raising eggs for the whole surrounding country, and that they may give up their whole time to this business, they are interdicted by law from producing silk. (20/16. M. Simon in 'Bull. de la Soc. d'Acclimat.' tome 9 1862 page 221.)

The care which successful breeders take in matching their birds is surprising. Sir John Sebright, whose fame is perpetuated by the "Sebright Bantam," used to spend "two and three days in examining, consulting, and disputing with a friend which were the best of five or six birds." (20/17. 'The Poultry Chronicle' volume 1 1854 page 607.) Mr. Bult, whose pouter-pigeons won so many prizes, and were exported to North America under the charge of a man sent on purpose, told me that he always deliberated for several days before he matched each pair. Hence we can understand the advice of an eminent fancier, who writes (20/18. J.M. Eaton 'A Treatise on Fancy Pigeons' 1852 page 14 and 'A Treatise on the Almond Tumbler' 1851 page 11.) "I would here particularly guard you against having too great a variety of pigeons, otherwise you will know a little of all, but nothing about one as it ought to be known." Apparently it transcends the power of the human intellect to breed all kinds: "it is possible that there may be a few fanciers that have a good general knowledge of fancy pigeons; but there are many more who labour under the delusion of supposing they know what they do not." The excellence of one sub- variety, the Almond Tumbler, lies in the plumage, carriage, head, beak, and eye; but it is too presumptuous in the beginner to try for all these points. The great judge above quoted says, "There are some young fanciers who are over-covetous, who go for all the above five properties at once; they have their reward by getting nothing." We thus see that breeding even fancy pigeons is no simple art: we may smile at the solemnity of these precepts, but he who laughs will win no prizes.

What methodical selection has effected for our animals is sufficiently proved, as already remarked, by our Exhibitions. So greatly were the sheep belonging to some of the earlier breeders, such as Bakewell and Lord Western, changed, that many persons could not be persuaded that they had not been crossed. Our pigs, as Mr. Corringham remarks (20/19. 'Journal Royal Agricultural Soc.' volume 6 page 22.) during the last twenty years have undergone, through rigorous selection together with crossing, a complete metamorphosis. The first exhibition for poultry was held in the Zoological Gardens in 1845; and the improvement effected since that time has been great. As Mr. Bailey, the great judge, remarked to me, it was formerly ordered that the comb of the Spanish cock should be upright, and in four or five years all good birds had upright combs; it was ordered that the Polish cock should have no comb or wattles, and now a bird thus furnished would be at once disqualified; beards were ordered, and out of fifty-seven pens lately (1860) exhibited at the Crystal Palace, all had beards. So it has been in many other cases. But in all cases the judges order only what is occasionally produced and what can be improved and rendered constant by selection. The steady increase in weight during the last few years in our fowls, turkeys, ducks, and geese is notorious; "six-pound ducks are now common, whereas four pounds was formerly the average." As the time required to make a change has not often been recorded, it may be worth mentioning that it took Mr. Wicking thirteen years to put a clean white head on an almond tumbler's body, "a triumph," says another fancier, "of which he may be justly proud." (20/20. 'Poultry Chronicle' volume 2 1855 page 596.)

Mr. Tollet, of Betley Hall, selected cows, and especially bulls, descended from good milkers, for the sole purpose of improving his cattle for the production of cheese; he steadily tested the milk with the lactometer, and in eight years he increased, as I was informed by him, the product in proportion of four to three. Here is a curious case (20/21. Isid. Geoffroy St.-Hilaire 'Hist. Nat. Gen.' tome 3 page 254.) of steady but slow progress, with the end not as yet fully attained: in 1784 a race of silkworms was introduced into France, in which one hundred in the thousand failed to produce white cocoons; but now after careful selection during sixty-five generations, the proportion of yellow cocoons has been reduced to thirty-five in the thousand.

With plants selection has been followed with the same good result as with animals. But the process is simpler, for plants in the great majority of cases bear both sexes. Nevertheless, with most kinds it is necessary to take as much care to prevent crosses as with animals or unisexual plants; but with some plants, such as peas, this care is not necessary. With all improved plants, excepting of course those which are propagated by buds, cuttings, etc., it is almost indispensable to examine the seedlings and destroy those which depart from the proper type. This is called "roguing," and is, in fact, a form of selection, like the rejection of inferior animals. Experienced horticulturists and agriculturists incessantly urge every one to preserve the finest plants for the production of seed.

Although plants often present much more conspicuous variations than animals, yet the closest attention is generally requisite to detect each slight and favourable change. Mr. Masters relates (20/22. 'Gardener's Chronicle' 1850 page 198.) how "many a patient hour was devoted," whilst he was young, to the detection of differences in peas intended for seed. Mr. Barnet (20/23. 'Transact. Hort. Soc.' volume 6 page 152.) remarks that the old scarlet American strawberry was cultivated for more than a century without producing a single variety; and another writer observes how singular it was that when gardeners first began to attend to this fruit it began to vary; the truth no doubt being that it had always varied, but that, until slight variations were selected and propagated by seed, no conspicuous result was obtained. The finest shades of difference in wheat have been discriminated and selected with almost as much care as, in the case of the higher animals, for instance by Col. Le Couteur and more especially by Major Hallett.

It may be worth while to give a few examples of methodical selection with plants; but in fact the great improvement of all our anciently cultivated plants may be attributed to selection long carried on, in part methodically, and in part unconsciously. I have shown in a former chapter how the weight of the gooseberry has been increased by systematic selection and culture. The flowers of the Heartsease have been similarly increased in size and regularity of outline. With the Cineraria, Mr. Glenny (20/24. 'Journal of Horticulture' 1862 page 369.) "was bold enough when the flowers were ragged and starry and ill defined in colour, to fix a standard which was then considered outrageously high and impossible, and which, even if reached, it was said, we should be no gainers by, as it would spoil the beauty of the flowers. He maintained that he was right; and the event has proved it to be so." The doubling of flowers has several times been effected by careful selection: the Rev. W. Williamson (20/25 'Transact. Hort. Soc.' volume 4 page 381.), after sowing during several years seed of Anemone coronaria, found a plant with one additional petal; he sowed the seed of this, and by perseverance in the same course obtained several varieties with six or seven rows of petals. The single Scotch rose was doubled, and yielded eight good varieties in nine or ten years. (20/26. 'Transact. Hort. Soc.' volume 4 page 285.) The Canterbury bell (Campanula medium) was doubled by careful selection in four generations. (20/27. Rev. W. Bromehead in 'Gardener's Chronicle' 1857 page 550.) In four years Mr. Buckman (20/28. 'Gardener's Chronicle' 1862 page 721.), by culture and careful selection, converted parsnips, raised from wild seed, into a new and good variety. By selection during a long course of years, the early maturity of peas has been hastened by between ten and twenty-one days. (20/29. Dr. Anderson in 'The Bee' volume 6 page 96; Mr. Barnes in 'Gardener's Chronicle' 1844 page 476.) A more curious case is offered by the beet plant, which since its cultivation in France, has almost exactly doubled its yield of sugar. This has been effected by the most careful selection; the specific gravity of the roots being regularly tested, and the best roots saved for the production of seed. (20/30. Godron 'De l'Espece' 1859 tome 2 page 69; 'Gardener's Chronicle' 1854 page 258.)

## SELECTION BY ANCIENT AND SEMI-CIVILISED PEOPLE.

In attributing so much importance to the selection of animals and plants, it may be objected, that methodical selection would not have been carried on during ancient times. A distinguished naturalist considers it as absurd to suppose that semi-civilised people should have practised selection of any kind. Undoubtedly the principle has been systematically acknowledged and followed to a far greater extent within the last hundred years than at any former period, and a corresponding result has been gained; but it would be a greater error to suppose, as we shall immediately see, that its importance was not recognised and acted on during the most ancient times, and by semi-civilised people. I should premise that many facts now to be given only show that care was taken in breeding; but when this is the case, selection is almost sure to be practised to a certain extent. We shall hereafter be enabled better to judge how far selection, when only occasionally carried on, by a few of the inhabitants of a country, will slowly produce a great effect.

In a well-known passage in the thirtieth chapter of Genesis, rules are given for influencing, as was then thought possible, the colour of sheep; and speckled and dark breeds are spoken of as being kept separate. By the time of David the fleece was likened to snow. Youatt (20/31. 'On Sheep' page 18.), who has discussed all the passages in relation to breeding in the Old Testament, concludes that at this early period "some of the best principles of breeding must have been steadily and long pursued." It was ordered, according to Moses, that "Thou shalt not let thy cattle gender with a diverse kind;" but mules were purchased (20/32. Volz 'Beitrage zur Kulturgeschichte' 1852 s. 47.) so that at this early period other nations must have crossed the horse and ass. It is said (20/33. Mitford 'History of Greece' volume 1 page 73.) that Erichthonius, some generations before the Trojan war, had many brood-mares, "which by his care and judgment in the choice of stallions produced a breed of horses superior to any in the surrounding countries." Homer (Book 5) speaks of Aeneas' horses as bred from mares which were put to the steeds of Laomedon. Plato, in his 'Republic' says to Glaucus, "I see that you raise at your house a great many dogs for the chase. Do you take care about breeding and pairing them? Among animals of good blood, are there not always some which are superior to the rest?" To which Glaucus answers in the affirmative. (20/34. Dr. Dally translated in

'Anthropological Review' May 1864 page 101.) Alexander the Great selected the finest Indian cattle to send to Macedonia to improve the breed. (20/35. Volz 'Beitrage' etc. 1852 s. 80.) According to Pliny (20/36 'History of the World' chapter 45.), King Pyrrhus had an especially valuable breed of oxen: and he did not suffer the bulls and cows to come together till four years old, that the breed might not degenerate. Virgil, in his Georgics (lib. 3), gives as strong advice as any modern agriculturist could do, carefully to select the breeding stock; "to note the tribe, the lineage, and the sire; whom to reserve for husband of the herd;"—to brand the progeny;—to select sheep of the purest white, and to examine if their tongues are swarthy. We have seen that the Romans kept pedigrees of their pigeons, and this would have been a senseless proceeding had not great care been taken in breeding them. Columella gives detailed instructions about breeding fowls: "Let the breeding hens therefore be of a choice colour, a robust body, square-built, full-breasted, with large heads, with upright and bright-red combs. Those are believed to be the best bred which have five toes." (20/37. 'Gardener's Chronicle' 1848 page 323.) According to Tacitus, the Celts attended to the races of their domestic animals; and Caesar states that they paid high prices to merchants for fine imported horses. (20/38. Reynier 'De l'Economie des Celtes' 1818 pages 487, 503.) In regard to plants, Virgil speaks of yearly culling the largest seeds; and Celsus says, "where the corn and crop is but small, we must pick out the best ears of corn, and of them lay up our seed separately by itself." (20/39. Le Couteur on 'Wheat' page 15.)

Coming down the stream of time, we may be brief. At about the beginning of the ninth century Charlemagne expressly ordered his officers to take great care of his stallions; and if any proved bad or old, to forewarn him in good time before they were put to the mares. (20/40. Michel 'Des Haras' 1861 page 84.) Even in a country so little civilised as Ireland during the ninth century, it would appear from some ancient verses (20/41. Sir W. Wilde an 'Essay on Unmanufactured Animal Remains' etc. 1860 page 11.), describing a ransom demanded by Cormac, that animals from particular places, or having a particular character, were valued. Thus it is said,—

Two pigs of the pigs of Mac Lir,
A ram and ewe both round and red,
I brought with me from Aengus.
I brought with me a stallion and a mare
From the beautiful stud of Manannan,
A bull and a white cow from Druim Cain.

Athelstan, in 930, received running-horses as a present from Germany; and he prohibited the exportation of English horses. King John imported "one hundred chosen stallions from Flanders." (20/42. Col. Hamilton Smith 'Nat. Library' volume 12 Horses, pages 135, 140.) On June 16th, 1305, the Prince of Wales wrote to the Archbishop of Canterbury, begging for the loan of any choice stallion, and promising its return at the end of the season. (20/43. Michel 'Des Haras' page 90.) There are numerous records at ancient periods in English history of the importation of choice animals of various kinds, and of foolish laws against their exportation. In the reigns of Henry VII. and VIII. it was ordered that the magistrates, at Michaelmas, should scour the heaths and commons, and destroy all mares beneath a certain size. (20/44. Mr. Baker 'History of the Horse' 'Veterinary' volume 13 page 423.) Some of our earlier kings passed laws against the slaughtering rams of any good breed before they were seven years old, so that they might have time to breed. In Spain Cardinal Ximenes issued, in 1509, regulations on the SELECTION of good rams for breeding. (20/45. M. l'Abbe Carlier in 'Journal de Physique' volume 24 1784 page 181; this memoir contains much information on the ancient selection of sheep; and is my authority for rams not being killed young in England.)

The Emperor Akbar Khan before the year 1600 is said to have "wonderfully improved" his pigeons by crossing the breeds; and this necessarily implies careful selection. About the same period the Dutch attended with the greatest care to the breeding of these birds. Belon in 1555 says that good managers in France examined the colour of their goslings in order to get geese of a white colour and better kinds. Markham in 1631 tells the breeder "to elect the largest and goodliest conies," and enters into minute details. Even with respect to seeds of plants for the flower-garden, Sir J. Hanmer writing about the year 1660 (20/46. 'Gardener's Chronicle' 1843 page 389.) says, in "choosing seed, the best seed is the most weighty, and is had from the lustiest and most vigorous stems;" and he then gives rules about leaving only a few flowers on plants for seed; so that even such details were attended to in our flower-gardens two hundred years ago. In order to show that selection has been silently carried on in places where it would not have been expected, I may add that in the middle of the last century, in a remote part of North America, Mr. Cooper improved by careful selection all his vegetables, "so that they were greatly superior to those of any other person. When his radishes, for instance, are fit for use, he takes ten or twelve that he most approves, and plants them at least 100 yards from others that blossom at the same time. In the same manner he treats all his other plants, varying the circumstances according to their nature." (20/47. 'Communications to Board of Agriculture' quoted in Dr. Darwin 'Phytologia' 1800 page 451.)

In the great work on China published in the last century by the Jesuits, and which is chiefly compiled from ancient Chinese encyclopaedias, it is said that with sheep "improving the breed consists in choosing with particular care the lambs which are destined for propagation, in nourishing them well, and in keeping the flocks separate." The same principles were applied by the

Chinese to various plants and fruit-trees. (20/48. 'Memoire sur les Chinois' 1786 tome 11 page 55; tome 5 page 507.) An imperial edict recommends the choice of seed of remarkable size; and selection was practised even by imperial hands, for it is said that the Ya-mi, or imperial rice, was noticed at an ancient period in a field by the Emperor Khang-hi, was saved and cultivated in his garden, and has since become valuable from being the only kind which will grow north of the Great Wall. (20/49. 'Recherches sur l'Agriculture des Chinois' par L. D'Hervey Saint-Denys 1850 page 229. With respect to Khang-hi see Huc's 'Chinese Empire' page 311.) Even with flowers, the tree paeony (P. moutan) has been cultivated, according to Chinese traditions, for 1400 years; between 200 and 300 varieties have been raised, which are cherished like tulips formerly were by the Dutch. (20/50. Anderson in 'Linn. Transact.' volume 12 page 253.)

Turning now to semi-civilised people and to savages: it occurred to me, from what I had seen of several parts of South America, where fences do not exist, and where the animals are of little value, that there would be absolutely no care in breeding or selecting them; and this to a large extent is true. Roulin (20/51. 'Mem. de l'Acad.' (divers savants), tome 6 1835 page 333.), however, describes in Columbia a naked race of cattle, which are not allowed to increase, on account of their delicate constitution. According to Azara (20/52. 'Des Quadrupedes du Paraguay' 1801 tome 2 pages 333, 371.) horses are often born in Paraguay with curly hair; but, as the natives do not like them, they are destroyed. On the other hand, Azara states that a hornless bull, born in 1770, was preserved and propagated its race. I was informed of the existence in Banda Oriental of a breed with reversed hair; and the extraordinary niata cattle first appeared and have since been kept distinct in La Plata. Hence certain conspicuous variations have been preserved, and others have been habitually destroyed, in these countries, which are so little favourable for careful selection. We have also seen that the inhabitants sometimes introduce fresh cattle on their estates to prevent the evil effects of close interbreeding. On the other hand, I have heard on reliable authority that the Gauchos of the Pampas never take any pains in selecting the best bulls or stallions for breeding; and this probably accounts for the cattle and horses being remarkably uniform in character throughout the immense range of the Argentine republic.

Looking to the Old World, in the Sahara Desert "The Touareg is as careful in the selection of his breeding Mahari (a fine race of the dromedary) as the Arab is in that of his horse. The pedigrees are handed down, and many a dromedary can boast a genealogy far longer than the descendants of the Darley Arabian." (20/53. 'The Great Sahara' by the Rev. H.B. Tristram 1860 page 238.) According to Pallas the Mongolians endeavour to breed the Yaks or horse-tailed buffaloes with white tails, for these are sold to the Chinese mandarins as fly-flappers; and Moorcroft, about seventy years after Pallas, found that white-tailed animals were still selected for breeding. (20/54. Pallas 'Act. Acad. St. Petersburg' 1777 page 249. Moorcroft and Trebeck 'Travels in the Himalayan Provinces' 1841.)

We have seen in the chapter on the Dog that savages in different parts of North America and in Guiana cross their dogs with wild Canidae, as did the ancient Gauls, according to Pliny. This was done to give their dogs strength and vigour, in the same way as the keepers in large warrens now sometimes cross their ferrets (as I have been informed by Mr. Yarrell) with the wild polecat, "to give them more devil." According to Varro, the wild ass was formerly caught and crossed with the tame animal to improve the breed, in the same manner as at the present day the natives of Java sometimes drive their cattle into the forests to cross with the wild Banteng (Bos sondaicus). (20/55. Quoted from Raffles in the 'Indian Field' 1859 page 196: for Varro see Pallas ut supra.) In Northern Siberia, among the Ostyaks, the dogs vary in markings in different districts, but in each place they are spotted black and white in a remarkably uniform manner (20/56. Erman 'Travels in Siberia' English translation volume 1 page 453.); and from this fact alone we may infer careful breeding, more especially as the dogs of one locality are famed throughout the country for their superiority. I have heard of certain tribes of Esquimaux who take pride in their teams of dogs being uniformly coloured. In Guiana, as Sir H. Schomburgk informs me (20/57. See also 'Journal of R. Geograph. Soc.' volume 13 part 1 page 65.), the dogs of the Turuma Indians are highly valued and extensively bartered: the price of a good one is the same as that given for a wife: they are kept in a sort of cage, and the Indians "take great care when the female is in season to prevent her uniting with a dog of an inferior description." The Indians told Sir Robert that, if a dog proved bad or useless, he was not killed, but was left to die from sheer neglect. Hardly any nation is more barbarous than the Fuegians, but I hear from Mr. Bridges, the Catechist to the Mission, that, "when these savages have a large, strong, and active bitch, they take care to put her to a fine dog, and even take care to feed her well, that her young may be strong and well favoured."

In the interior of Africa, negroes, who have not associated with white men, show great anxiety to improve their animals; they "always choose the larger and stronger males for stock;" the Malakolo were much pleased at Livingstone's promise to send them a bull, and some Bakalolo carried a live cock all the way from Loanda into the interior. (20/58. Livingstone 'First Travels' pages 191, 439, 565; see also 'Expedition to the Zambesi' 1865 page 495, for an analogous case respecting a good breed of goats.) At Falaba Mr. Winwood Reade noticed an unusually fine horse, and the negro King informed him that "the owner was noted for his skill in breeding horses." Further south on the same continent, Andersson states that he has known a

Damara give two fine oxen for a dog which struck his fancy. The Damaras take great delight in having whole droves of cattle of the same colour, and they prize their oxen in proportion to the size of their horns. "The Namaquas have a perfect mania for a uniform team; and almost all the people of Southern Africa value their cattle next to their women, and take a pride in possessing animals that look high-bred. They rarely or never make use of a handsome animal as a beast of burden." (20/59. Andersson 'Travels in South Africa' pages 232, 318, 319.) The power of discrimination which these savages possess is wonderful, and they can recognise to which tribe any cattle belong. Mr. Andersson further informs me that the natives frequently match a particular bull with a particular cow.

The most curious case of selection by semi-civilised people, or indeed by any people, which I have found recorded, is that given by Garcilazo de la Vega, a descendant of the Incas, as having been practised in Peru before the country was subjugated by the Spaniards. (20/60. Dr. Vavasseur in 'Bull. de La Soc. d'Acclimat.' tome 8 1861 page 136.) The Incas annually held great hunts, when all the wild animals were driven from an immense circuit to a central point. The beasts of prey were first destroyed as injurious. The wild Guanacos and Vicunas were sheared; the old males and females killed, and the others set at liberty. The various kinds of deer were examined; the old males and females were likewise killed, "but the young females, with a certain number of males, selected from the most beautiful and strong," were given their freedom. Here, then, we have selection by man aiding natural selection. So that the Incas followed exactly the reverse system of that which our Scottish sportsman are accused of following, namely, of steadily killing the finest stags, thus causing the whole race to degenerate. (20/61. 'The Natural History of Dee Side' 1855 page 476.) In regard to the domesticated llamas and alpacas, they were separated in the time of the Incas according to colour: and if by chance one in a flock was born of the wrong colour, it was eventually put into another flock.

In the genus Auchenia there are four forms,—the Guanaco and Vicuna, found wild and undoubtedly distinct species; the Llama and Alpaca, known only in a domesticated condition. These four animals appear so different, that most naturalists, especially those who have studied these animals in their native country, maintain that they are specifically distinct, notwithstanding that no one pretends to have seen a wild llama or alpaca. Mr. Ledger, however, who has closely studied these animals both in Peru and during their exportation to Australia, and who has made many experiments on their propagation, adduces arguments (20/62. 'Bull. de la Soc. d'Acclimat.' tome 7 1860 page 457.) which seem to me conclusive, that the llama is the domesticated descendant of the guanaco, and the alpaca of the vicuna. And now that we know that these animals were systematically bred and selected many centuries ago, there is nothing surprising in the great amount of change which they have undergone.

It appeared to me at one time probable that, though ancient and semi-civilised people might have attended to the improvement of their more useful animals in essential points, yet that they would have disregarded unimportant characters. But human nature is the same throughout the world: fashion everywhere reigns supreme, and man is apt to value whatever he may chance to possess. We have seen that in South America the niata cattle, which certainly are not made useful by their shortened faces and upturned nostrils, have been preserved. The Damaras of South Africa value their cattle for uniformity of colour and enormously long horns. And I will now show that there is hardly any peculiarity in our most useful animals which, from fashion, superstition, or some other motive, has not been valued, and consequently preserved. With respect to cattle, "an early record," according to Youatt (20/63. 'Cattle' page 48.) "speaks of a hundred white cows with red ears being demanded as a compensation by the princes of North and South Wales. If the cattle were of a dark or black colour, 150 were to be presented." So that colour was attended to in Wales before its subjugation by England. In Central Africa, an ox that beats the ground with its tail is killed; and in South Africa some of the Damaras will not eat the flesh of a spotted ox. The Kaffirs value an animal with a musical voice; and "at a sale in British Kaffraria the low "of a heifer excited so much admiration that a sharp competition sprung up for her possession, and she realised a considerable price." (20/64. Livingstone 'Travels' page 576; Andersson 'Lake Ngami' 1856 page 222. With respect to the sale in Kaffraria see 'Quarterly Review' 1860 page 139.) With respect to sheep, the Chinese prefer rams without horns; the Tartars prefer them with spirally wound horns, because the hornless are thought to lose courage. (20/65. 'Memoire sur les Chinois' by the Jesuits 1786 tome 11 page 57.) Some of the Damaras will not eat the flesh of hornless sheep. In regard to horses, at the end of the fifteenth century animals of the colour described as liart pomme were most valued in France. The Arabs have a proverb, "Never buy a horse with four white feet, for he carries his shroud with him" (20/66. F. Michel 'Des Haras' pages 47, 50.); the Arabs also, as we have seen, despise dun- coloured horses. So with dogs, Xenophon and others at an ancient period were prejudiced in favour of certain colours; and "white or slate-coloured hunting dogs were not esteemed." (20/67. Col. Hamilton Smith 'Dogs' in 'Nat. Lib.' volume 10 page 103.)

Turning to poultry, the old Roman gourmands thought that the liver of a white goose was the most savoury. In Paraguay black-skinned fowls are kept because they are thought to be more productive, and their flesh the most proper for invalids. (20/68. Azara 'Quadrupedes du Paraguay' tome 2 page 324.) In Guiana, as I am informed by Sir R. Schomburgk, the aborigines will not eat the flesh or eggs of the fowl, but two races are kept distinct merely for ornament. In the Philippines, no less than nine sub-varieties of the game-cock are kept and named, so that they must be separately bred.

At the present time in Europe, the smallest peculiarities are carefully attended to in our most useful animals, either from fashion, or as a mark of purity of blood. Many examples could be given; two will suffice. "In the Western counties of England the prejudice against a white pig is nearly as strong as against a black one in Yorkshire." In one of the Berkshire sub- breeds, it is said, "the white should be confined to four white feet, a white spot between the eyes, and a few white hairs behind each shoulder." Mr. Saddler possessed three hundred pigs, every one of which was marked in this manner." (20/69. Sidney's edition of Youatt 1860 pages 24, 25.) Marshall, towards the close of the last century, in speaking of a change in one of the Yorkshire breeds of cattle, says the horns have been considerably modified, as "a clean, small, sharp horn has been FASHIONABLE for the last twenty years." (20/70. 'Rural Economy of Yorkshire' volume 2 page 182.) In a part of Germany the cattle of the Race de Gfoehl are valued for many good qualities, but they must have horns of a particular curvature and tint, so much so that mechanical means are applied if they take a wrong direction; but the inhabitants "consider it of the highest importance that the nostrils of the bull should be flesh-coloured, and the eyelashes light; this is an indispensable condition. A calf with blue nostrils would not be purchased, or purchased at a very low price." (20/71. Moll et Gayot 'Du Boeuf' 1860 page 547.) Therefore let no man say that any point or character is too trifling to be methodically attended to and selected by breeders.

**UNCONSCIOUS SELECTION.**

By this term I mean, as already more than once explained, the preservation by man of the most valued, and the destruction of the least valued individuals, without any conscious intention on his part of altering the breed. It is difficult to offer direct proofs of the results which follow from this kind of selection; but the indirect evidence is abundant. In fact, except that in the one case man acts intentionally, and in the other unintentionally, there is little difference between methodical and unconscious selection. In both cases man preserves the animals which are most useful or pleasing to him, and destroys or neglects the others. But no doubt a far more rapid result follows from methodical than from unconscious selection. The "roguing" of plants by gardeners, and the destruction by law in Henry VIII.'s reign of all under- sized mares, are instances of a process the reverse of selection in the ordinary sense of the word, but leading to the same general result. The influence of the destruction of individuals having a particular character is well shown by the necessity of killing every lamb with a trace of black about it, in order to keep the flock white; or again, by the effects on the average height of the men of France of the destructive wars of Napoleon, by which many tall men were killed, the short ones being left to be the fathers of families. This at least is the conclusion of some of those who have closely studied the effects of the conscription; and it is certain that since Napoleon's time the standard for the army has been lowered two or three times.

Unconscious selection blends with methodical, so that it is scarcely possible to separate them. When a fancier long ago first happened to notice a pigeon with an unusually short beak, or one with the tail-feathers unusually developed, although he bred from these birds with the distinct intention of propagating the variety, yet he could not have intended to make a short-faced tumbler or a fantail, and was far from knowing that he had made the first step towards this end. If he could have seen the final result, he would have been struck with astonishment, but, from what we know of the habits of fanciers, probably not with admiration. Our English carriers, barbs, and short-faced tumblers have been greatly modified in the same manner, as we may infer both from the historical evidence given in the chapters on the Pigeon, and from the comparison of birds brought from distant countries.

So it has been with dogs; our present fox-hounds differ from the old English hound; our greyhounds have become lighter: the Scotch deer-hound has been modified, and is now rare. Our bulldogs differ from those which were formerly used for baiting bulls. Our pointers and Newfoundlands do not closely resemble any native dog now found in the countries whence they were brought. These changes have been effected partly by crosses; but in every case the result has been governed by the strictest selection. Nevertheless, there is no reason to suppose that man intentionally and methodically made the breeds exactly what they now are. As our horses became fleeter, and the country more cultivated and smoother, fleeter fox-hounds were desired and produced, but probably without any one distinctly foreseeing what they would become. Our pointers and setters, the latter almost certainly descended from large spaniels, have been greatly modified in accordance with fashion and the desire for increased speed. Wolves have become extinct, and so has the wolf-dog; deer have become rarer, bulls are no longer baited, and the corresponding breeds of the dog have answered to the change. But we may feel almost sure that when, for instance, bulls were no longer baited, no man said to himself, I will now breed my dogs of smaller size, and thus create the present race. As circumstances changed, men unconsciously and slowly modified their course of selection.

With racehorses selection for swiftness has been followed methodically, and our horses now easily surpass their progenitors. The increased size and different appearance of the English racehorse led a good observer in India to ask," Could any one in this year of 1856, looking at our racehorses, conceive that they were the result of the union of the Arab horse and the

African mare?" (20/72. 'The India Sporting Review' volume 2 page 181; 'The Stud Farm' by Cecil page 58.) This change has, it is probable, been largely effected through unconscious selection, that is, by the general wish to breed as fine horses as possible in each generation, combined with training and high feeding, but without any intention to give to them their present appearance. According to Youatt (20/73. 'The Horse' page 22.), the introduction in Oliver Cromwell's time of three celebrated Eastern stallions speedily affected the English breed; "so that Lord Harleigh, one of the old school, complained that the great horse was fast disappearing." This is an excellent proof how carefully selection must have been attended to; for without such care, all traces of so small an infusion of Eastern blood would soon have been absorbed and lost. Notwithstanding that the climate of England has never been esteemed particularly favourable to the horse, yet long-continued selection, both methodical and unconscious, together with that practised by the Arabs during a still longer and earlier period, has ended in giving us the best breed of horses in the world. Macaulay (20/74. 'History of England' volume 1 page 316.) remarks, "Two men whose authority on such subjects was held in great esteem, the Duke of Newcastle and Sir John Fenwick, pronounced that the meanest hack ever imported from Tangier would produce a finer progeny than could be expected from the best sire of our native breed. They would not readily have believed that a time would come when the princes and nobles of neighbouring lands would be as eager to obtain horses from England as ever the English had been to obtain horses from Barbary."

The London dray-horse, which differs so much in appearance from any natural species, and which from its size has so astonished many Eastern princes, was probably formed by the heaviest and most powerful animals having been selected during many generations in Flanders and England, but without the least intention or expectation of creating a horse such as we now see. If we go back to an early period of history, we behold in the antique Greek statues, as Schaaffhausen has remarked (20/75. 'Ueber Bestandigkeit der Arten.'), a horse equally unlike a race or dray horse, and differing from any existing breed.

The results of unconscious selection, in an early stage, are well shown in the difference between the flocks descended from the same stock, but separately reared by careful breeders. Youatt gives an excellent instance of this fact in the sheep belonging to Messrs. Buckley and Burgess, which "have been purely bred from the original stock of Mr. Bakewell for upwards of fifty years. There is not a suspicion existing in the mind of any one at all acquainted with the subject that the owner of either flock has deviated in any one instance from the pure blood of Mr. Bakewell's flock; yet the difference between the sheep possessed by these two gentlemen is so great, that they have the appearance of being quite different varieties." (20/76. 'Youatt on Sheep' page 315.) I have seen several analogous and well marked cases with pigeons: for instance, I had a family of barbs descended from those long bred by Sir J. Sebright, and another family long bred by another fancier, and the two families plainly differed from each other. Nathusius—and a more competent witness could not be cited—observes that, though the Shorthorns are remarkably uniform in appearance (except in colour), yet the individual character and wishes of each breeder become impressed on his cattle, so that different herds differ slightly from one another. (20/77. 'Ueber Shorthorn Rindvieh' 1857 s. 51.) The Hereford cattle assumed their present well-marked character soon after the year 1769, through careful selection by Mr. Tomkins (20/78. Low 'Domesticated Animals' 1845 page 363.) and the breed has lately split into two strains—one strain having a white face, and differing slightly, it is said (20/79. 'Quarterly Review' 1849 page 392.), in some other points: but there is no reason to believe that this split, the origin of which is unknown, was intentionally made; it may with much more probability be attributed to different breeders having attended to different points. So again, the Berkshire breed of swine in the year 1810 had greatly changed from what it was in 1780; and since 1810 at least two distinct sub-breeds have arisen bearing the same name. (20/80. H. von Nathusius 'Vorstudien...Schweineschadel' 1864 s 140.) Keeping in mind how rapidly all animals increase, and that some must be annually slaughtered and some saved for breeding, then, if the same breeder during a long course of years deliberately settles which shall be saved and which shall be killed, it is almost inevitable that his individual turn of mind will influence the character of his stock, without his having had any intention to modify the breed.

Unconscious selection in the strictest sense of the word, that is, the saving of the more useful animals and the neglect or slaughter of the less useful, without any thought of the future, must have gone on occasionally from the remotest period and amongst the most barbarous nations. Savages often suffer from famines, and are sometimes expelled by war from their own homes. In such cases it can hardly be doubted that they would save their most useful animals. When the Fuegians are hard pressed by want, they kill their old women for food rather than their dogs; for, as we were assured, "old women no use—dogs catch otters." The same sound sense would surely lead them to preserve their more useful dogs when still harder pressed by famine. Mr. Oldfield, who has seen so much of the aborigines of Australia, informs me that "they are all very glad to get a European kangaroo dog, and several instances have been known of the father killing his own infant that the mother might suckle the much-prized puppy." Different kinds of dogs would be useful to the Australian for hunting opossums and kangaroos, and to the Fuegian for catching fish and otters; and the occasional preservation in the two countries of the most useful animals would ultimately lead to the formation of two widely distinct breeds.

With plants, from the earliest dawn of civilisation, the best variety which was known would generally have been cultivated at each period and its seeds occasionally sown; so that there will have been some selection from an extremely remote period, but without any prefixed standard of excellence or thought of the future. We at the present day profit by a course of selection occasionally and unconsciously carried on during thousands of years. This is proved in an interesting manner by Oswald Heer's researches on the lake-inhabitants of Switzerland, as given in a former chapter; for he shows that the grain and seed of our present varieties of wheat, barley, oats, peas, beans, lentils, and poppy, exceed in size those which were cultivated in Switzerland during the Neolithic and Bronze periods. These ancient people, during the Neolithic period, possessed also a crab considerably larger than that now growing wild on the Jura. (20/81. See also Dr. Christ in Rutimeyer's 'Pfahlbauten' 1861 s. 226.) The pears described by Pliny were evidently extremely inferior in quality to our present pears. We can realise the effects of long-continued selection and cultivation in another way, for would any one in his senses expect to raise a first-rate apple from the seed of a truly wild crab, or a luscious melting pear from the wild pear? Alphonse de Candolle informs me that he has lately seen on an ancient mosaic at Rome a representation of the melon; and as the Rotnans, who were such gourmands, are silent on this fruit, he infers that the melon has been greatly ameliorated since the classical period.

Coming to later times, Buffon (20/82. The passage is given 'Bull. Soc. d'Acclimat.' 1858 page 11.) on comparing the flowers, fruit, and vegetables which were then cultivated with some excellent drawings made a hundred and fifty years previously, was struck with surprise at the great improvement which had been effected; and remarks that these ancient flowers and vegetables would now be rejected, not only by a florist but by a village gardener. Since the time of Buffon the work of improvement has steadily and rapidly gone on. Every florist who compares our present flowers with those figured in books published not long since, is astonished at the change. A well-known amateur (20/83. 'Journal of Horticulture' 1862 page 394.), in speaking of the varieties of Pelargonium raised by Mr. Garth only twenty-two years before, remarks, "What a rage they excited: surely we had attained perfection, it was said; and now not one of the flowers of those days will be looked at. But none the less is the debt of gratitude which we owe to those who saw what was to be done, and did it." Mr. Paul, the well-known horticulturist, in writing of the same flower (20/84. 'Gardener's Chronicle' 1857 page 85.), says he remembers when young being delighted with the portraits in Sweet's work; "but what are they in point of beauty compared with the Pelargoniums of this day? Here again nature did not advance by leaps; the improvement was gradual, and if we had neglected those very gradual advances, we must have foregone the present grand results." How well this practical horticulturist appreciates and illustrates the gradual and accumulative force of selection! The Dahlia has advanced in beauty in a like manner; the line of improvement being guided by fashion, and by the successive modifications which the flower slowly underwent. (20/85. See Mr. Wildman's address to the Floricult. Soc. in 'Gardener's Chronicle' 1843 page 86.) A steady and gradual change has been noticed in many other flowers: thus an old florist (20/86. 'Journal of Horticulture' October 24, 1865 page 239.), after describing the leading varieties of the Pink which were grown in 1813 adds, "the pinks of those days would now be scarcely grown as border- flowers." The improvement of so many flowers and the number of the varieties which have been raised is all the more striking when we hear that the earliest known flower-garden in Europe, namely at Padua, dates only from the year 1545. (20/87. Prescott 'Hist. of Mexico' volume 2 page 61.)

## EFFECTS OF SELECTION, AS SHOWN BY THE PARTS MOST VALUED BY MAN PRESENTING THE GREATEST AMOUNT OF DIFFERENCE.

The power of long-continued selection, whether methodical or unconscious, or both combined, is well shown in a general way, namely, by the comparison of the differences between the varieties of distinct species, which are valued for different parts, such as for the leaves, or stems, or tubers, the seed, or fruit, or flowers. Whatever part man values most, that part will be found to present the greatest amount of difference. With trees cultivated for their fruit, Sageret remarks that the fruit is larger than in the parent-species, whilst with those cultivated for the seed, as with nuts, walnuts, almonds, chestnuts, etc., it is the seed itself which is larger; and he accounts for this fact by the fruit in the one case, and by the seed in the other, having been carefully attended to and selected during many ages. Gallesio has made the same observation. Godron insists on the diversity of the tuber in the potato, of the bulb in the onion, and of the fruit in the melon; and on the close similarity of the other parts in these same plants. (20/88. Sagaret 'Pomologie Physiologique' 1830 page 47; Gallesio 'Teoria della Riproduzione' 1816 page 88; Godron 'De l'Espece' 1859 tome 2 pages 63, 67, 70. In my tenth and eleventh chapters I have given details on the potato; and I can confirm similar remarks with respect to the onion. I have also shown how far Naudin concurs in regard to the varieties of the melon.)

In order to judge how far my own impression on this subject was correct, I cultivated numerous varieties of the same species close to one another. The comparison of the amount of difference between widely different organs is necessarily vague; I will therefore give the results in only a few cases. We have previously seen in the ninth chapter how greatly the varieties of the cabbage differ in their foliage and stems, which are the selected parts, and how closely they resemble one another in their flowers, capsules, and seeds. In seven varieties of the radish, the roots differed greatly in colour and shape, but no difference whatever could be detected in their foliage, flowers, or seeds. Now what a contrast is presented, if we compare the flowers of the varieties of these two plants with those of any species cultivated in our flower-gardens for ornament; or if we compare their seeds with those of the varieties of maize, peas, beans, etc., which are valued and cultivated for their seeds. In the ninth chapter it was shown that the varieties of the pea differ but little except in the tallness of the plant, moderately in the shape of the pod, and greatly in the pea itself, and these are all selected points. The varieties, however, of the Pois sans parchemin differ much more in their pods, and these are eaten and valued. I cultivated twelve varieties of the common bean; one alone, the Dwarf Fan, differed considerably in general appearance; two differed in the colour of their flowers, one being an albino, and the other being wholly instead of partially purple; several differed considerably in the shape and size of the pod, but far more in the bean itself, and this is the valued and selected part. Toker's bean, for instance, is twice-and-a-half as long and broad as the horse-bean, and is much thinner and of a different shape.

The varieties of the gooseberry, as formerly described, differ much in their fruit, but hardly perceptibly in their flowers or organs of vegetation. With the plum, the differences likewise appear to be greater in the fruit than in the flowers or leaves. On the other hand, the seed of the strawberry, which corresponds with the fruit of the plum, differs hardly at all; whilst every one knows how greatly the fruit—that is, the enlarged receptacle—differs in several varieties. In apples, pears, and peaches the flowers and leaves differ considerably, but not, as far as I can judge, in proportion with the fruit. The Chinese double-flowering peaches, on the other hand, show that varieties of this tree have been formed, which differ more in flower than in fruit. If, as is highly probable, the peach is the modified descent of the almond, a surprising amount of change has been effected in the same species, in the fleshy covering of the former and in the kernels of the latter.

When parts stand in close relationship to each other, such as the seed and the fleshy covering of the fruit (whatever its homological nature may be), changes in the one are usually accompanied by modifications in the other, though not necessarily to the same degree. With the plum-tree, for instance, some varieties produce plums which are nearly alike, but include stones extremely dissimilar in shape; whilst conversely other varieties produce dissimilar fruit with barely distinguishable stones; and generally the stones, though they have never been subjected to selection, differ greatly in the several varieties of the plum. In other cases organs which are not manifestly related, through some unknown bond vary together, and are consequently liable, without any intention on man's part, to be simultaneously acted on by selection. Thus the varieties of the stock (Matthiola) have been selected solely for the beauty of their flowers, but the seeds differ greatly in colour and somewhat in size. Varieties of the lettuce have been selected solely on account of their leaves, yet produce seeds which likewise differ in colour. Generally, through the law of correlation, when a variety differs greatly from its fellow-varieties in any one character, it differs to a certain extent in several other characters. I observed this fact when I cultivated together many varieties of the same species, for I used first to make a list of the varieties which differed most from each other in their foliage and manner of growth, afterwards of those that differed most in their flowers, then in their seed-capsules, and lastly in their mature seed; and I found that the same names generally occurred in two, three, or four of the successive lists. Nevertheless the greatest amount of difference between the varieties was always exhibited, as far as I could judge, by that part or organ for which the plant was cultivated.

When we bear in mind that each plant was at first cultivated because useful to man, and that its variation was a subsequent, often a long subsequent, event, we cannot explain the greater amount of diversity in the valuable parts by supposing that species endowed with an especial tendency to vary in any particular manner were originally chosen. We must attribute the result to the variations in these parts having been successively preserved, and thus continually augmented; whilst other variations, excepting such as inevitably appeared through correlation, were neglected and lost. We may therefore infer that most plants might be made, through long-continued selection, to yield races as different from one another in any character as they now are in those parts for which they are valued and cultivated.

With animals we see nothing of the same kind; but a sufficient number of species have not been domesticated for a fair comparison. Sheep are valued for their wool, and the wool differs much more in the several races than the hair in cattle. Neither sheep, goats, European cattle, nor pigs are valued for their fleetness or strength; and we do not possess breeds differing in these respects like the racehorse and dray-horse. But fleetness and strength are valued in camels and dogs; and we have with the former the swift dromedary and heavy camel; with the latter the greyhound and mastiff. But dogs are valued even in a higher degree for their mental qualities and senses; and every one knows how greatly the races differ in these respects. On the other hand, where the dog is kept solely to serve for food, as in the Polynesian islands and China, it is described as an extremely stupid animal. (20/89. Godron 'De l'Espece' tome 2 page 27.) Blumenbach remarks that "many dogs, such as the badger-dog, have a build so marked and so appropriate for particular purposes, that I should find it very difficult to persuade myself that this astonishing figure was an accidental consequence of degeneration." (20/90. 'The Anthropological Treatises of Blumenbach' 1856 page 292.) Had Blumenbach reflected on the great principle of selection, he would not have used the term degeneration, and he would not have been astonished that dogs and other animals should become excellently adapted for the service of man.

On the whole we may conclude that whatever part or character is most valued— whether the leaves, stems, tubers, bulbs, flowers, fruit, or seed of plants, or the size, strength, fleetness, hairy covering, or intellect of animals— that character will almost invariably be found to present the greatest amount of difference both in kind and degree. And this result may be safely attributed to man having preserved during a long course of generations the variations which were useful to him, and neglected the others.

I will conclude this chapter by some remarks on an important subject. With animals such as the giraffe, of which the whole structure is admirably co-ordinated for certain purposes, it has been supposed that all the parts must have been simultaneously modified; and it has been argued that, on the principle of natural selection, this is scarcely possible. But in thus arguing, it has been tacitly assumed that the variations must have been abrupt and great. No doubt, if the neck of a ruminant were suddenly to become greatly elongated, the fore limbs and back would have to be simultaneously strengthened and modified; but it cannot be denied that an animal might have its neck, or head, or tongue, or fore-limbs elongated a very little without any corresponding modification in other parts of the body; and animals thus slightly modified would, during a dearth, have a slight advantage, and be enabled to browse on higher twigs, and thus survive. A few mouthfuls more or less every day would make all the difference between life and death. By the repetition of the same process, and by the occasional intercrossing of the survivors, there would be some progress, slow and fluctuating though it would be, towards the admirably coordinated structure of the giraffe. If the short- faced tumbler-pigeon, with its small conical beak, globular head, rounded body, short wings, and small feet—characters which appear all in harmony—had been a natural species, its whole structure would have been viewed as well fitted for its life; but in this case we know that inexperienced breeders are urged to attend to point after point, and not to attempt improving the whole structure at the same time. Look at the greyhound, that perfect image of grace, symmetry, and vigour; no natural species can boast of a more admirably co-ordinated structure, with its tapering head, slim body, deep chest, tucked- up abdomen, rat-like tail, and long muscular limbs, all adapted for extreme fleetness, and for running down weak prey. Now, from what we see of the variability of animals, and from what we know of the method which different men follow in improving their stock—some chiefly attending to one point, others to another point, others again correcting defects by crosses, and so forth—we may feel assured that if we could see the long line of ancestors of a first-rate greyhound up to its wild wolf-like progenitor, we should behold an infinite number of the finest gradations, sometimes in one character and sometimes in another, but all leading towards our present perfect type. By small and doubtful steps such as these, nature, as we may confidently believe, has progressed, on her grand march of improvement and development.

A similar line of reasoning is as applicable to separate organs as to the whole organisation. A writer (20/91. Mr. J.J. Murphy in his opening address to the Belfast Nat. Hist. Soc. as given in the 'Belfast Northern Whig' November 19, 1866. Mr. Murphy here follows the line of argument against my views previously and more cautiously given by the Rev. C. Pritchard, Pres. Royal Astronomical Soc., in his sermon Appendix page 33 preached before the British Association at Nottingham 1866.) has recently maintained that "it is probably no exaggeration to suppose that in order to improve such an organ as the eye at all, it must be improved in ten different ways at once. And the improbability of any complex organ being produced and brought to perfection in any such way is an improbability of the same kind and degree as that of producing a poem or a mathematical demonstration by throwing letters at random on a table." If the eye were abruptly and greatly modified, no doubt many parts would have to be simultaneously altered, in order that the organ should remain serviceable.

But is this the case with smaller changes? There are persons who can see distinctly only in a dull light, and this condition depends, I believe, on the abnormal sensitiveness of the retina, and is known to be inherited. Now if a bird, for instance, receive some great advantage from seeing well in the twilight, all the individuals with the most sensitive retina would succeed best and be the most likely to survive; and why should not all those which happened to have the eye itself a little larger, or the pupil capable of greater dilatation, be likewise preserved, whether or not these modifications were strictly simultaneous? These individuals would subsequently intercross and blend their respective advantages. By such slight successive changes, the eye of a diurnal bird would be brought into the condition of that of an owl, which has often been advanced as an excellent instance of adaptation. Short-sight, which is often inherited, permits a person to see distinctly a minute object at so near a distance that it would be indistinct to ordinary eyes; and here we have a capacity which might be serviceable under certain conditions, abruptly gained. The Fuegians on board the Beagle could certainly see distant objects more distinctly than our sailors with all their long practice; I do not know whether this depends upon sensitiveness or on the power of adjustment in the focus; but this capacity for distant vision might, it is probable, be slightly augmented by successive modifications of either kind. Amphibious animals which are enabled to see both in the water and in the air, require and possess, as M. Plateau has shown (20/92. On the Vision of Fishes and Amphibia, translated in 'Annals and Mag. of Nat. Hist.' volume 18 1866 page 469.), eyes constructed on the following plan: "the cornea is always flat, or at least much flattened in the front of the crystalline and over a space equal to the diameter of that lens, whilst the lateral portions may be much curved." The crystalline is very nearly a sphere, and the humours have nearly the same density as water. Now as a terrestrial animal became more and more aquatic in its habits, very slight changes, first in the curvature of the cornea or crystalline, and then in the density of the humours, or conversely, might successively occur, and would be advantageous to the animal whilst under water, without serious detriment to its power of vision in the air. It is of course impossible to conjecture by what steps the fundamental structure of the eye in the Vertebrata was originally acquired, for we know nothing about this organ in the first progenitors of the class. With respect to the lowest animals in the scale, the transitional states through which the eye at first probably passed, can by the aid of analogy be indicated, as I have attempted to show in my 'Origin of Species.' (20/93. Sixth edition 1872 page 144.)

# CHAPTER 2.XXI.

SELECTION, continued.

**NATURAL SELECTION AS AFFECTING DOMESTIC PRODUCTIONS. CHARACTERS WHICH APPEAR OF TRIFLING VALUE OFTEN OF REAL IMPORTANCE. CIRCUMSTANCES FAVOURABLE TO SELECTION BY MAN. FACILITY IN PREVENTING CROSSES, AND THE NATURE OF THE CONDITIONS. CLOSE ATTENTION AND PERSEVERANCE INDISPENSABLE. THE PRODUCTION OF A LARGE NUMBER OF INDIVIDUALS ESPECIALLY FAVOURABLE. WHEN NO SELECTION IS APPLIED, DISTINCT RACES ARE NOT FORMED. HIGHLY-BRED ANIMALS LIABLE TO DEGENERATION. TENDENCY IN MAN TO CARRY THE SELECTION OF EACH CHARACTER TO AN EXTREME POINT, LEADING TO DIVERGENCE OF CHARACTER, RARELY TO CONVERGENCE. CHARACTERS CONTINUING TO VARY IN THE SAME DIRECTION IN WHICH THEY HAVE ALREADY VARIED. DIVERGENCE OF CHARACTER, WITH THE EXTINCTION OF INTERMEDIATE VARIETIES, LEADS TO DISTINCTNESS IN OUR DOMESTIC RACES. LIMIT TO THE POWER OF SELECTION. LAPSE OF TIME IMPORTANT. MANNER IN WHICH DOMESTIC RACES HAVE ORIGINATED. SUMMARY.**

**NATURAL SELECTION, OR THE SURVIVAL OF THE FITTEST, AS AFFECTING DOMESTIC PRODUCTIONS.**

We know little on this head. But as animals kept by savages have to provide throughout the year their own food either entirely or to a large extent, it can hardly be doubted that in different countries, varieties differing in constitution and in various characters would succeed best, and so be naturally selected. Hence perhaps it is that the few domesticated animals kept by savages partake, as has been remarked by more than one writer, of the wild appearance of their masters, and likewise resemble natural species. Even in long-civilised countries, at least in the wilder parts, natural selection must act on our domestic races. It is obvious that varieties having very different habits, constitution, and structure, would succeed best on mountains and on rich lowland pastures. For example, the improved Leicester sheep were formerly taken to the Lammermuir Hills; but an intelligent sheep-master reported that "our coarse lean pastures were unequal to the task of supporting such heavy-bodied sheep; and they gradually dwindled away into less and less bulk: each generation was inferior to the preceding one; and when the spring was severe, seldom more than two-thirds of the lambs survived the ravages of the storms." (21/1. Quoted by Youatt on 'Sheep' page 325. See also Youatt on 'Cattle' pages 62, 69.) So with the mountain cattle of North Wales and the Hebrides, it has been found that they could not withstand being crossed with the larger and more delicate lowland breeds. Two French naturalists, in describing the horses of Circassia, remark that, subjected as they are to extreme vicissitudes of climate, having to search for scanty pasture, and exposed to constant danger from wolves, the strongest and most vigorous alone survive. (21/2. MM. Lherbette and De Quatrefages in 'Bull. Soc. d'Acclimat.' tome 8 1861 page 311.)

Every one must have been struck with the surpassing grace, strength, and vigour of the Game-cock, with its bold and confident air, its long, yet firm neck, compact body, powerful and closely pressed wings, muscular thighs, strong beak massive at the base, dense and sharp spurs set low on the legs for delivering the fatal blow, and its compact, glossy, and mail-like plumage serving as a defence. Now the English game-cock has not only been improved during many years by man's careful selection, but in addition, as Mr. Tegetmeier has remarked (21/3. 'The Poultry Book' 1866 page 123. Mr. Tegetmeier, 'The Homing or Carrier Pigeon' 1871 pages 45-58.), by a kind of natural selection, for the strongest, most active and courageous birds have stricken down their antagonists in the cockpit, generation after generation, and have subsequently served as the progenitors of their race. The same kind of double selection has come into play with the carrier pigeon, for during their training the inferior birds fail to return home and are lost, so that even without selection by man only the superior birds propagate their race.

In Great Britain, in former times, almost every district had its own breed of cattle and sheep; "they were indigenous to the soil, climate, and pasturage of the locality on which they grazed: they seemed to have been formed for it and by it." (21/4. 'Youatt on Sheep' page 312.) But in this case we are quite unable to disentangle the effects of the direct action of the conditions of life,—of use or habit—of natural selection—and of that kind of selection which we have seen is occasionally and unconsciously followed by man even during the rudest periods of history.

Let us now look to the action of natural selection on special characters. Although nature is difficult to resist, yet man often strives against her power, and sometimes with success. From the facts to be given, it will also be seen that natural selection would powerfully affect many of our domestic productions if left unprotected. This is a point of much interest, for we thus learn that differences apparently of very slight importance would certainly determine the survival of a form when forced to struggle for its own existence. It may have occurred to some naturalists, as it formerly did to me, that, though selection acting under natural conditions would determine the structure of all important organs, yet that it could not affect characters which are esteemed by us of little importance; but this is an error to which we are eminently liable, from our ignorance of what characters are of real value to each living creature.

When man attempts to make a breed with some serious defect in structure, or in the mutual relation of the several parts, he will partly or completely fail, or encounter much difficulty; he is in fact resisted by a form of natural selection. We have seen that an attempt was once made in Yorkshire to breed cattle with enormous buttocks, but the cows perished so often in bringing forth their calves, that the attempt had to be given up. In rearing short-faced tumblers, Mr. Eaton says (21/5. 'Treatise on the Almond Tumbler' 1851 page 33.), "I am convinced that better head and beak birds have perished in the shell than ever were hatched; the reason being that the amazingly short-faced bird cannot reach and break the shell with its beak, and so perishes." Here is a more curious case, in which natural selection comes into play only at long intervals of time: during ordinary seasons the Niata cattle can graze as well as others, but occasionally, as from 1827 to 1830 the plains of La Plata suffer from long-continued droughts and the pasture is burnt up; at such times common cattle and horses perish by the thousand, but many survive by browsing on twigs, reeds, etc.; this the Niata cattle cannot so well effect from their upturned jaws and the shape of their lips; consequently, if not attended to, they perish before the other cattle. In Columbia, according to Roulin, there is a breed of nearly hairless cattle, called Pelones; these succeed in their native hot district, but are found too tender for the Cordillera; in this case, however, natural selection determines only the range of the variety. It is obvious that a host of artificial races could never survive in a state of nature;—such as Italian greyhounds,—hairless and almost toothless Turkish dogs,—fantail pigeons, which cannot fly well against a strong wind,—barbs and Polish fowls, with their vision impeded by their eye wattles and great topknots,—hornless bulls and rams, which consequently cannot cope with other males, and thus have a poor chance of leaving offspring,—seedless plants, and many other such cases.

Colour is generally esteemed by the systematic naturalist as unimportant: let us, therefore, see how far it indirectly affects our domestic productions, and how far it would affect them if they were left exposed to the full force of natural selection. In a future chapter I shall have to show that constitutional peculiarities of the strangest kind, entailing liability to the action of certain poisons, are correlated with the colour of the skin. I will here give a single case, on the high authority of Professor Wyman; he informs me that, being surprised at all the pigs in a part of Virginia being black, he made inquiries, and ascertained that these animals feed on the roots of the Lachnanthes tinctoria, which colours their bones pink, and, excepting in the case of the black varieties, causes the hoofs to drop off. Hence, as one of the squatters remarked, "we select the black members of the litter for raising, as they alone have a good chance of living." So that here we have artificial and natural selection working hand in hand. I may add that in the Tarentino the inhabitants keep black sheep alone, because the Hypericum crispum abounds there; and this plant does not injure black sheep, but kills the white ones in about a fortnight's time. (21/6. Dr. Heusinger 'Wochenschrift fur die Heilkunde' Berlin 1846 s. 279.)

Complexion, and liability to certain diseases, are believed to run together in man and the lower animals. Thus white terriers suffer more than those of any other colour from the fatal distemper. (21/7. Youatt on the 'Dog' page 232.) In North America plum-trees are liable to a disease which Downing (21/8. 'The Fruit-trees of America' 1845 page 270: for peaches page 466.) believes is not caused by insects; the kinds bearing purple fruit are most affected, "and we have never known the green or yellow fruited varieties infected until the other sorts had first become filled with the knots." On the other hand, peaches in North America suffer much from a disease called the "yellows," which seems to be peculiar to that continent, and more than nine-tenths of the victims, "when the disease first appeared, were the yellow-fleshed peaches. The white-fleshed kinds are much more rarely attacked; in some parts of the country never." In Mauritius, the white sugar-canes have of late years been so severely attacked by a disease, that many planters have been compelled to give up growing this variety (although fresh plants were imported from China for trial), and cultivate only red canes. (21/9. 'Proc. Royal Soc. of Arts and Sciences of Mauritius' 1852 page 135.) Now, if these plants had been forced to struggle with other competing plants and enemies, there cannot be a doubt that the colour of the flesh or skin of the fruit, unimportant as these characters are considered, would have rigorously determined their existence.

Liability to the attacks of parasites is also connected with colour. White chickens are certainly more subject than dark-coloured chickens to the "gapes," which is caused by a parasitic worm in the trachea. (21/10. 'Gardener's Chronicle' 1856 page 379.) On the other hand, experience has shown that in France the caterpillars which produce white cocoons resist the deadly fungus better than those producing yellow cocoons. (21/11. Quatrefages 'Maladies Actuelles du Ver a Soie' 1859 pages 12, 214.) Analogous facts have been observed with plants: a new and beautiful white onion, imported from France, though planted close to other kinds, was alone attacked by a parasitic fungus. (21/12. 'Gardener's Chronicle' 1851 page 595.) White verbenas are especially liable to mildew. (21/13. 'Journal of Horticulture' 1862 page 476.) Near Malaga, during an early period of the vine-disease, the green sorts suffered most; "and red and black grapes, even when interwoven with the sick plants, suffered not at all." In France whole groups of varieties were comparatively free, and others, such as the Chasselas, did not afford a single fortunate exception; but I do not know whether any correlation between colour and liability to disease was here observed. (21/14. 'Gardener's Chronicle' 1852 pages 435, 691.) In a former chapter it was shown how curiously liable one variety of the strawberry is to mildew.

It is certain that insects regulate in many cases the range and even the existence of the higher animals, whilst living under their natural conditions. Under domestication light-coloured animals suffer most: in Thuringia (21/15. Bechstein 'Naturgesch. Deutschlands' 1801 b. 1 s. 310.) the inhabitants do not like grey, white, or pale cattle, because they are much more troubled by various kinds of flies than the brown, red, or black cattle. An Albino negro, it has been remarked (21/16. Prichard 'Phys. Hist. of Mankind' 1851 volume 1 page 224.), was peculiarly sensitive to the bites of insects. In the West Indies (21/17. G. Lewis 'Journal of Residence in West Indies' 'Home and Col. Library' page 100.) it is said that "the only horned cattle fit for work are those which have a good deal of black in them. The white are terribly tormented by the insects; and they are weak and sluggish in proportion to the white."

In Devonshire there is a prejudice against white pigs, because it is believed that the sun blisters them when turned out (21/18. Sidney's edition of Youatt on the 'Pig' page 24. I have given analogous facts in the case of mankind in my 'Descent of Man' 2nd edition page 195.); and I knew a man who would not keep white pigs in Kent, for the same reason. The scorching of flowers by the sun seems likewise to depend much on colour; thus, dark pelargoniums suffer most; and from various accounts it is clear that the cloth-of-gold variety will not withstand a degree of exposure to sunshine which other varieties enjoy. Another amateur asserts that not only all dark-coloured verbenas, but likewise scarlets, suffer from the sun: "the paler kinds stand better, and pale blue is perhaps the best of all." So again with the heartsease (Viola tricolor); hot weather suits the blotched sorts, whilst it destroys the beautiful markings of some other kinds. (21/19. 'Journal of Horticulture' 1862 pages 476, 498; 1865 page 460. With respect to the heartsease 'Gardener's Chronicle' 1863 page 628.) During one extremely cold season in Holland all red-flowered hyacinths were observed to be very inferior in quality. It is believed by many agriculturists that red wheat is hardier in northern climates than white wheat. (21/20. 'Des Jacinthes, de leur Culture' 1768 page 53: on wheat 'Gardener's Chronicle' 1846 page 653.)

With animals, white varieties from being conspicuous are the most liable to be attacked by beasts and birds of prey. In parts of France and Germany where hawks abound, persons are advised not to keep white pigeons; for, as Parmentier says, "it is certain that in a flock the white always first fall victims to the kite." In Belgium, where so many societies have been established for the flight of carrier-pigeons, white is the one colour which for the same reason is disliked. (20/21. W.B. Tegetmeier 'The Field' February 25, 1865. With respect to black fowls see a quotation in Thompson 'Nat. Hist. of Ireland' 1849 volume 1 page 22.) Prof. G. Jaeger (21/22. 'In Sachen Darwin's contra Wigand' 1874 page 70.) whilst fishing found four pigeons which had been killed by hawks, and all were white; on another occasion he examined the eyrie of a hawk, and the feathers of the pigeons which had been caught were all of a white or yellow colour. On the other hand, it is said that the sea-eagle (Falco ossifragus, Linn.) on the west coast of Ireland picks out the black fowls, so that "the villagers avoid as much as possible rearing birds of that colour." M. Daudin (20/23. 'Bull. de la Soc. d'Acclimat.' tome 7 1860 page 359.), speaking of white rabbits kept in warrens in Russia, remarks that their colour is a great disadvantage, as they are thus more exposed to attack, and can be seen during bright nights from a distance. A gentleman in Kent, who failed to stock his woods with a nearly white and hardy kind of rabbit, accounted in the same manner for their early disappearance. Any one who will watch a white cat prowling after her prey will soon perceive under what a disadvantage she lies.

The white Tartarian cherry, "owing either to its colour being so much like that of the leaves, or to the fruit always appearing from a distance unripe," is not so readily attacked by birds as other sorts. The yellow-fruited raspberry, which generally comes nearly true by seed, "is very little molested by birds, who evidently are not fond of it; so that nets may be dispensed with in places where nothing else will protect the red fruit." (21/24. 'Transact. Hort. Soc.' volume 1 2nd series 1835 page 275. For raspberries see 'Gardener's Chronicle' 1855 page 154 and 1863 page 245.) This immunity, though a benefit to the gardener, would be a disadvantage in a state of nature both to the cherry and raspberry, as dissemination depends on birds. I noticed during several winters that some trees of the yellow-berried holly, which were raised from seed from a tree found wild by my father remained covered with fruit, whilst not a scarlet berry could be seen on the adjoining trees of the common kind. A friend informs me that a mountain-ash (Pyrus aucuparia) growing in his garden bears berries which, though not differently coloured, are always devoured by birds before those on the other trees. This variety of the mountain-ash would thus be more freely disseminated, and the yellow-berried variety of the holly less freely, than the common varieties of these two trees.

Independently of colour, trifling differences are sometimes found to be of importance to plants under cultivation, and would be of paramount importance if they had to fight their own battle and to struggle with many competitors. The thin-shelled peas, called pois sans parchemin, are attacked by birds (21/25. 'Gardener's Chronicle' 1843 page 806.) much more commonly than ordinary peas. On the other hand, the purple-podded pea, which has a hard shell, escaped the attacks of tomtits (Parus major) in my garden far better than any other kind. The thin-shelled walnut likewise suffers greatly from the tomtit. (21/26. Ibid 1850 page 732.) These same birds have been observed to pass over and thus favour the filbert, destroying only the other kinds of nuts which grew in the same orchard. (21/27. Ibid 1860 page 956.)

Certain varieties of the pear have soft bark, and these suffer severely from wood-boring beetles; whilst other varieties are known to resist their attacks much better. (21/28. J. De Jonghe in 'Gardener's Chronicle' 1860 page 120.) In North America the smoothness, or absence of down on the fruit, makes a great difference in the attacks of the weevil, "which is the uncompromising foe of all smooth stone-fruits;" and the cultivator "has the frequent mortification of seeing nearly all, or indeed often the whole crop, fall from the trees when half or two-thirds grown." Hence the nectarine suffers more than the peach. A particular variety of the Morello cherry, raised in North America, is, without any assignable cause, more liable to be injured by this same insect than other cherry-trees. (21/29. Downing 'Fruit-trees of North America' pages 266, 501: in regard to the cherry page 198.) From some unknown cause, certain varieties of the apple enjoy, as we have seen, the great advantage in various parts of the world of not being infested by the coccus. On the other hand, a particular case has been recorded in which aphides confined themselves to the Winter Nelis pear and touched no other kind in an extensive orchard. (21/30. 'Gardener's Chronicle' 1849 page 755.) The existence of minute glands on the leaves of peaches, nectarines, and apricots, would not be esteemed by botanists as a character of the least importance for they are present or absent in closely-related sub-varieties, descended from the same parent-tree; yet there is good evidence (21/31. 'Journal of Horticulture' September 26, 1865 page 254; see other references given in chapter 10.) that the absence of glands leads to mildew, which is highly injurious to these trees.

A difference either in flavour or in the amount of nutriment in certain varieties causes them to be more eagerly attacked by various enemies than other varieties of the same species. Bullfinches (Pyrrhula vulgaris) injure our fruit-trees by devouring the flower-buds, and a pair of these birds have been seen "to denude a large plum-tree in a couple of days of almost every bud;" but certain varieties (21/32. Mr. Selby in 'Mag. of Zoology and Botany' Edinburgh volume 2 1838 page 393.) of the apple and thorn (Crataegus oxyacantha) are more especially liable to be attacked. A striking instance of this was observed in Mr. Rivers's garden, in which two rows of a particular variety of plum (21/33. The Reine Claude de Bavay 'Journal of Horticulture'

December 27, 1864 page 511.) had to be carefully protected, as they were usually stripped of all their buds during the winter, whilst other sorts growing near them escaped. The root (or enlarged stem) of Laing's Swedish turnip is preferred by hares, and therefore suffers more than other varieties. Hares and rabbits eat down common rye before St. John's-day-rye, when both grow together. (21/34. Mr. Pusey in 'Journal of R. Agricult. Soc.' volume 6 page 179. For Swedish turnips see 'Gardener's Chronicle' 1847 page 91.) In the south of France, when an orchard of almond-trees is formed, the nuts of the bitter variety are sown, "in order that they may not be devoured by field-mice" (21/35. Godron 'De l'Espece' tome 2 page 98.); so we see the use of the bitter principle in almonds.

Other slight differences, which would be thought quite unimportant, are no doubt sometimes of great service both to plants and animals. The Whitesmith's gooseberry, as formerly stated, produces its leaves later than other varieties, and, as the flowers are thus left unprotected, the fruit often fails. In one variety of the cherry, according to Mr. Rivers (21/36. 'Gardener's Chronicle' 1866 page 732.), the petals are much curled backwards, and in consequence of this the stigmas were observed to be killed by a severe frost; whilst at the same time, in another variety with incurved petals, the stigmas were not in the least injured. The straw of the Fenton wheat is remarkably unequal in height; and a competent observer believes that this variety is highly productive, partly because the ears from being distributed at various heights above the ground are less crowded together. The same observer maintains that in the upright varieties the divergent awns are serviceable by breaking the shocks when the ears are dashed together by the wind. (21/37. 'Gardener's Chronicle' 1862 pages 820, 821.) If several varieties of a plant are grown together, and the seed is indiscriminately harvested, it is clear that the hardier and more productive kinds will, by a sort of natural selection, gradually prevail over the others; this takes place, as Colonel Le Couteur believes (21/38. 'On the Varieties of Wheat' page 59.), in our wheat-fields, for, as formerly shown, no variety is quite uniform in character. The same thing, as I am assured by nurserymen, would take place in our flower-gardens, if the seed of the different varieties were not separately saved. When the eggs of the wild and tame duck are hatched together, the young wild ducks almost invariably perish, from being of smaller size and not getting their fair share of food. (21/39. Mr. Hewitt and others, in 'Journal of Hort.' 1862 page 773.)

Facts in sufficient number have now been given showing that natural selection often checks, but occasionally favours, man's power of selection. These facts teach us, in addition, a valuable lesson, namely, that we ought to be extremely cautious in judging what characters are of importance in a state of nature to animals and plants, which have to struggle for existence from the hour of their birth to that of their death,—their existence depending on conditions, about which we are profoundly ignorant.

## CIRCUMSTANCES FAVOURABLE TO SELECTION BY MAN.

The possibility of selection rests on variability, and this, as we shall see in the following chapters, mainly depends on changed conditions of life, but is governed by infinitely complex and unknown laws. Domestication, even when long continued, occasionally causes but a small amount of variability, as in the case of the goose and turkey. The slight differences, however, which characterise each individual animal and plant would in most, probably in all, cases suffice for the production of distinct races through careful and prolonged selection. We see what selection, though acting on mere individual differences, can effect when families of cattle, sheep, pigeons, etc., of the same race, have been separately bred during a number of years by different men without any wish on their part to modify the breed. We see the same fact in the difference between hounds bred for hunting in different districts (21/40. 'Encyclop. of Rural Sports' page 405.), and in many other such cases.

In order that selection should produce any result, it is manifest that the crossing of distinct races must be prevented; hence facility in pairing, as with the pigeon, is highly favourable for the work; and difficulty in pairing, as with cats, prevents the formation of distinct breeds. On nearly the same principle the cattle of the small island of Jersey have been improved in their milking qualities "with a rapidity that could not have been obtained in a widely extended country like France." (21/41. Col. Le Couteur 'Journal Roy. Agricult. Soc.' volume 4 page 43.) Although free crossing is a danger on the one side which every one can see, too close interbreeding is a hidden danger on the other side. Unfavourable conditions of life overrule the power of selection. Our improved heavy breeds of cattle and sheep could not have been formed on mountainous pastures; nor could dray-horses have been raised on a barren and inhospitable land, such as the Falkland Islands, where even the light horses of La Plata rapidly decrease in size. It seems impossible to preserve several English breeds of sheep in France; for as soon as the lambs are weaned their vigour decays as the heat of the summer increases (21/42. Malingie-Nouel 'Journal R. Agricult. Soc.' volume 14 1853 pages 215, 217.): it would be impossible to give great length of wool to sheep within the tropics; yet selection has kept the Merino breed nearly true under diversified and unfavourable conditions. The power of selection is so great, that breeds of the dog, sheep, and poultry, of the largest and smallest size, long and short beaked pigeons, and other breeds with opposite characters, have had their characteristic qualities augmented, though treated in every way alike, being exposed to the same climate and fed on the same food. Selection, however, is either checked or favoured by the effects of use or habit. Our wonderfully-improved pigs could never have been formed if they had been forced to search for their own food; the English racehorse and greyhound could not have been improved up to their present high standard of excellence without constant training.

As conspicuous deviations of structure occur rarely, the improvement of each breed is generally the result of the selection of slight individual differences. Hence the closest attention, the sharpest powers of observation, and indomitable perseverance, are indispensable. It is, also, highly important that many individuals of the breed which is to be improved should be raised; for thus there will be a better chance of the appearance of variations in the right direction, and individuals varying in an unfavourable manner may be freely rejected or destroyed. But that a large number of individuals should be raised, it is necessary that the conditions of life should favour the propagation of the species. Had the peacock been reared as easily as the fowl, we should probably ere this have had many distinct races. We see the importance of a large number of plants, from the fact of nursery gardeners almost always beating amateurs in the exhibition of new varieties. In 1845 it was estimated (21/43. Gardener's Chronicle' 1845 page 273.) that between 4000 and 5000 pelargoniums were annually raised from seed in England, yet a decidedly improved variety is rarely obtained. At Messrs. Carter's grounds, in Essex, where such flowers as the Lobelia, Nemophila, Mignonette, etc., are grown by the acre for seed, "scarcely a season passes without some new kinds being raised, or some improvement effected on old kinds." (21/44. 'Journal of Horticulture' 1862 page 157.) At Kew, as Mr. Beaton remarks, where many seedlings of common plants are raised, "you see new forms of Laburnums, Spiraeas, and other shrubs." (21/45. 'Cottage Gardener' 1860 page 368.) So with animals: Marshall (21/46. 'A Review of Reports' 1808 page 406.), in speaking of the sheep in one part of Yorkshire, remarks, "as they belong to poor people, and are mostly in small lots, they never can be improved." Lord Rivers, when asked how he succeeded in always having first-rate greyhounds, answered, "I breed many, and hang many." This, as another man remarks, "was the secret of his success; and the same will be found in exhibiting fowls,— successful competitors breed largely, and keep the best." (21/47. 'Gardener's Chronicle' 1853 page 45.)

It follows from this that the capacity of breeding at an early age and at short intervals, as with pigeons, rabbits, etc., facilitates selection; for the result is thus soon made visible, and perseverance in the work encouraged. It can hardly be an accident that the great majority of the culinary and agricultural plants which have yielded numerous races are annuals or biennials, which therefore are capable of rapid propagation, and thus of improvement. Sea-kale, asparagus, common and Jerusalem artichokes, potatoes, and onions, must be excepted, as they are perennials: but onions are propagated like annuals, and of the other plants just specified, none, with the exception of the potato, have yielded in this country more than one or two varieties. In the Mediterranean region, where artichokes are often raised from seed, there are several kinds, as I hear from Mr. Bentham. No doubt fruit-trees, which cannot be propagated quickly by seed, have yielded a host of varieties, though not permanent races; but these, judging from prehistoric remains, have been produced at a comparatively late period.

A species may be highly variable, but distinct races will not be formed, if from any cause selection be not applied. It would be difficult to select slight variations in fishes from their place of habitation; and though the carp is extremely variable and is much attended to in Germany, only one well-marked race has been formed, as I hear from Lord A. Russell, namely the spiegel-carpe; and this is carefully secluded from the common scaly kind. On the other hand, a closely allied species, the gold-fish, from being reared in small vessels, and from having been carefully attended to by the Chinese, has yielded many races. Neither the bee, which has been semi-domesticated from an extremely remote period, nor the cochineal insect, which was cultivated by the aboriginal Mexicans (21/48. Isidore Geoffroy Saint-Hilaire 'Hist. Nat. Gen.' tome 3 page 49. 'On the Cochineal Insect' page 46.), has yielded races; and it would be impossible to match the queen-bee with any particular drone, and most difficult to match cochineal insects. Silk-moths, on the other hand, have been subjected to rigorous selection, and have produced a host of races. Cats, which from their nocturnal habits cannot be selected for breeding, do not, as formerly remarked, yield distinct races within the same country. Dogs are held in abomination in the East, and their breeding is neglected; consequently, as Prof. Moritz Wagner (21/49. 'Die Darwin'sche Theorie und das Migrationsgesetz der Organismen' 1868 page 19.) remarks, one kind alone exists there. The ass in England varies much in colour and size; but as it is an animal of little value and bred by poor people, there has been no selection, and distinct races have not been formed. We must not attribute the inferiority of our asses to climate, for in India they are of even smaller size than in Europe. But when selection is brought to bear on the ass, all is changed. Near Cordova, as I am informed (February 1860) by Mr. W.E. Webb, C.E., they are carefully bred, as much as 200 pounds having been paid for a stallion ass, and they have been immensely improved. In Kentucky, asses have been imported (for breeding mules) from Spain, Malta, and France; these "seldom averaged more than fourteen hands high: but the Kentuckians, by great care, have raised them up to fifteen hands, and sometimes even to sixteen. The prices paid for these splendid animals, for such they really are, will prove how much they are in request. One male, of great celebrity, was sold for upwards of one thousand pounds sterling." These choice asses are sent to cattle-shows, a day being given for their exhibition. (21/50. Capt. Marryat quoted by Blyth in 'Journ. Asiatic Soc. of Bengal' volume 28 page 229.)

Analogous facts have been observed with plants: the nutmeg-tree in the Malay archipelago is highly variable, but there has been no selection, and there are no distinct races. (21/51. Mr. Oxley 'Journal of the Indian Archipelago' volume 2 1848 page 645.) The common mignonette (Reseda odorata), from bearing inconspicuous flowers, valued solely for their fragrance, "remains in the same unimproved condition as when first introduced." (21/52. Mr. Abbey 'Journal of Horticulture' December 1, 1863 page 430.) Our common forest-trees are very variable, as may be seen in every extensive nursery-ground; but as they are not valued like fruit-trees, and as they seed late in life, no selection has been applied to them; consequently, as Mr. Patrick Matthews remarks (21/53. 'On Naval Timber' 1831 page 107.), they have not yielded distinct races, leafing at different periods, growing to different sizes, and producing timber fit for different purposes. We have gained only some fanciful and semi- monstrous varieties, which no doubt appeared suddenly as we now see them.

Some botanists have argued that plants cannot have so strong a tendency to vary as is generally supposed, because many species long grown in botanic gardens, or unintentionally cultivated year after year mingled with our corn crops, have not produced distinct races; but this is accounted for by slight variations not having been selected and propagated. Let a plant which is now grown in a botanic garden, or any common weed, be cultivated on a large scale, and let a sharp-sighted gardener look out for each slight variety and sow the seed, and then, if distinct races are not produced, the argument will be valid.

The importance of selection is likewise shown by considering special characters. For instance, with most breeds of fowls the form of the comb and the colour of the plumage have been attended to, and are eminently characteristic of each race; but in Dorkings fashion has never demanded uniformity of comb or colour; and the utmost diversity in these respects prevails. Rose-combs, double-combs, cup-combs, etc., and colours of all kinds, may be seen in purely bred and closely related Dorking fowls, whilst other points, such as the general form of body, and the presence of an additional toe, have been attended to, and are invariably present. It has also been ascertained that colour can be fixed in this breed, as well as in any other. (21/54. Mr. Baily in 'The Poultry Chronicle' volume 2 1854 page 150. Also volume 1 page 342; volume 3 page 245.)

During the formation or improvement of a breed, its members will always be found to vary much in those characters to which especial attention is directed, and of which each slight improvement is eagerly sought and selected. Thus, with short-faced tumbler-pigeons, the shortness of the beak, shape of head and plumage,—with carriers, the length of the beak and wattle,—with fantails, the tail and carriage,—with Spanish fowls, the white face and comb,—with long-eared rabbits, the length of ear, are all points which are eminently variable. So it is in every case; and the large price paid for first-rate animals proves the difficulty of breeding them up to the highest standard of excellence. This subject has been discussed by fanciers (21/55. 'Cottage Gardener' 1855 December page 171; 1856 January pages 248, 323.), and the greater prizes given for highly improved breeds, in comparison with those given for old breeds which are not now undergoing rapid improvement, have been fully justified. Nathusius makes (21/56. 'Ueber Shorthorn Rindvieh' 1857 s. 51.) a similar remark when discussing the less uniform character of improved Shorthorn cattle and of the English horse, in comparison, for example, with the unennobled cattle of Hungary, or with the horses of the Asiatic steppes. This want of uniformity in the parts which at the time are undergoing selection chiefly depends on the strength of the principle of reversion; but it likewise depends to a certain extent on the continued variability of the parts which have recently varied. That the same parts do continue varying in the same manner we must admit, for if it were not so, there could be no improvement beyond an early standard of excellence, and we know that such improvement is not only possible, but is of general occurrence.

As a consequence of continued variability, and more especially of reversion, all highly improved races, if neglected or not subjected to incessant selection, soon degenerate. Youatt gives a curious instance of this in some cattle formerly kept in Glamorganshire; but in this case the cattle were not fed with sufficient care. Mr. Baker, in his memoir on the Horse, sums up: "It must have been observed in the preceding pages that, whenever there has been neglect, the breed has proportionally deteriorated." (21/57. 'The Veterinary' volume 13 page 720. For the Glamorganshire cattle see Youatt on 'Cattle' page 51.) If a considerable number of improved cattle, sheep, or other animals of the same race, were allowed to breed freely together, with no selection, but with no change in their condition of life, there can be no doubt that after a score or hundred generations they would be very far from excellent of their kind; but, from what we see of the many common races of dogs, cattle, fowls, pigeons, etc., which without any particular care have long retained nearly the same character, we have no grounds for believing that they would altogether depart from their type.

It is a general belief amongst breeders that characters of all kinds become fixed by long-continued inheritance. But I have attempted to show in the fourteenth chapter that this belief apparently resolves itself into the following proposition, namely, that all characters whatever, whether recently acquired or ancient, tend to be transmitted, but that those which have already long withstood all counteracting influences, will, as a general rule, continue to withstand them, and consequently be faithfully transmitted.

# TENDENCY IN MAN TO CARRY THE PRACTICE OF SELECTION TO AN EXTREME POINT.

It is an important principle that in the process of selection man almost invariably wishes to go to an extreme point. Thus, there is no limit to his desire to breed certain kinds of horses and dogs as fleet as possible, and others as strong as possible; certain kinds of sheep for extreme fineness, and others for extreme length of wool; and he wishes to produce fruit, grain, tubers, and other useful parts of plants, as large and excellent as possible. With animals bred for amusement, the same principle is even more powerful; for fashion, as we see in our dress, always runs to extremes. This view has been expressly admitted by fanciers. Instances were given in the chapters on the pigeon, but here is another: Mr. Eaton, after describing a comparatively new variety, namely, the Archangel, remarks, "What fanciers intend doing with this bird I am at a loss to know, whether they intend to breed it down to the tumbler's head and beak, or carry it out to the carrier's head and beak; leaving it as they found it, is not progressing." Ferguson, speaking of fowls, says, "their peculiarities, whatever they may be, must necessarily be fully developed: a little peculiarity forms nought but ugliness, seeing it violates the existing laws of symmetry." So Mr. Brent, in discussing the merits of the sub-varieties of the Belgian canary-bird, remarks, "Fanciers always go to extremes; they do not admire indefinite properties." (21/58. J.M. Eaton 'A Treatise on Fancy Pigeons' page 82; Ferguson on 'Rare and Prize Poultry' page 162; Mr. Brent in 'Cottage Gardener' October 1860 page 13.)

This principle, which necessarily leads to divergence of character, explains the present state of various domestic races. We can thus see how it is that racehorses and dray-horses, greyhounds and mastiffs, which are opposed to each other in every character,—how varieties so distinct as Cochin-china fowls and bantams, or carrier-pigeons with very long beaks, and tumblers with excessively short beaks, have been derived from the same stock. As each breed is slowly improved, the inferior varieties are first neglected and finally lost. In a few cases, by the aid of old records, or from intermediate varieties still existing in countries where other fashions have prevailed, we are enabled partially to trace the graduated changes through which certain breeds have passed. Selection, whether methodical or unconscious, always tending towards an extreme point, together with the neglect and slow extinction of the intermediate and less-valued forms, is the key which unlocks the mystery of how man has produced such wonderful results.

In a few instances selection, guided by utility for a single purpose, has led to convergence of character. All the improved and different races of the pig, as Nathusius has well shown (21/59. 'Die Racen des Schweines' 1860 s. 48.), closely approach each other in character, in their shortened legs and muzzles, their almost hairless, large, rounded bodies, and small tusks. We see some degree of convergence in the similar outline of the body in well-bred cattle belonging to distinct races. (21/60. See some good remarks on this head by M. de Quatrefages 'Unite de l'Espece Humaine' 1861 page 119.) I know of no other such cases.

Continued divergence of character depends on, and is indeed a clear proof, as previously remarked, of the same parts continuing to vary in the same direction. The tendency to mere general variability or plasticity of organisation can certainly be inherited, even from one parent, as has been shown by Gartner and Kolreuter, in the production of varying hybrids from two species, of which one alone was variable. It is in itself probable that, when an organ has varied in any manner, it will again vary in the same manner, if the conditions which first caused the being to vary remain, as far as can be judged, the same. This is either tacitly or expressly admitted by all horticulturists: if a gardener observes one or two additional petals in a flower, he feels confident that in a few generations he will be able to raise a double flower, crowded with petals. Some of the seedlings from the weeping Moccas oak were so prostrate that they only crawled along the ground. A seedling from the fastigiate or upright Irish yew is described as differing greatly from the parent-form "by the exaggeration of the fastigiate habit of its branches." (21/61. Verlot 'Des Varietes' 1865 page 94.) Mr. Shirreff, who has been highly successful in raising new kinds of wheat, remarks, "A good variety may safely be regarded as the forerunner of a better one." (21/62. Mr. Patrick Shirreff 'Gardener's Chronicle' 1858 page 771.) A great rose-grower, Mr. Rivers, has made the same remark with respect to roses. Sageret (21/63. 'Pomologie Physiolog.' 1830 page 106.), who had large experience, in speaking of the future progress of fruit-trees, observes that the most important principle is "that the more plants have departed from their original type, the more they tend to depart from it." There is apparently much truth in this remark; for we can in no other way understand the surprising amount of difference between varieties in the parts or qualities which are valued, whilst other parts retain nearly their original character.

The foregoing discussion naturally leads to the question, what is the limit to the possible amount of variation in any part or quality, and, consequently, is there any limit to what selection can effect? Will a racehorse ever be reared fleeter than Eclipse? Can our prize-cattle and sheep be still further improved? Will a gooseberry ever weigh more than that produced by "London" in 1852? Will the beet-root in France yield a greater percentage of sugar? Will future varieties of wheat and other grain produce heavier crops than our present varieties? These questions cannot be positively answered; but it is certain that we ought to be cautious in answering them by a negative. In some lines of variation the limit has probably been reached. Youatt believes that the reduction of bone in some of our sheep has already been carried so far that it entails great delicacy of constitution. (21/64. Youatt on 'Sheep' page 521.) But seeing the great improvement within recent times in our cattle and sheep, and especially in our pigs; seeing the wonderful increase in weight in our poultry of all kinds during the last few years; he would be a bold man who would assert that perfection has been reached. It has often been said that Eclipse never was, and never will be, beaten in speed by any other horse; but on making inquiries I find that the best judges believe that our present racehorses are fleeter. (21/65. See also Stonehenge 'British Rural Sports' edition of 1871 page 384.) The attempt to raise a new variety of wheat more productive than the many old kinds, might have been thought until lately quite hopeless; but this has been effected by Major Hallett, by careful selection. With respect to almost all our animals and plants, those who are best qualified to judge do not believe that the extreme point of perfection has yet been reached even in the characters which have already been carried to a high standard. For instance, the short-faced tumbler-pigeon has been greatly modified; nevertheless, according to Mr. Eaton (21/66. 'A Treatise on the Almond Tumbler' page 1.) "the field is still as open for fresh competitors as it was one hundred years ago." Over and over again it has been said that perfection had been attained with our flowers, but a higher standard has soon been reached. Hardly any fruit has been more improved than the strawberry, yet a great authority remarks (21/67. M. J. de Jonghe in 'Gardener's Chronicle' 1858 page 173.), "it must not be concealed that we are far from the extreme limits at which we may arrive."

No doubt there is a limit beyond which the organisation cannot be modified compatibly with health or life. The extreme degree of fleetness, for instance, of which a terrestrial animal is capable, may have been acquired by our present racehorses; but as Mr. Wallace has well remarked (21/68. 'Contributions to the Theory of Natural Selection' 2nd edition 1871 page 292.), the question that interests us, "is not whether indefinite and unlimited change in any or all directions is possible, but whether such differences as do occur in nature could have been produced by the accumulation of varieties by selection." And in the case of our domestic productions, there can be no doubt that many parts of the organisation, to which man has attended, have been thus modified to a greater degree than the corresponding parts in the natural species of the same genera or even families. We see this in the form and size of our light and heavy dogs or horses,—in the beak and many other characters of our pigeons,—in the size and quality of many fruits,—in comparison with the species belonging to the same natural groups.

Time is an important element in the formation of our domestic races, as it permits innumerable individuals to be born, and these when exposed to diversified conditions are rendered variable. Methodical selection has been occasionally practised from an ancient period to the present day, even by semi-civilised people, and during former times will have produced some effect. Unconscious selection will have been still more effective; for during a lengthened period the more valuable individual animals will occasionally have been saved, and the less valuable neglected. In the course of time, different varieties, especially in the less civilised countries, will also have been more or less modified through natural selection. It is generally believed, though on this head we have little or no evidence, that new characters in time become fixed; and after having long remained fixed it seems possible that under new conditions they might again be rendered variable.

How great the lapse of time has been since man first domesticated animals and cultivated plants, we begin dimly to see. When the lake-dwellings of Switzerland were inhabited during the Neolithic period, several animals were already domesticated and various plants cultivated. The science of language tells us that the art of ploughing and sowing the land was followed, and the chief animals had been already domesticated, at an epoch so immensely remote, that the Sanskrit, Greek, Latin, Gothic, Celtic, and Sclavonic languages had not as yet diverged from their common parent-tongue. (21/69. Max Muller 'Science of Language' 1861 page 223.)

It is scarcely possible to overrate the effects of selection occasionally carried on in various ways and places during thousands of generations. All that we know, and, in a still stronger degree, all that we do not know (21/70. 'Youatt on Cattle' pages 116, 128.), of the history of the great majority of our breeds, even of our more modern breeds, agrees with the view that their production, through the action of unconscious and methodical selection, has been almost insensibly slow. When a man attends rather more closely than is usual to the breeding of his animals, he is almost sure to improve them to a slight extent. They are in consequence valued in his immediate neighbourhood, and are bred by others; and their characteristic features, whatever these may be, will then slowly but steadily be increased, sometimes by methodical and almost always by unconscious selection. At last a strain, deserving to be called a sub-variety, becomes a little more widely known, receives a local name, and spreads. The spreading will have been extremely slow during ancient and less civilised times, but now is rapid. By the time that the new breed had assumed a somewhat distinct character, its history, hardly noticed at the time, will have been completely forgotten; for, as Low remarks (21/71. 'Domesticated Animals' page 188.), "we know how quickly the memory of such events is effaced."

As soon as a new breed is thus formed, it is liable through the same process to break up into new strains and sub-varieties. For different varieties are suited for, and are valued under, different circumstances. Fashion changes, but, should a fashion last for even a moderate length of time, so strong is the principle of inheritance, that some effect will probably be impressed on the breed. Thus varieties go on increasing in number, and history shows us how wonderfully they have increased since the earliest records. (21/72. Volz 'Beitrage zur Kulturgeschichte' 1852 s. 99 et passim.) As each new variety is produced, the earlier, intermediate, and less valuable forms will be neglected, and perish. When a breed, from not being valued, is kept in small numbers, its extinction almost inevitably follows sooner or later, either from accidental causes of destruction or from close interbreeding; and this is an event which, in the case of well-marked breeds, excites attention. The birth or production of a new domestic race is so slow a process that it escapes notice; its death or destruction is comparatively sudden, is often recorded, and when too late sometimes regretted.

Several authors have drawn a wide distinction between artificial and natural races. The latter are more uniform in character, possessing in a high degree the appearance of natural species, and are of ancient origin. They are generally found in less civilised countries, and have probably been largely modified by natural selection, and only to a small extent by man's unconscious and methodical selection. They have, also, during a long period, been directly acted on by the physical conditions of the countries which they inhabit. The so-called artificial races, on the other hand, are not so uniform in character; some have a semi-monstrous character, such as "the wry-legged terriers so useful in rabbit-shooting" (21/73. Blaine 'Encyclop. of Rural Sports' page 213.), turnspit dogs, ancon sheep, niata oxen, Polish fowls, fantail-pigeons, etc.; their characteristic features have generally been acquired suddenly, though subsequently increased by careful selections in many cases. Other races, which certainly must be called artificial, for they have been largely modified by methodical selection and by crossing, as the English racehorse, terrier-dogs, the English game-cock, Antwerp carrier-pigeons, etc., nevertheless cannot be said to have an unnatural appearance; and no distinct line, as it seems to me, can be drawn between natural and artificial races.

It is not surprising that domestic races should generally present a different aspect from natural species. Man selects and propagates modifications solely for his own use or fancy, and not for the creature's own good. His attention is struck by strongly marked modifications, which have appeared suddenly, due to some great disturbing cause in the organisation. He attends almost exclusively to external characters; and when he succeeds in modifying internal organs,—when for instance he reduces the bones and offal, or loads the viscera with fat, or gives early maturity, etc.-the chances are strong that he will at the same time weaken the constitution. On the other hand, when an animal has to struggle throughout its life with many competitors and enemies, under circumstances inconceivably complex and liable to change, modifications of the most varied nature in the internal organs as well as in external characters, in the functions and mutual relations of parts, will be rigorously tested, preserved, or rejected. Natural selection often checks man's comparatively feeble and capricious attempts at improvement; and if it were not so, the result of his work, and of nature's work, would be even still more different. Nevertheless, we must not overrate the amount of difference between natural species and domestic races; the most experienced naturalists have often disputed whether the latter are descended from one or from several aboriginal stocks, and this clearly shows that there is no palpable difference between species and races.

Domestic races propagate their kind far more truly, and endure for munch longer periods, than most naturalists are willing to admit. Breeders feel no doubt on this head: ask a man who has long reared Shorthorn or Hereford cattle, Leicester or Southdown sheep, Spanish or Game poultry, tumbler or carrier-pigeons, whether these races may not have been derived from common progenitors, and he will probably laugh you to scorn. The breeder admits that he may hope to produce sheep with finer or longer wool and with better carcases, or handsomer fowls, or carrier-pigeons with beaks just perceptibly longer to the practised eye, and thus be successful at an exhibition. Thus far he will go, but no farther. He does not reflect on what follows from adding up during a long course of time many slight, successive modifications; nor does he reflect on the former existence of numerous varieties, connecting the links in each divergent line of descent. He concludes, as was shown in the earlier chapters, that all the chief breeds to which he has long attended are aboriginal productions. The systematic naturalist, on the other hand, who generally knows nothing of the art of breeding, who does not pretend to know how and when the several domestic races were formed, who cannot have seen the intermediate gradations, for they do not now exist, nevertheless feels no doubt that these races are sprung from a single source. But ask him whether the closely allied natural species which he has studied may not have descended from a common progenitor, and he in his turn will perhaps reject the notion with scorn. Thus the naturalist and breeder may mutually learn a useful lesson from each other.

## SUMMARY ON SELECTION BY MAN.

There can be no doubt that methodical selection has effected and will effect wonderful results. It was occasionally practised in ancient times, and is still practised by semi-civilised people. Characters of the highest importance, and others of trifling value, have been attended to, and modified. I need not here repeat what has been so often said on the part which unconscious selection has played: we see its power in the difference between flocks which have been separately bred, and in the slow changes, as circumstances have slowly changed, which many animals have undergone in the same country, or when transported into a foreign land. We see the combined effects of methodical and unconscious selection, in the great amount of difference in those parts or qualities which are valued by man in comparison with the parts which are not valued, and consequently have not been attended to. Natural selection often determines man's power of selection. We sometimes err in imagining that characters, which are considered as unimportant by the systematic naturalist, could not be affected by the struggle for existence, and could not be acted on by natural selection; but striking cases have been given, showing how great an error this is.

The possibility of selection coming into action rests on variability; and this is mainly caused, as we shall hereafter see, by changes in the conditions of life. Selection is sometimes rendered difficult, or even impossible, by the conditions being opposed to the desired character or quality. It is sometimes checked by the lessened fertility and weakened constitution which follow from long-continued close interbreeding. That methodical selection may be successful, the closest attention and discernment, combined with unwearied patience, are absolutely necessary; and these same qualities, though not indispensable, are highly serviceable in the case of unconscious selection. It is almost necessary that a large number of individuals should be reared; for thus there will be a fair chance of variations of the desired nature arising, and of every individual with the slightest blemish or in any degree inferior being freely rejected. Hence length of time is an important element of success. Thus, also, reproduction at an early age and at short intervals favours the work. Facility in pairing animals, or their inhabiting a confined area, is advantageous as a check to free crossing. Whenever and wherever selection is not practised, distinct races are not formed within the same country. When any one part of the body or one quality is not attended to, it remains either unchanged or varies in a fluctuating manner, whilst at the same time other parts and other qualities may become permanently and greatly modified. But from the tendency to reversion and to continued variability, those parts or organs which are now undergoing rapid improvement through selection, are likewise found to vary much. Consequently highly-bred animals when neglected soon degenerate; but we have no reason to believe that the effects of long-continued selection would, if the conditions of life remained the same, be soon and completely lost.

Man always tends to go to an extreme point in the selection, whether methodical or unconscious, of all useful and pleasing qualities. This is an important principle, as it leads to continued divergence, and in some rare cases to convergence of character. The possibility of continued divergence rests on the tendency in each part or organ to go on varying in the same manner in which it has already varied; and that this occurs, is proved by the steady and gradual improvement of many animals and plants during lengthened periods. The principle of divergence of character, combined with the neglect and final extinction of all previous, less-valued, and intermediate varieties, explains the amount of difference and the distinctness of our several races. Although we may have reached the utmost limit to which certain characters can be modified, yet we are far from having reached, as we have good reason to believe, the limit in the majority of cases. Finally, from the difference between selection as carried on by man and by nature, we can understand how it is that domestic races often, though by no means always, differ in general aspect from closely allied natural species.

Throughout this chapter and elsewhere I have spoken of selection as the paramount power, yet its action absolutely depends on what we in our ignorance call spontaneous or accidental variability. Let an architect be compelled to build an edifice with uncut stones, fallen from a precipice. The shape of each fragment may be called accidental; yet the shape of each has been determined by the force of gravity, the nature of the rock, and the slope of the precipice,—events and circumstances, all of which depend on natural laws; but there is no relation between these laws and the purpose for which each fragment is used by the builder. In the same manner the variations of each creature are determined by fixed and immutable laws; but these bear no relation to the living structure which is slowly built up through the power of selection, whether this be natural or artificial selection.

If our architect succeeded in rearing a noble edifice, using the rough wedge-shaped fragments for the arches, the longer stones for the lintels, and so forth, we should admire his skill even in a higher degree than if he had used stones shaped for the purpose. So it is with selection, whether applied by man or by nature; for although variability is indispensably necessary, yet, when we look at some highly complex and excellently adapted organism, variability sinks to a quite subordinate position in importance in comparison with selection, in the same manner as the shape of each fragment used by our supposed architect is unimportant in comparison with his skill.

# CHAPTER 2.XXII.

CAUSES OF VARIABILITY.

VARIABILITY DOES NOT NECESSARILY ACCOMPANY REPRODUCTION. CAUSES ASSIGNED BY VARIOUS AUTHORS. INDIVIDUAL DIFFERENCES. VARIABILITY OF EVERY KIND DUE TO CHANGED CONDITIONS OF LIFE. ON THE NATURE OF SUCH CHANGES. CLIMATE, FOOD, EXCESS OF NUTRIMENT. SLIGHT CHANGES SUFFICIENT. EFFECTS OF GRAFTING ON THE VARIABILITY OF SEEDLING-TREES. DOMESTIC PRODUCTIONS BECOME HABITUATED TO CHANGED CONDITIONS. ON THE ACCUMULATIVE ACTION OF CHANGED CONDITIONS. CLOSE INTERBREEDING AND THE IMAGINATION OF THE MOTHER SUPPOSED TO CAUSE VARIABILITY. CROSSING AS A CAUSE OF THE APPEARANCE OF NEW CHARACTERS. VARIABILITY FROM THE COMMINGLING OF CHARACTERS AND FROM REVERSION. ON THE MANNER AND PERIOD OF ACTION OF THE CAUSES WHICH EITHER DIRECTLY, OR INDIRECTLY THROUGH THE REPRODUCTIVE SYSTEM, INDUCE VARIABILITY.

We will now consider, as far as we can, the causes of the almost universal variability of our domesticated productions. The subject is an obscure one; but it may be useful to probe our ignorance. Some authors, for instance Dr. Prosper Lucas, look at variability as a necessary contingent on reproduction, and as much an aboriginal law as growth or inheritance. Others have of late encouraged, perhaps unintentionally, this view by speaking of inheritance and variability as equal and antagonistic principles. Pallas maintained, and he has had some followers, that variability depends exclusively on the crossing of primordially distinct forms. Other authors attribute variability to an excess of food, and with animals to an excess relatively to the amount of exercise taken, or again to the effects of a more genial climate. That these causes are all effective is highly probable. But we must, I think, take a broader view, and conclude that organic beings, when subjected during several generations to any change whatever in their conditions, tend to vary; the kind of variation which ensues depending in most cases in a far higher degree on the nature or constitution of the being, than on the nature of the changed conditions.

Those authors who believe that it is a law of nature that each individual should differ in some slight degree from every other, may maintain, apparently with truth, that this is the fact, not only with all domesticated animals and cultivated plants, but likewise with all organic beings in a state of nature. The Laplander by long practice knows and gives a name to each reindeer, though, as Linnaeus remarks, "to distinguish one from another among such multitudes was beyond my comprehension, for they were like ants on an anthill." In Germany shepherds have won wagers by recognising each sheep in a flock of a hundred, which they had never seen until the previous fortnight. This power of discrimination, however, is as nothing compared to that which some florists have acquired. Verlot mentions a gardener who could distinguish 150 kinds of camellia, when not in flower; and it has been positively asserted that the famous old Dutch florist Voorhelm, who kept above 1200 varieties of the hyacinth, was hardly ever deceived in knowing each variety by the bulb alone. Hence we must conclude that the bulbs of the hyacinth and the branches and leaves of the camellia, though appearing to an unpractised eye absolutely undistinguishable, yet really differ. (22/1. 'Des Jacinthes' etc. Amsterdam 1768 page 43; Verlot 'Des Varietes' etc. page 86. On the reindeer see Linnaeus 'Tour in Lapland' translated by Sir J.E. Smith volume 1 page 314. The statement in regard to German shepherds is given on the authority of Dr. Weinland.)

As Linnaeus has compared the reindeer in number to ants, I may add that each ant knows its fellow of the same community. Several times I carried ants of the same species (Formica rufa) from one ant-hill to another, inhabited apparently by tens of thousands of ants; but the strangers were instantly detected and killed. I then put some ants taken from a very large nest into a bottle strongly perfumed with assafoetida, and after an interval of twenty-four hours returned them to their home; they were at first threatened by their fellows, but were soon recognised and allowed to pass. Hence each ant certainly recognised, independently of odour, its fellow; and if all the ants of the same community have not some countersign or watchword, they must present to each other's senses some distinguishable character.

The dissimilarity of brothers or sisters of the same family, and of seedlings from the same capsule, may be in part accounted for by the unequal blending of the characters of the two parents, and by the more or less complete recovery through reversion of ancestral characters on either side; but we thus only push the difficulty further back in time, for what made the parents or their progenitors different? Hence the belief (22/2. Muller 'Physiology' English translation, volume 2 page 1662. With respect to the similarity of twins in constitution, Dr. William Ogle has given me the following extract from Professor Trousseau's Lectures 'Clinique Medicale' tome 1 page 523, in which a curious case is recorded:—"J'ai donne mes soins a deux freres jumeaux, tous deux si extraordinairement ressemblants qu'il m'etait impossible de les reconnaitre, a moin de les voir l'un a cote de l'autre. Cette ressemblance physique s'etendait plus loin: ils avaient, permettez-moi l'expression, une similitude pathologique plus remarquable encore. Ainsi l'un d'eux que je voyais aux neothermes a Paris malade d'une ophthalmie rhumatismale me disait, 'En ce moment mon frere doit avoir une ophthalmie comme la mienne;' et comme je m'etais recrie, il me montrait quelques jours apres une lettre qu'il venait de recevoir de ce frere alors a Vienne, et qui lui ecrivait en effet—'J'ai mon ophthalmie, tu dois avoir la tienne.' Quelque singulier que ceci puisse paraitre, le fait n'en est pas moins exact: on ne me l'a pas raconte, je l'ai vu, et j'en ai vu d'autres analogues dans ma pratique. Ces deux jumeaux etaient aussi tous deux asthmatiques, et asthmatiques a un effroyable degre. Originaires de Marseille, ils n'ont jamais pu demeurer dans cette ville, ou leurs interets les appelaient souvent, sans etre pris de leurs acces; jamais ils n'en eprouvaient a Paris. Bien mieux, il leur suffisait de gagner Toulon pour etre gueris de leurs attaques de Marseille. Voyageant sans cesse et dans tous pays pour leurs affaires, ils avaient remarque que certaines localites leur etaient funestes, que dans d'autres ils etaient exempts de tout phenomene d'oppression.") that an innate tendency to vary exists, independently of external differences, seems at first sight probable. But even the seeds nurtured in the same capsule are not subjected to absolutely uniform conditions, as they draw their nourishment from different points; and we shall see in a future chapter that this difference sometimes suffices to affect the character of the future plant. The greater dissimilarity of the successive children of the same family in comparison with twins, which often resemble each other in external appearance, mental disposition, and constitution, in so extraordinary a manner, apparently proves that the state of the parents at the exact period of conception, or the nature of the subsequent embryonic development, has a direct and powerful influence on the character of the offspring. Nevertheless, when we reflect on the individual differences between organic beings in a state of nature, as shown by every wild animal knowing its mate; and when we reflect on the infinite diversity of the many varieties of our domesticated productions, we may well be inclined to exclaim, though falsely as I believe, that Variability must be looked at as an ultimate fact, necessarily contingent on reproduction.

Those authors who adopt this latter view would probably deny that each separate variation has its own proper exciting cause. Although we can seldom trace the precise relation between cause and effect, yet the considerations presently to be given lead to the conclusion that each modification must have its own distinct cause, and is not the result of what we blindly call accident. The following striking case has been communicated to me by Dr. William Ogle. Two girls, born as twins, and in all respects extremely alike, had their little fingers on both hands crooked; and in both children the second bicuspid tooth of the second dentition on the right side in the upper jaw was misplaced; for, instead of standing in a line with the others, it grew from the roof of the mouth behind the first bicuspid. Neither the parents nor any other members of the family were known to have exhibited any similar peculiarity;

but a son of one of these girls had the same tooth similarly misplaced. Now, as both the girls were affected in exactly the same manner, the idea of accident is at once excluded: and we are compelled to admit that there must have existed some precise and sufficient cause which, if it had occurred a hundred times, would have given crooked fingers and misplaced bicuspid teeth to a hundred children. It is of course possible that this case may have been due to reversion to some long-forgotten progenitor, and this would much weaken the value of the argument. I have been led to think of the probability of reversion, from having been told by Mr. Galton of another case of twin girls born with their little fingers slightly crooked, which they inherited from their maternal grandmother.

We will now consider the general arguments, which appear to me to have great weight, in favour of the view that variations of all kinds and degrees are directly or indirectly caused by the conditions of life to which each being, and more especially its ancestors, have been exposed.

No one doubts that domesticated productions are more variable than organic beings which have never been removed from their natural conditions. Monstrosities graduate so insensibly into mere variations that it is impossible to separate them; and all those who have studied monstrosities believe that they are far commoner with domesticated than with wild animals and plants (22/3. Isid. Geoffroy St.-Hilaire 'Hist. des Anomalies' tome 3 page 352; Moquin-Tandon 'Teratologie Vegetale' 1841 page 115.); and in the case of plants, monstrosities would be equally noticeable in the natural as in the cultivated state. Under nature, the individuals of the same species are exposed to nearly uniform conditions, for they are rigorously kept to their proper places by a host of competing animals and plants; they have, also, long been habituated to their conditions of life; but it cannot be said that they are subject to quite uniform conditions, and they are liable to a certain amount of variation. The circumstances under which our domestic productions are reared are widely different: they are protected from competition; they have not only been removed from their natural conditions and often from their native land, but they are frequently carried from district to district, where they are treated differently, so that they rarely remain during any considerable length of time exposed to closely similar conditions. In conformity with this, all our domesticated productions, with the rarest exceptions, vary far more than natural species. The hive-bee, which feeds itself and follows in most respects its natural habits of life, is the least variable of all domesticated animals, and probably the goose is the next least variable; but even the goose varies more than almost any wild bird, so that it cannot be affiliated with perfect certainty to any natural species. Hardly a single plant can be named, which has long been cultivated and propagated by seed, that is not highly variable; common rye (Secale cereale) has afforded fewer and less marked varieties than almost any other cultivated plant (22/4. Metzger 'Die Getreidarten' 1841 s. 39.); but it may be doubted whether the variations of this, the least valuable of all our cereals, have been closely observed.

Bud-variation, which was fully discussed in a former chapter, shows us that variability may be quite independent of seminal reproduction, and likewise of reversion to long-lost ancestral characters. No one will maintain that the sudden appearance of a moss-rose on a Provence-rose is a return to a former state, for mossiness of the calyx has been observed in no natural species; the same argument is applicable to variegated and laciniated leaves; nor can the appearance of nectarines on peach-trees be accounted for on the principle of reversion. But bud-variations more immediately concern us, as they occur far more frequently on plants which have been highly cultivated during a length of time, than on other and less highly cultivated plants; and very few well-marked instances have been observed with plants growing under strictly natural conditions. I have given one instance of an ash-tree growing in a gentleman's pleasure-grounds; and occasionally there may be seen, on beech and other trees, twigs leafing at a different period from the other branches. But our forest trees in England can hardly be considered as living under strictly natural conditions; the seedlings are raised and protected in nursery-grounds, and must often be transplanted into places where wild trees of the kind would not naturally grow. It would be esteemed a prodigy if a dog-rose growing in a hedge produced by bud-variation a moss-rose, or a wild bullace or wild cherry- tree yielded a branch bearing fruit of a different shape and colour from the ordinary fruit. The prodigy would be enhanced if these varying branches were found capable of propagation, not only by grafts, but sometimes by seed; yet analogous cases have occurred with many of our highly cultivated trees and herbs.

These several considerations alone render it probable that variability of every kind is directly or indirectly caused by changed conditions of life. Or, to put the case under another point of view, if it were possible to expose all the individuals of a species during many generations to absolutely uniform conditions of life, there would be no variability.

## ON THE NATURE OF THE CHANGES IN THE CONDITIONS OF LIFE WHICH INDUCE VARIABILITY.

From a remote period to the present day, under climates and circumstances as different as it is possible to conceive, organic beings of all kinds, when domesticated or cultivated, have varied. We see this with the many domestic races of quadrupeds and birds belonging to different orders, with goldfish and silkworms, with plants of many kinds, raised in various quarters of the world. In the deserts of northern Africa the date-palm has yielded thirty-eight varieties; in the fertile plains of India it is notorious how many varieties of rice and of a host of other plants exist; in a single Polynesian island, twenty-four varieties of the bread-fruit, the same number of the banana, and twenty-two varieties of the arum, are cultivated by the natives; the mulberry-tree in India and Europe has yielded many varieties serving as food for the silkworm; and in China sixty-three varieties of the bamboo are used for various domestic purposes. (22/5. On the date-palm see Vogel 'Annals and Mag. of Nat. Hist.' 1854 page 460. On Indian varieties Dr. F. Hamilton 'Transact. Linn. Soc.' volume 14 page 296. On the varieties cultivated in Tahiti see Dr. Bennett in Loudon's 'Mag. of N. Hist.' volume 5 1832 page 484. Also Ellis 'Polynesian Researches' volume 1 pages 370, 375. On twenty varieties of the Pandanus and other trees in the Marianne Island see 'Hooker's Miscellany' volume 1 page 308. On the bamboo in China see Huc 'Chinese Empire' volume 2 page 307.) These facts, and innumerable others which could be added, indicate that a change of almost any kind in the conditions of life suffices to cause variability—different changes acting on different organisms.

Andrew Knight (22/6. 'Treatise on the Culture of the Apple' etc. page 3.) attributed the variation of both animals and plants to a more abundant supply of nourishment, or to a more favourable climate, than that natural to the species. A more genial climate, however, is far from necessary; the kidney-bean, which is often injured by our spring frosts, and peaches, which require the protection of a wall, have varied much in England, as has the orange-tree in northern Italy, where it is barely able to exist. (22/7. Gallesio 'Teoria della Riproduzione Veg.' page 125.) Nor can we overlook the fact, though not immediately connected with our present subject, that the plants and shells of the Arctic regions are eminently variable. (22/8. See Dr. Hooker's Memoir on Arctic Plants in 'Linn. Transact.' volume 23 part 2. Mr. Woodward, and a higher authority cannot be quoted, speaks of the Arctic mollusca in his 'Rudimentary Treatise' 1856 page 355 as remarkably subject to variation.) Moreover, it does not appear that a change of climate, whether more or less genial, is one of the most potent causes of variability; for in regard to plants Alph. De Candolle, in his 'Geographie Botanique' repeatedly shows that the native country of a plant, where in most cases it has been longest cultivated, is that where it has yielded the greatest number of varieties.

It is doubtful whether a change in the nature of the food is a potent cause of variability. Scarcely any domesticated animal has varied more than the pigeon or the fowl, but their food, especially that of highly-bred pigeons, is generally the same. Nor can our cattle and sheep have been subjected to any great change in this respect. But in all these cases the food probably is much less varied in kind than that which was consumed by the species in its natural state. (22/9. Bechstein in his 'Naturgeschichte der Stubenvogel' 1840 s. 238, has some good remarks on this subject. He states that his canary-birds varied in colour, though kept on uniform food.)

Of all the causes which induce variability, excess of food, whether or not changed in nature, is probably the most powerful. This view was held with regard to plants by Andrew Knight, and is now held by Schleiden, more especially in reference to the inorganic elements of the food. (22/10. 'The Plant' by Schleiden translated by Henfrey 1848 page 169. See also Alex. Braun in 'Bot. Memoirs' Ray Soc. 1853 page 313.) In order to give a plant more food it suffices in most cases to grow it separately, and thus prevent other plants robbing its roots. It is surprising, as I have often seen, how vigorously our common wild species flourish when planted by themselves, though not in highly manured land; separate growth is, in fact, the first step in cultivation. We see the converse of the belief that excess of food induces variability in the following statement by a great raiser of seeds of all kinds (22/11. Messrs. Hardy and Son of Maldon in 'Gardener's Chronicle' 1856 page 458. Carriere 'Production et Fixation des Varietes' 1865 page 31.): "It is a rule invariably with us, when we desire to keep a true stock of any one kind of seed, to grow it on poor land without dung; but when we grow for quantity, we act contrary, and sometimes have dearly to repent of it." According also to Carriere, who has had great experience with flower-garden seeds, "On remarque en general les plantes de vigeur moyenne sont celles qui conservent le mieux leurs caracteres."

In the case of animals the want of a proper amount of exercise, as Bechstein remarked, has perhaps played, independently of the direct effects of the disuse of any particular organ, an important part in causing variability. We can see in a vague manner that, when the organised and nutrient fluids of the body are not used during growth, or by the wear and tear of the tissues, they will be in excess; and as growth, nutrition, and reproduction are intimately allied processes, this superfluity might disturb the due and proper action of the reproductive organs, and consequently affect the character of the future offspring. But it may be argued that neither an excess of food nor a superfluity in the organised fluids of the body necessarily induces variability. The goose and the turkey have been well fed for many generations, yet have varied very little. Our fruit-trees and culinary plants, which are so variable, have been cultivated from an ancient period, and, though they probably still receive more nutriment than in their natural state, yet they must have received during many generations nearly the same amount; and it might be thought that they would have become habituated to the excess. Nevertheless, on the whole, Knight's view, that excess of food is one of the most potent causes of variability, appears, as far as I can judge, probable.

Whether or not our various cultivated plants have received nutriment in excess, all have been exposed to changes of various kinds. Fruit-trees are grafted on different stocks, and grown in various soils. The seeds of culinary and agricultural plants are carried from place to place; and during the last century the rotation of our crops and the manures used have been greatly changed.

Slight changes of treatment often suffice to induce variability. The simple fact of almost all our cultivated plants and domesticated animals having varied in all places and at all times, leads to this conclusion. Seeds taken from common English forest-trees, grown under their native climate, not highly manured or otherwise artificially treated, yield seedlings which vary much, as may be seen in every extensive seed-bed. I have shown in a former chapter what a number of well-marked and singular varieties the thorn (Crataegus oxycantha) has produced: yet this tree has been subjected to hardly any cultivation. In Staffordshire I carefully examined a large number of two British plants, namely Geranium phaeum and pyrenaicum, which have never been highly cultivated. These plants had spread spontaneously by seed from a common garden into an open plantation; and the seedlings varied in almost every single character, both in their flower and foliage, to a degree which I have never seen exceeded; yet they could not have been exposed to any great change in their conditions.

With respect to animals, Azara has remarked with much surprise (22/12. 'Quadrupedes du Paraguay' 1801 tome 2 page 319.) that, whilst the feral horses on the Pampas are always of one of three colours, and the cattle always of a uniform colour, yet these animals, when bred on the unenclosed estancias, though kept in a state which can hardly be called domesticated, and apparently exposed to almost identically the same conditions as when they are feral, nevertheless display a great diversity of colour. So again in India several species of fresh-water fish are only so far treated artificially, that they are reared in great tanks; but this small change is sufficient to induce much variability. (22/13. M'Clelland on Indian Cyprinidae 'Asiatic Researches' volume 19 part 2 1839 pages 266, 268, 313.)

Some facts on the effects of grafting, in regard to the variability of trees, deserve attention. Cabanis asserts that when certain pears are grafted on the quince, their seeds yield a greater number of varieties than do the seeds of the same variety of pear when grafted on the wild pear. (22/14. Quoted by Sageret 'Pom. Phys.' 1830 page 43. This statement, however, is not believed by Decaisne.) But as the pear and quince are distinct species, though so closely related that the one can be readily grafted and succeeds admirably on the other, the fact of variability being thus caused is not surprising; as we are here enabled to see the cause, namely, the very different nature of the stock and graft. Several North American varieties of the plum and peach are well known to reproduce themselves truly by seed; but Downing asserts (22/15. 'The Fruits of America' 1845 page 5.), "that when a graft is taken from one of these trees and placed upon another stock, this grafted tree is found to lose its singular property of producing the same variety by seed, and becomes like all other worked trees;"—that is, its seedlings become highly variable. Another case is worth giving: the Lalande variety of the walnut-tree leafs between April 20th and May 15th, and its seedlings invariably inherit the same habit; whilst several other varieties of the walnut leaf in June. Now, if seedlings are raised from the May-leafing Lalande variety, grafted on another May-leafing variety, though both stock and graft have the same early habit of leafing, yet the seedlings leaf at various times, even as late as the 5th of June. (22/16. M. Cardan in 'Comptes Rendus' December 1848 quoted in 'Gardener's Chronicle' 1849 page 101.) Such facts as these are well fitted to show on what obscure and slight causes variability depends.

[I may here just allude to the appearance of new and valuable varieties of fruit-trees and of wheat in woods and waste places, which at first sight seems a most anomalous circumstance. In France a considerable number of the best pears have been discovered in woods; and this has occurred so frequently, that Poiteau asserts that "improved varieties of our cultivated fruits rarely originate with nurserymen." (22/17. M. Alexis Jordan mentions four excellent pears found in woods in France, and alludes to others ('Mem. Acad. de Lyon' tome 2 1852 page 159). Poiteau's remark is quoted in 'Gardener's Mag.' volume 4 1828 page 385. See 'Gardener's Chronicle' 1862 page 335, for another case of a new variety of the pear found in a hedge in France. Also for another case, see Loudon's 'Encyclop. of Gardening' page 901. Mr. Rivers has given me similar information.) In England, on the other hand, no instance of a good pear having been found wild has been recorded; and Mr. Rivers informs me that he knows of only one instance with apples, namely, the Bess Poole, which was discovered in a wood in Nottinghamshire. This difference between the two countries may be in part accounted for by the more favourable climate of France, but chiefly from the great number of seedlings which spring up there in the woods. I infer that this is the case from a remark made by a French gardener (22/18. Duval 'Hist. du Poirier' 1849 page 2.), who regards it as a national calamity that such a number of pear-trees are periodically cut down for firewood, before they have borne fruit. The new varieties which thus spring up in the woods, though they cannot have received any excess of nutriment, will have been exposed to abruptly changed conditions, but whether this is the cause of their production is very doubtful. These varieties, however, are probably all descended (22/19. I infer that this is the fact from Van Mons' statement ('Arbres Fruitiers' 1835 tome 1 page 446) that he finds in the woods seedlings resembling all the chief cultivated races of both the pear and apple. Van Mons, however, looked at these wild varieties as aboriginal species.) from old cultivated kinds growing in adjoining orchards— a circumstance which will account for their variability; and out of a vast number of varying trees there will always be a good chance of the appearance of a valuable kind. In North America, where fruit-trees frequently spring up in waste places, the Washington pear was found in a hedge, and the Emperor peach in a wood. (22/20. Downing 'Fruit-trees of North America' page 422; Foley in 'Transact. Hort. Soc.' volume 6 page 412.)

With respect to wheat, some writers have spoken (22/21. 'Gardener's Chronicle' 1847 page 244.) as if it were an ordinary event for new varieties to be found in waste places; the Fenton wheat was certainly discovered growing on a pile of basaltic detritus in a quarry, but in such a situation the plant would probably receive a sufficient amount of nutriment. The Chidham wheat was raised from an ear found ON a hedge; and Hunter's wheat was discovered BY the roadside in Scotland, but it is not said that this latter variety grew where it was found. (22/22. 'Gardener's Chronicle' 1841 page 383; 1850 page 700; 1854 page 650.)]

Whether our domestic productions would ever become so completely habituated to the conditions under which they now live, as to cease varying, we have no sufficient means for judging. But, in fact, our domestic productions are never exposed for a great length of time to uniform conditions, and it is certain that our most anciently cultivated plants, as well as animals, still go on varying, for all have recently undergone marked improvement. In some few cases, however, plants have become habituated to new conditions. Thus, Metzger, who cultivated in Germany during many years numerous varieties of wheat, brought from different countries (22/23. 'Die Getreidearten' 1843 s. 66, 116, 117.), states that some kinds were at first extremely variable, but gradually, in one instance after an interval of twenty-five years, became constant; and it does not appear that this resulted from the selection of the more constant forms.

## ON THE ACCUMULATIVE ACTION OF CHANGED CONDITIONS OF LIFE.

We have good grounds for believing that the influence of changed conditions accumulates, so that no effect is produced on a species until it has been exposed during several generations to continued cultivation or domestication. Universal experience shows us that when new flowers are first introduced into our gardens they do not vary; but ultimately all, with the rarest exceptions, vary to a greater or less extent. In a few cases the requisite number of generations, as well as the successive steps in the progress of variation, have been recorded, as in the often quoted instance of the Dahlia. (22/24. Sabine in 'Hort. Transact.' volume 3 page 225; Bronn 'Geschichte der Natur' b. 2 s. 119.) After several years' culture the Zinnia has only lately (1860) begun to vary in any great degree. "In the first seven or eight years of high cultivation, the Swan River daisy (Brachycome iberidifolia) kept to its original colour; it then varied into lilac and purple and other minor shades." (22/25. 'Journal of Horticulture' 1861 page 112; on Zinnia 'Gardener's

Chronicle' 1860 page 852.) Analogous facts have been recorded with the Scotch rose. In discussing the variability of plants several experienced horticulturists have spoken to the same general effect. Mr. Salter (22/26. 'The Chrysanthemum, its History, etc.' 1865 page 3.) remarks, "Every one knows that the chief difficulty is in breaking through the original form and colour of the species, and every one will be on the look-out for any natural sport, either from seed or branch; that being once obtained, however trifling the change may be, the result depends upon himself." M. de Jonghe, who has had so much success in raising new varieties of pears and strawberries (22/27. 'Gardener's Chronicle' 1855 page 54; 'Journal of Horticulture' May 9, 1865 page 363.), remarks with respect to the former, "There is another principle, namely, that the more a type has entered into a state of variation, the greater is its tendency to continue doing so; and the more it has varied from the original type, the more it is disposed to vary still farther." We have, indeed, already discussed this latter point when treating of the power which man possesses, through selection, of continually augmenting in the same direction each modification; for this power depends on continued variability of the same general kind. The most celebrated horticulturist in France, namely, Vilmorin (22/28. Quoted by Verlot 'Des Varietes' etc. 1865 page 28.), even maintains that, when any particular variation is desired, the first step is to get the plant to vary in any manner whatever, and to go on selecting the most variable individuals, even though they vary in the wrong direction; for the fixed character of the species being once broken, the desired variation will sooner or later appear.

As nearly all our animals were domesticated at an extremely remote epoch, we cannot, of course, say whether they varied quickly or slowly when first subjected to new conditions. But Dr. Bachman (22/29. 'Examination of the Characteristics of Genera and Species' Charleston 1855 page 14.) states that he has seen turkeys raised from the eggs of the wild species lose their metallic tints and become spotted with white in the third generation. Mr. Yarrell many years ago informed me that the wild ducks bred on the ponds in St. James's Park, which had never been crossed, as it is believed, with domestic ducks, lost their true plumage after a few generations. An excellent observer (22/30. Mr. Hewitt 'Journal of Hort.' 1863 page 39.), who has often reared ducks from the eggs of the wild bird, and who took precautions that there should be no crossing with domestic breeds, has given, as previously stated, full details on the changes which they gradually undergo. He found that he could not breed these wild ducks true for more than five or six generations, "as they then proved so much less beautiful. The white collar round the neck of the mallard became much broader and more irregular, and white feathers appeared in the ducklings' wings." They increased also in size of body; their legs became less fine, and they lost their elegant carriage. Fresh eggs were then procured from wild birds; but again the same result followed. In these cases of the duck and turkey we see that animals, like plants, do not depart from their primitive type until they have been subjected during several generations to domestication. On the other hand, Mr. Yarrell informed me that the Australian dingos, bred in the Zoological Gardens, almost invariably produced in the first generation puppies marked with white and other colours; but, these introduced dingos had probably been procured from the natives, who keep them in a semi-domesticated state. It is certainly a remarkable fact that changed conditions should at first produce, as far as we can see, absolutely no effect; but that they should subsequently cause the character of the species to change. In the chapter on pangenesis I shall attempt to throw a little light on this fact.

Returning now to the causes which are supposed to induce variability. Some authors (22/31. Devay 'Mariages Consanguins' pages 97, 125. In conversation I have found two or three naturalists of the same opinion.) believe that close interbreeding gives this tendency, and leads to the production of monstrosities. In the seventeenth chapter some few facts were advanced, showing that monstrosities are, as it appears, occasionally thus induced; and there can be no doubt that close interbreeding causes lessened fertility and a weakened constitution; hence it may lead to variability: but I have not sufficient evidence on this head. On the other hand, close interbreeding, if not carried to an injurious extreme, far from causing variability, tends to fix the character of each breed.

It was formerly a common belief, still held by some persons, that the imagination of the mother affects the child in the womb. (22/32. Muller has conclusively argued against this belief, 'Elements of Phys.' English translation volume 2 1842 page 1405.) This view is evidently not applicable to the lower animals, which lay unimpregnated eggs, or to plants. Dr. William Hunter, in the last century, told my father that during many years every woman in a large London Lying-in Hospital was asked before her confinement whether anything had specially affected her mind, and the answer was written down; and it so happened that in no one instance could a coincidence be detected between the woman's answer and any abnormal structure; but when she knew the nature of the structure, she frequently suggested some fresh cause. The belief in the power of the mother's imagination may perhaps have arisen from the children of a second marriage resembling the previous father, as certainly sometimes occurs, in accordance with the facts given in the eleventh chapter.

## CROSSING AS A CAUSE OF VARIABILITY.

In an early part of this chapter it was stated that Pallas (22/33. 'Act. Acad. St. Petersburg' 1780 part 2 page 84 etc.) and a few other naturalists maintain that variability is wholly due to crossing. If this means that new characters never spontaneously appear in our domestic races, but that they are all directly derived from certain aboriginal species, the doctrine is little less than absurd; for it implies that animals like Italian greyhounds, pug-dogs, bull-dogs, pouter and fantail pigeons, etc., were able to exist in a state of nature. But the doctrine may mean something widely different, namely, that the crossing of distinct species is the sole cause of the first appearance of new characters, and that without this aid man could not have formed his various breeds. As, however, new characters have appeared in certain cases by bud-variation, we may conclude with certainty that crossing is not necessary for variability. It is, moreover, certain that the breeds of various animals, such as of the rabbit, pigeon, duck, etc., and the varieties of several plants, are the modified descendants of a single wild species. Nevertheless, it is probable that the crossing of two forms, when one or both have long been domesticated or cultivated, adds to the variability of the offspring, independently of the commingling of the characters derived from the two parent-forms; and this implies that new characters actually arise. But we must not forget the facts advanced in the thirteenth chapter, which clearly prove that the act of crossing often leads to the reappearance or reversion of long-lost characters; and in most cases it would be impossible to distinguish between the reappearance of ancient characters and the first appearance of absolutely new characters. Practically, whether new or old, they would be new to the breed in which they reappeared.

[Gartner declares (22/34. 'Bastarderzeugung' s. 249, 255, 295.), and his experience is of the highest value on such a point, that, when he crossed native plants which had not been cultivated, he never once saw in the offspring any new character; but that from the odd manner in which the characters derived from the parents were combined, they sometimes appeared as if new. When, on the other hand, he crossed cultivated plants, he admits that new characters occasionally appeared, but he is strongly inclined to attribute their appearance to ordinary variability, not in any way to the cross. An opposite conclusion, however, appears to me the more probable. According to Kolreuter, hybrids in the genus Mirabilis vary almost infinitely, and he describes new and singular characters in the form of the seeds, in the colour of the anthers, in the cotyledons being of immense size, in new and highly peculiar odours, in the flowers expanding early in the season, and in their closing at night. With respect to one lot of these hybrids, he remarks that they presented characters exactly the reverse of what might have been expected from their parentage. (22/35. 'Nova Acta, St. Petersburg' 1794 page 378; 1795 pages 307, 313, 316; 1787 page 407.)

Prof. Lecoq (22/36. 'De la Fecondation' 1862 page 311.) speaks strongly to the same effect in regard to this same genus, and asserts that many of the hybrids from Mirabilis jalapa and multiflora might easily be mistaken for distinct species, and adds that they differed in a greater degree than the other species of the genus, from M. jalapa. Herbert, also, has described (22/37. 'Amaryllidaceae' 1837 page 362.) certain hybrid Rhododendrons as being "as UNLIKE ALL OTHERS in foliage, as if they had been a separate species." The common experience of floriculturists proves that the crossing and recrossing of distinct but allied plants, such as the species of Petunia, Calceolaria, Fuchsia, Verbena, etc., induces excessive variability; hence the appearance of quite new characters is probable. M. Carriere (22/38. Abstracted in 'Gardener's Chronicle' 1860 page 1081.) has lately discussed this subject: he states that Erythrina cristagalli had been multiplied by seed for many years, but had not yielded any varieties: it was then crossed with the allied E. herbacea, and "the resistance was now overcome, and varieties were produced with flowers of extremely different size, form, and colour."

From the general and apparently well-founded belief that the crossing of distinct species, besides commingling their characters, adds greatly to their variability, it has probably arisen that some botanists have gone so far as to maintain (22/39. This was the opinion of the elder De Candolle, as quoted in 'Dic. Class. d'Hist. Nat.' tome 8 page 405. Puvis in his work 'De la Degeneration' 1837 page 37, has discussed this same point.) that, when a genus includes only a single species, this when cultivated never varies. The proposition made so broadly cannot be admitted; but it is probably true that the variability of monotypic genera when cultivated is generally less than that of genera including numerous species, and this quite independently of the effects of crossing. I have shown in my 'Origin of Species' that the species belonging to small genera generally yield a less number of varieties in a state of nature than those belonging to large genera. Hence the species of small genera would, it is probable, produce fewer varieties under cultivation than the already variable species of larger genera.

Although we have not at present sufficient evidence that the crossing of species, which have never been cultivated, leads to the appearance of new characters, this apparently does occur with species which have been already rendered in some degree variable through cultivation. Hence crossing, like any other change in the conditions of life, seems to be an element, probably a potent one, in causing variability. But we seldom have the means of distinguishing, as previously remarked, between the appearance of really new characters and the reappearance of long-lost characters, evoked through the act of crossing. I will give an instance of the difficulty in distinguishing such cases. The species of Datura may be divided into two sections, those having white flowers with green stems, and those having purple flowers with brown stems: now Naudin (22/40. 'Comptes Rendus' Novembre 21, 1864 page 838.) crossed Datura laevis and ferox, both of which belong to the white section, and raised from them 205 hybrids. Of these hybrids, every one had brown stems and bore purple flowers; so that they resembled the species of the other section of the genus, and not their own two parents. Naudin was so much astonished at this fact, that he was led carefully to observe both parent- species, and he discovered that the pure seedlings of D. ferox, immediately after germination, had dark purple stems, extending from the young roots up to the cotyledons, and that this tint remained ever afterwards as a ring round the base of the stem of the plant when old. Now I have shown in the thirteenth chapter that the retention or exaggeration of an early character is so intimately related to reversion, that it evidently comes under the same principle. Hence probably we ought to look at the purple flowers and brown stems of these hybrids, not as new characters due to variability, but as a return to the former state of some ancient progenitor.

Independently of the appearance of new characters from crossing, a few words may be added to what has been said in former chapters on the unequal combination and transmission of the characters proper to the two parent-forms. When two species or races are crossed, the offspring of the first generation are generally uniform, but those subsequently produced display an almost infinite diversity of character. He who wishes, says Kolreuter (22/41. 'Nova Acta, St. Petersburg' 1794 page 391.), to obtain an endless number of varieties from hybrids should cross and recross them. There is also much variability when hybrids or mongrels are reduced or absorbed by repeated crosses with either pure parent-form: and a still higher degree of variability when three distinct species, and most of all when four species, are blended together by successive crosses. Beyond this point Gartner (22/42. 'Bastarderzeugung' s. 507, 516, 572.), on whose authority the foregoing statements are made, never succeeded in effecting a union; but Max Wichura (22/43. 'Die Bastardbefruchtung' etc. 1865 s. 24.) united six distinct species of willows into a single hybrid. The sex of the parent species affects in an inexplicable manner the degree of variability of hybrids; for Gartner (22/44. 'Bastarderzeugung' s. 452, 507.) repeatedly found that when a hybrid was used as a father and either one of the pure parent-species, or a third species, was used as the mother, the offspring were more variable than when the same hybrid was used as the mother, and either pure parent or the same third species as the father: thus seedlings from Dianthus barbatus crossed by the hybrid D. chinensi-barbatus were more variable than those raised from this latter hybrid fertilised by the pure D. barbatus. Max Wichura (22/45. 'Die Bastardbefruchtung' s. 56.) insists strongly on an analogous result with his hybrid willows. Again Gartner (22/46. 'Bastarderzeugung' s. 423.) asserts that the degree of variability sometimes differs in hybrids raised from reciprocal crosses between the same two species; and here the sole difference is, that the one species is first used as the father and then as the mother. On the whole we see that, independently of the appearance of new characters, the variability of successive crossed generations is extremely complex, partly from the offspring partaking unequally of the characters of the two parent-forms, and more especially from their unequal tendency to revert to such characters or to those of more ancient progenitors.]

## ON THE MANNER AND ON THE PERIOD OF ACTION OF THE CAUSES WHICH INDUCE VARIABILITY.

This is an extremely obscure subject, and we need here only consider, whether inherited variations are due to certain parts being acted on after they have been formed, or through the reproductive system being affected before their formation; and in the former case at what period of growth or development the effect is produced. We shall see in the two following chapters that various agencies, such as an abundant supply of food, exposure to a different climate, increased use or disuse of parts, etc., prolonged during several generations, certainly modify either the whole organisation or certain organs; and it is clear at least in the case of bud-variation that the action cannot have been through the reproductive system.

[With respect to the part which the reproductive system takes in causing variability, we have seen in the eighteenth chapter that even slight changes in the conditions of life have a remarkable power in causing a greater or less degree of sterility. Hence it seems not improbable that beings generated through a system so easily affected should themselves be affected, or should fail to inherit, or inherit in excess, characters proper to their parents. We know that certain groups of organic beings, but with exceptions in each group, have their reproductive systems much more easily affected by changed conditions than other groups; for instance, carnivorous birds, more readily than carnivorous mammals, and parrots more readily than pigeons; and this fact harmonises with the apparently capricious manner and degree in which various groups of animals and plants vary under domestication.

Kolreuter (22/47. 'Dritte Fortsetzung' etc. 1766 s. 85.) was struck with the parallelism between the excessive variability of hybrids when crossed and recrossed in various ways,—these hybrids having their reproductive powers more or less affected,—and the variability of anciently cultivated plants. Max Wichura (22/48. 'Die Bastardbefruchtung' etc. 1865 s. 92: see also the Rev. M.J. Berkeley on the same subject in 'Journal of Royal Hort. Soc.' 1866 page 80.) has gone one step farther, and shows that with many of our highly cultivated plants, such as the hyacinth, tulip, auricula, snapdragon, potato, cabbage, etc., which there is no reason to believe have been hybridised, the anthers contain many irregular pollen-grains in the same state as in hybrids. He finds also in certain wild forms, the same coincidence between the state of the pollen and a high degree of variability, as in many species of Rubus; but in R. caesius and idaeus, which are not highly variable species, the pollen is sound. It is also notorious that many cultivated plants, such as the banana, pineapple, bread-fruit, and others previously mentioned, have their reproductive organs so seriously affected as to be generally quite sterile; and when they do yield seed, the seedlings, judging from the large number of cultivated races which exist, must be variable in an extreme degree. These facts indicate that there is some relation between the state of the reproductive organs and a tendency to variability; but we must not conclude that the relation is strict. Although many of our highly cultivated plants may have their pollen in a deteriorated condition, yet, as we have previously seen, they yield more seeds, and our anciently domesticated animals are more prolific, than the corresponding species in a state of nature. The peacock is almost the only bird which is believed to be less fertile under domestication than in its native state, and it has varied in a remarkably small degree. From these considerations it would seem that changes in the conditions of life lead either to sterility or to variability, or to both; and not that sterility induces variability. On the whole it is probable that any cause affecting the organs of reproduction would likewise affect their product,—that is, the offspring thus generated.

The period of life at which the causes that induce variability act, is likewise an obscure subject, which has been discussed by various authors. (22/49. Dr. P. Lucas has given a history of opinion on this subject 'Hered. Nat.' 1847 tome 1 page 175.) In some of the cases, to be given in the following chapter, of modifications from the direct action of changed conditions, which are inherited, there can be no doubt that the causes have acted on the mature or nearly mature animal. On the other hand, monstrosities, which cannot be distinctly separated from lesser variations, are often caused by the embryo being injured whilst in the mother's womb or in the egg. Thus I. Geoffroy Saint-Hilaire (22/50. 'Hist. des Anomalies' tome 3 page 499.) asserts that poor women who work hard during their pregnancy, and the mothers of illegitimate children troubled in their minds and forced to conceal their state, are far more liable to give birth to monsters than women in easy circumstances. The eggs of the fowl when placed upright or otherwise treated unnaturally frequently produce monstrous chickens. It would, however, appear that complex monstrosities are induced more frequently during a rather late than during a very early period of embryonic life; but this may partly result from some one part, which has been injured during an early period, affecting by its abnormal growth other parts subsequently developed; and this would be less likely to occur with parts injured at a later period. (22/51. Ibid tome 3 pages 392, 502. The several memoirs by M. Dareste hereafter referred to are of special value on this whole subject.) When any part or organ becomes monstrous through abortion, a rudiment is generally left, and this likewise indicates that its development had already commenced.

Insects sometimes have their antennae or legs in a monstrous condition, the larvae of which do not possess either antennae or legs; and in these cases, as Quatrefages (22/52. See his interesting work 'Metamorphoses de l'Homme' etc. 1862 page 129.) believes, we are enabled to see the precise period at which the normal progress of development was troubled. But the nature of the food given to a caterpillar sometimes affects the colours of the moth, without the caterpillar itself being affected; therefore it seems possible that other characters in the mature insect might be indirectly modified through the larvae. There is no reason to suppose that organs which have been rendered monstrous have always been acted on during their development; the cause may have acted on the organisation at a much earlier stage. It is even probable that either the male or female sexual elements, or both, before their union, may be affected in such a manner as to lead to modifications in organs developed at a late period of life; in nearly the same manner as a child may inherit from his father a disease which does not appear until old age.

In accordance with the facts above given, which prove that in many cases a close relation exists between variability and the sterility following from changed conditions, we may conclude that the exciting cause often acts at the earliest possible period, namely, on the sexual elements, before impregnation has taken place. That an affection of the female sexual element may induce variability we may likewise infer as probable from the occurrence of bud-variations; for a bud seems to be the analogue of an ovule. But the male element is apparently much oftener affected by changed conditions, at least in a visible manner, than the female element or ovule and we know from Gartner's and Wichura's statements that a hybrid used as the father and crossed with a pure species gives a greater degree of variability to the offspring, than does the same hybrid when used as the mother. Lastly, it is certain that variability may be transmitted through either sexual element, whether or not originally excited in them, for Kolreuter and Gartner (22/53. 'Dritte Fortsetzung' etc. s. 123; 'Bastarderzeugung' s. 249.) found that when two species were crossed, if either one was variable, the offspring were rendered variable.]

## SUMMARY.

From the facts given in this chapter, we may conclude that the variability of organic beings under domestication, although so general, is not an inevitable contingent on life, but results from the conditions to which the parents have been exposed. Changes of any kind in the conditions of life, even extremely slight changes, often suffice to cause variability. Excess of nutriment is perhaps the most efficient single exciting cause. Animals and plants continue to be variable for an immense period after their first domestication; but the conditions to which they are exposed never long remain quite constant. In the course of time they can be habituated to certain changes, so as to become less variable; and it is possible that when first domesticated they may have been even more variable than at present. There is good evidence that the power of changed conditions accumulates; so that two, three, or more generations must be exposed to new conditions before any effect is visible. The crossing of distinct forms, which have already become variable, increases in the offspring the tendency to further variability, by the unequal commingling of the characters of the two parents, by the reappearance of long-lost characters, and by the appearance of absolutely new characters. Some variations are induced by the direct action of the surrounding conditions on the whole organisation, or on certain parts alone; other variations appear to be induced indirectly through the reproductive system being affected, as we know is often the case with various beings, which when removed from their natural conditions become sterile. The causes which induce variability act on the mature organism, on the embryo, and, probably, on the sexual elements before impregnation has been effected.

# CHAPTER 2.XXIII.

## DIRECT AND DEFINITE ACTION OF THE EXTERNAL CONDITIONS OF LIFE.

SLIGHT MODIFICATIONS IN PLANTS FROM THE DEFINITE ACTION OF CHANGED CONDITIONS, IN SIZE, COLOUR, CHEMICAL PROPERTIES, AND IN THE STATE OF THE TISSUES. LOCAL DISEASES. CONSPICUOUS MODIFICATIONS FROM CHANGED CLIMATE OR FOOD, ETC. PLUMAGE OF BIRDS AFFECTED BY PECULIAR NUTRIMENT, AND BY THE INOCULATION OF POISON. LAND-SHELLS. MODIFICATIONS OF ORGANIC BEINGS IN A STATE OF NATURE THROUGH THE DEFINITE ACTION OF EXTERNAL CONDITIONS. COMPARISON OF AMERICAN AND EUROPEAN TREES. GALLS. EFFECTS OF PARASITIC FUNGI. CONSIDERATIONS OPPOSED TO THE BELIEF IN THE POTENT INFLUENCE OF CHANGED EXTERNAL CONDITIONS. PARALLEL SERIES OF VARIETIES. AMOUNT OF VARIATION DOES NOT CORRESPOND WITH THE DEGREE OF CHANGE IN THE CONDITIONS. BUD-VARIATION. MONSTROSITIES PRODUCED BY UNNATURAL TREATMENT. SUMMARY.

If we ask ourselves why this or that character has been modified under domestication, we are, in most cases, lost in utter darkness. Many naturalists, especially of the French school, attribute every modification to the "monde ambiant," that is, to changed climate, with all its diversities of heat and cold, dampness and dryness, light and electricity, to the nature of the soil, and to varied kinds and amount of food. By the term definite action, as used in this chapter, I mean an action of such a nature that, when many individuals of the same variety are exposed during several generations to any particular change in their conditions of life, all, or nearly all the individuals, are modified in the same manner. The effects of habit, or of the increased use and disuse of various organs, might have been included under this head; but it will be convenient to discuss this subject in a separate chapter. By the term indefinite action I mean an action which causes one individual to vary in one way and another individual in another way, as we often see with plants and animals after they have been subjected for some generations to changed conditions of life. But we know far too little of the causes and laws of variation to make a sound classification. The action of changed conditions, whether leading to definite or indefinite results, is a totally distinct consideration from the effects of selection; for selection depends on the preservation by man of certain individuals, or on their survival under various and complex natural circumstances, and has no relation whatever to the primary cause of each particular variation.

I will first give in detail all the facts which I have been able to collect, rendering it probable that climate, food, etc., have acted so definitely and powerfully on the organisation of our domesticated productions, that new sub- varieties or races have been thus formed without the aid of selection by man or nature. I will then give the facts and considerations opposed to this conclusion, and finally we will weigh, as fairly as we can, the evidence on both sides.

When we reflect that distinct races of almost all our domesticated animals exist in each kingdom of Europe, and formerly even in each district of England, we are at first strongly inclined to attribute their origin to the definite action of the physical conditions of each country; and this has been the conclusion of many authors. But we should bear in mind that man annually has to choose which animals shall be preserved for breeding, and which shall be slaughtered. We have also seen that both methodical and unconscious selection were formerly practised, and are now occasionally practised by the most barbarous races, to a much greater extent than might have been anticipated. Hence it is difficult to judge how far differences in the conditions between, for instance, the several districts in England, have sufficed to modify the breeds which have been reared in each. It may be argued that, as numerous wild animals and plants have ranged during many ages throughout Great Britain, and still retain the same character, the difference in conditions between the several districts could not have modified in a marked manner the various native races of cattle, sheep, pigs, and horses. The same difficulty of distinguishing between the effects of natural selection and the definite action of external conditions is encountered in a still higher degree when we compare closely allied species inhabiting two countries, such as North America and Europe, which do not differ greatly in climate, nature of soil, etc., for in this case natural selection will inevitably and rigorously have acted during a long succession of ages.

Prof. Weismann has suggested (23/1. 'Ueber den Einfluss der Isolirung auf die Artbildung' 1872.) that when a variable species enters a new and isolated country, although the variations may be of the same general nature as before, yet it is improbable that they should occur in the same proportional numbers. After a longer or shorter period, the species will tend to become nearly uniform in character from the incessant crossing of the varying individuals; but owing to the proportion of the individuals varying in different ways not being the same in the two cases, the final result will be the production of two forms somewhat different from one another. In cases of this kind it would falsely appear as if the conditions had induced certain definite modifications, whereas they had only excited indefinite variability, but with the variations in slightly different proportional numbers. This view may throw some light on the fact that the domestic animals which formerly inhabited the several districts in Great Britain, and the half wild cattle lately kept in several British parks, differed slightly from one another; for these animals were prevented from wandering over the whole country and intercrossing, but would have crossed freely within each district or park.

[From the difficulty of judging how far changed conditions have caused definite modifications of structure, it will be advisable to give as large a body of facts as possible, showing that extremely slight differences within the same country, or during different seasons, certainly produce an appreciable effect, at least on varieties which are already in an unstable condition. Ornamental flowers are good for this purpose, as they are highly variable, and are carefully observed. All floriculturists are unanimous that certain varieties are affected by very slight differences in the nature of the artificial compost in which they are grown, and by the natural soil of the district, as well as by the season. Thus, a skilful judge, in writing on Carnations and Picotees (23/2. 'Gardener's Chronicle' 1853 page 183.) asks "where can Admiral Curzon be seen possessing the colour, size, and strength which it has in Derbyshire? Where can Flora's Garland be found equal to those at Slough? Where do high-coloured flowers revel better than at Woolwich and Birmingham? Yet in no two of these districts do the same varieties attain an equal degree of excellence, although each may be receiving the attention of the most skilful cultivators." The same writer then recommends every cultivator to keep five different kinds of soil and manure, "and to endeavour to suit the respective appetites of the plants you are dealing with, for without such attention all hope of general success will be vain." So it is with the Dahlia (23/3. Mr. Wildman 'Floricultural Soc.' February 7, 1843 reported in 'Gardener's Chronicle' 1843 page 86.): the Lady Cooper rarely succeeds near London, but does admirably in other districts; the reverse holds good with other varieties; and again, there are others which succeed equally well in various situations. A skilful gardener (23/4. Mr. Robson in 'Journal of Horticulture' February 13, 1866 page 122.) states that he procured cuttings of an old and well-known variety (pulchella) of Verbena, which from having been propagated in a different situation presented a slightly different shade of colour; the two varieties were afterwards multiplied by cuttings, being carefully kept distinct; but in the second year they could hardly be distinguished, and in the third year no one could distinguish them.

The nature of the season has an especial influence on certain varieties of the Dahlia: in 1841 two varieties were pre-eminently good, and the next year these same two were pre-eminently bad. A famous amateur (23/5. 'Journal of Horticulture' 1861 page 24.) asserts that in 1861 many varieties of the Rose came so untrue in character, "that it was hardly possible to recognise them, and the thought was not seldom entertained that the grower had lost his tally." The same amateur (23/6. Ibid 1862 page 83.) states that in 1862 two-thirds of his Auriculas produced central trusses of flowers, and such trusses are liable not to keep true; and he adds that in some seasons certain varieties of this plant all prove good, and the next season all prove bad; whilst exactly the reverse happens with other varieties. In 1845 the editor of the 'Gardener's Chronicle' (23/7. 'Gardener's Chronicle' 1845 page 660.) remarked how singular it was that this year many Calceolarias tended to assume a tubular form. With Heartsease (23/8. Ibid 1863 page 628.) the blotched sorts do not acquire their proper character until hot weather sets in; whilst other varieties lose their beautiful marks as soon as this occurs.

Analogous facts have been observed with leaves: Mr. Beaton asserts (23/9. 'Journal of Hort.' 1861 pages 64, 309.) that he raised at Shrubland, during six years, twenty thousand seedlings from the Punch Pelargonium, and not one had variegated leaves; but at Surbiton, in Surrey, one-third, or even a greater proportion, of the seedlings from this same variety were more or less variegated. The soil of another district in Surrey has a strong tendency to cause variegation, as appears from information given me by Sir F. Pollock. Verlot (23/10. 'Des Varietes' etc. page 76.) states that the variegated strawberry retains its character as long as grown in a dryish soil, but soon loses it when planted in fresh and humid soil. Mr. Salter, who is well known for his success in cultivating variegated plants, informs me that rows of strawberries were planted in his garden in 1859, in the usual way; and at various distances in one row, several plants simultaneously became variegated; and what made the case more extraordinary, all were variegated in precisely the same manner. These plants were removed, but during the three succeeding years other plants in the same row became variegated, and in no instance were the plants in any adjoining row affected.

The chemical qualities, odours, and tissues of plants are often modified by a change which seems to us slight. The Hemlock is said not to yield conicine in Scotland. The root of the Aconitum napellus becomes innocuous in frigid climates. The medicinal properties of the Digitalis are easily affected by culture. As the Pistacia lentiscus grows abundantly in the South of France, the climate must suit it, but it yields no mastic. The Laurus sassafras in Europe loses the odour proper to it in North America. (23/11. Engel 'Sur les Prop. Medicales des Plantes' 1860 pages 10, 25. On changes in the odours of plants see Dalibert's Experiments quoted by Beckman 'Inventions' volume 2 page 344; and Nees in Ferussac 'Bull. des Sc. Nat.' 1824 tome 1 page 60. With respect to the rhubarb etc. see also 'Gardener's Chronicle' 1849 page 355; 1862 page 1123.) Many similar facts could be given, and they are remarkable because it might have been thought that definite chemical compounds would have been little liable to change either in quality or quantity.

The wood of the American Locust-tree (Robinia) when grown in England is nearly worthless, as is that of the Oak-tree when grown at the Cape of Good Hope. (23/12. Hooker 'Flora Indica' page 32.) Hemp and flax, as I hear from Dr. Falconer, flourish and yield plenty of seed on the plains of India, but their fibres are brittle and useless. Hemp, on the other hand, fails to produce in England that resinous matter which is so largely used in India as an intoxicating drug.

The fruit of the Melon is greatly influenced by slight differences in culture and climate. Hence it is generally a better plan, according to Naudin, to improve an old kind than to introduce a new one into any locality. The seed of the Persian Melon produces near Paris fruit inferior to the poorest market kinds, but at Bordeaux yields delicious fruit. (23/13. Naudin 'Annales des Sc. Nat.' 4th series, Bot. tome 11 1859 page 81. 'Gardener's Chronicle' 1859 page 464.) Seed is annually brought from Thibet to Kashmir (23/14. Moorcroft 'Travels' etc. volume 2 page 143.) and produces fruit weighing from four to ten pounds, but plants raised next year from seed saved in Kashmir give fruit weighing only from two to three pounds. It is well known that American varieties of the Apple produce in their native land magnificent and brightly-coloured fruit, but these in England are of poor quality and a dull colour. In Hungary there are many varieties of the kidney-bean, remarkable for the beauty of their seeds, but the Rev. M.J. Berkeley (23/15. 'Gardener's Chronicle' 1861 page 1113.) found that their beauty could hardly ever be preserved in England, and in some cases the colour was greatly changed. We have seen in the ninth chapter, with respect to wheat, what a remarkable effect transportal from the north to the south of France, and conversely, produced on the weight of the grain.]

When man can perceive no change in plants or animals which have been exposed to a new climate or to different treatment, insects can sometimes perceive a marked change. A cactus has been imported into India from Canton, Manilla Mauritius, and from the hot-houses of Kew, and there is likewise a so-called native kind which was formerly introduced from South America; all these plants belong to the same species and are alike in appearance, but the cochineal insect flourishes only on the native kind, on which it thrives prodigiously. (23/16. Royle 'Productive Resources of India' page 59.) Humboldt remarks (23/17. 'Personal Narrative' English translation volume 5 page 101. This statement has been confirmed by Karsten 'Beitrag zur Kenntniss der Rhynchoprion' Moscow 1864 s. 39 and by others.) that white men "born in the torrid zone walk barefoot with impunity in the same apartment where a European, recently landed, is exposed to the attacks of the Pulex penetrans." This insect, the too well-known chigoe, must therefore be able to perceive what the most delicate chemical analysis fails to discover, namely, a difference between the blood or tissues of a European and those of a white man born in the tropics. But the discernment of the chigoe is not so surprising as it at first appears; for according to Liebig (23/18. 'Organic Chemistry' English translation 1st edition page 369.) the blood of men with different complexions, though inhabiting the same country, emits a different odour.

[Diseases peculiar to certain localities, heights, or climates, may be here briefly noticed, as showing the influence of external circumstances on the human body. Diseases confined to certain races of man do not concern us, for the constitution of the race may play the more important part, and this may have been determined by unknown causes. The Plica Polonica stands, in this respect, in a nearly intermediate position; for it rarely affects Germans, who inhabit the neighbourhood of the Vistula, where so many Poles are grievously affected; neither does it affect Russians, who are said to belong to the same original stock as the Poles. (23/19. Prichard 'Phys. Hist. of Mankind' 1851 volume 1 page 155.) The elevation of a district often governs the appearance of diseases; in Mexico the yellow fever does not extend above 924 metres; and in Peru, people are affected with the verugas only between 600 and 1600 metres above the sea; many other such cases could be given. A peculiar cutaneous complaint, called the Bouton d'Alep, affects in Aleppo and some neighbouring districts almost every native infant, and some few strangers; and it seems fairly well established that this singular complaint depends on drinking certain waters. In the healthy little island of St. Helena the scarlet-fever is dreaded like the Plague; analogous facts have been observed in Chili and Mexico. (23/20. Darwin 'Journal of Researches' 1845 page 434.) Even in the different departments of France it is found that the various infirmities which render the conscript unfit for serving in the army, prevail with remarkable inequality, revealing, as Boudin observes, that many of them are endemic, which otherwise would never have been suspected. (23/21. These statements on disease are taken from Dr. Boudin 'Geographie et Statistique Medicale' 1857 tome 1 pages 44 and 52; tome 2 page 315.) Any one who will study the distribution of disease will be struck with surprise at what slight differences in the surrounding circumstances govern the nature and severity of the complaints by which man is at least temporarily affected.

The modifications as yet referred to are extremely slight, and in most cases have been caused, as far as we can judge, by equally slight differences in the conditions. But such conditions acting during a series of generations would perhaps produce a marked effect.

With plants, a considerable change of climate sometimes produces a conspicuous result. I have given in the ninth chapter the most remarkable case known to me, namely, that of varieties of maize, which were greatly modified in the course of only two or three generations when taken from a tropical country to a cooler one, or conversely. Dr. Falconer informs me that he has seen the English Ribston-pippin apple, a Himalayan oak, Prunus and Pyrus, all assume in the hotter parts of India a fastigiate or pyramidal habit; and this fact is the more interesting, as a Chinese tropical species of Pyrus naturally grows thus. Although in these cases the changed manner of growth seems to have been directly caused by the great heat, we know that many fastigiate trees have originated in their temperate homes. In the Botanic Gardens of Ceylon the apple-tree (23/22. 'Ceylon' by Sir J.E. Tennent volume 1 1859 page 89.) "sends out numerous runners under ground, which continually rise into small stems, and form a growth around the parent-tree.) The varieties of the cabbage which produce heads in Europe fail to do so in certain tropical countries (23/23. Godron 'De l'Espece' tome 2 page 52.) The Rhododendron ciliatum produced at Kew flowers so much larger and paler-coloured than those which it bears on its native Himalayan mountain, that Dr. Hooker (23/24. 'Journal of Horticultural Soc.' volume 7 1852 page 117.) would hardly have recognised the species by the flowers alone. Many similar facts with respect to the colour and size of flowers could be given.

The experiments of Vilmorin and Buckman on carrots and parsnips prove that abundant nutriment produces a definite and inheritable effect on the roots, with scarcely any change in other parts of the plant. Alum directly influences the colour of the flowers of the Hydrangea. (23/25. 'Journal of Hort. Soc.' volume 1 page 160.) Dryness seems generally to favour the hairiness or villosity of plants. Gartner found that hybrid Verbascums became extremely woolly when grown in pots. Mr. Masters, on the other hand, states that the Opuntia leucotricha "is well clothed with beautiful white hairs when grown in a damp heat, but in a dry heat exhibits none of this peculiarity." (23/26. See Lecoq on the Villosity of Plants 'Geograph. Bot.' tome 3 pages 287, 291; Gartner 'Bastarderz.' s. 261; Mr. Masters on the Opuntia in 'Gardener's Chronicle' 1846 page 444.) Slight variations of many kinds, not worth specifying in detail, are retained only as long as plants are grown in certain soils, of which Sageret (23/27. 'Pom. Phys.' page 136.) gives some instances from his own experience. Odart, who insists strongly on the permanence of the varieties of the grape, admits (23/28. 'Ampelographie' 1849 page 19.) that some varieties, when grown under a different climate or treated differently, vary in a slight degree, as in the tint of the fruit and in the period of ripening. Some authors have denied that grafting causes even the slightest difference in the scion; but there is sufficient evidence that the fruit is sometimes slightly affected in size and flavour, the leaves in duration, and the flowers in appearance. (23/29. Gartner 'Bastarderz.' s. 606, has collected nearly all recorded facts. Andrew Knight in 'Transact. Hort. Soc.' volume 2 page 160, goes so far as to maintain that few varieties are absolutely permanent in character when propagated by buds or grafts.)

There can be no doubt, from the facts given in the first chapter, that European dogs deteriorate in India, not only in their instincts but in structure; but the changes which they undergo are of such a nature, that they may be partly due to reversion to a primitive form, as in the case of feral animals. In parts of India the turkey becomes reduced in size, "with the pendulous appendage over the beak enormously developed." (23/30. Mr. Blyth 'Annals and Mag of Nat. Hist.' volume 20 1847 page 391.) We have seen how soon the wild duck, when domesticated, loses its true character, from the effects of abundant or changed food, or from taking little exercise. From the direct action of a humid climate and poor pasture the horse rapidly decreases in size in the Falkland Islands. From information which I have received, this seems likewise to be the case to a certain extent with sheep in Australia.

Climate definitely influences the hairy covering of animals; in the West Indies a great change is produced in the fleece of sheep, in about three generations. Dr. Falconer states (23/31. 'Natural History Review' 1862 page 113.) that the Thibet mastiff and goat, when brought down from the Himalaya to Kashmir, lose their fine wool. At Angora not only goats, but shepherd-dogs and cats, have fine fleecy hair, and Mr. Ainsworth (23/32. 'Journal of Roy. Geographical Soc.' volume 9 1839 page 275.) attributes the thickness of the fleece to the severe winters, and its silky lustre to the hot summers. Burnes states positively (23/33. 'Travels in Bokhara' volume 3 page 151.) that the Karakool sheep lose their peculiar black curled fleeces when removed into any other country. Even within the limits of England, I have been assured that the wool of two breeds of sheep was slightly changed by the flocks being pastured in different localities. (23/34. See also on the influence of marshy pastures on the wool Godron 'L'Espece' tome 2 page 22.) It has been asserted on good authority (23/35. Isidore Geoffroy Saint-Hilaire 'Hist. Nat. Gen.' tome 3 page 438.) that horses kept during several years in the deep coal-mines of Belgium become covered with velvety hair, almost like that on the mole. These cases probably stand in close relation to the natural change of coat in winter and summer. Naked varieties of several domestic animals have occasionally appeared; but there is no reason to believe that this is in any way related to the nature of the climate to which they have been exposed. (23/36. Azara has made some good remarks on this subject 'Quadrupedes du Paraguay' tome 2 page 337. See an account of a family of naked mice produced in England 'Proc. Zoolog. Soc.' 1856 page 38.)

It appears at first sight probable that the increased size, the tendency to fatten, the early maturity and altered forms of our improved cattle, sheep, and pigs, have directly resulted from their abundant supply of food. This is the opinion of many competent judges, and probably is to a great extent true. But as far as form is concerned, we must not overlook the more potent influence of lessened use on the limbs and lungs. We see, moreover, as far as size is concerned, that selection is apparently a more powerful agent than a large supply of food, for we can thus only account for the existence, as remarked to me by Mr. Blyth, of the largest and smallest breeds of sheep in the same country, of Cochin-China fowls and Bantams, of small Tumbler and large Runt pigeons, all kept together and supplied with abundant nourishment. Nevertheless there can be little doubt that our domesticated animals have been modified, independently of the increased or lessened use of parts, by the conditions to which they have been subjected, without the aid of selection. For instance, Prof. Rutimeyer (23/37. 'Die Fauna der Pfahlbauten' 1861 s. 15.) shows that the bones of domesticated quadrupeds can be distinguished from those of wild animals by the state of their surface and general appearance. It is scarcely possible to read Nathusius's excellent 'Vorstudien' (23/38. 'Schweineschadel' 1864 s. 99.) and doubt that, with the highly improved races of the pig, abundant food has produced a conspicuous effect on the general form of the body, on the breadth of the head and face, and even on the teeth. Nathusius rests much on the case of a purely bred Berkshire pig, which when two months old became diseased in its digestive organs, and was preserved for observation until nineteen months old; at this age it had lost several characteristic features of the breed, and had acquired a long, narrow head, of large size relatively to its small body, and elongated legs. But in this case and in some others we ought not to assume that, because certain characters are lost, perhaps through reversion, under one course of treatment, therefore that they were at first directly produced by an opposite treatment.

In the case of the rabbit, which has become feral on the island of Porto Santo, we are at first strongly tempted to attribute the whole change—the greatly reduced size, the altered tints of the fur, and the loss of certain characteristic marks—to the definite action of the new conditions to which it has been exposed. But in all such cases we have to consider in addition the tendency to reversion to progenitors more or less remote, and the natural selection of the finest shades of difference.

The nature of the food sometimes either definitely induces certain peculiarities, or stands in some close relation with them. Pallas long ago asserted that the fat-tailed sheep of Siberia degenerate and lose their enormous tails when removed from certain saline pastures; and recently Erman (23/39. 'Travels in Siberia' English translation volume 1 page 228.) states that this occurs with the Kirgisian sheep when brought to Orenburgh.

It is well known that hemp-seed causes bullfinches and certain other birds to become black. Mr. Wallace has communicated to me some much more remarkable facts of the same nature. The natives of the Amazonian region feed the common green parrot (Chrysotis festiva, Linn.) with the fat of large Siluroid fishes, and the birds thus treated become beautifully variegated with red and yellow feathers. In the Malayan archipelago, the natives of Gilolo alter in an analogous manner the colours of another parrot, namely, the Lorius garrulus, Linn., and thus produce the Lori rajah or King-Lory. These parrots in the Malay Islands and South America, when fed by the natives on natural vegetable food, such as rice and plaintains, retain their proper colours. Mr. Wallace has, also, recorded (23/40. A.R. Wallace 'Travels on the Amazon and Rio Negro' page 294.) a still more singular fact. "The Indians (of S. America) have a curious art by which they change the colours of the feathers of many birds. They pluck out those from the part they wish to paint, and inoculate the fresh wound with the milky secretion from the skin of a small toad. The feathers grow of a brilliant yellow colour, and on being plucked out, it is said, grow again of the same colour without any fresh operation."

Bechstein (23/41. 'Naturgeschichte der Stubenvogel' 1840 s. 262, 308.) does not entertain any doubt that seclusion from light affects, at least temporarily, the colours of cage-birds.

It is well known that the shells of land-mollusca are affected by the abundance of lime in different districts. Isidore Geoffroy Saint-Hilaire (23/42. 'Hist. Nat Gen.' tome 3 page 402.) gives the case of Helix lactea, which has recently been carried from Spain to the South of France and to the Rio Plata, and in both countries now presents a distinct appearance, but whether this has resulted from food or climate is not known. With respect to the common oyster, Mr. F. Buckland informs me that he can generally distinguish the shells from different districts; young oysters brought from Wales and laid down in beds where "natives" are indigenous, in the short space of two months begin to assume the "native" character. M. Costa (23/43. 'Bull. de La Soc. Imp. d'Acclimat.' tome 8 page 351.) has recorded a much more remarkable case of the same nature, namely, that young shells taken from the shores of England and placed in the Mediterranean, at once altered their manner of growth and formed prominent diverging rays, like those on the shells of the proper Mediterranean oyster. The same individual shell, showing both forms of growth, was exhibited before a society in Paris. Lastly, it is well known that caterpillars fed on different food sometimes either themselves acquire a different colour or produce moths differing in colour. (23/44. See an account of Mr. Gregson's experiments on the Abraxus grossulariata 'Proc. Entomolog. Soc.' January 6, 1862: these experiments have been confirmed by Mr. Greening in 'Proc. of the Northern Entomolog. Soc.' July 28, 1862. For the effects of food on caterpillars see a curious account by M. Michely in 'Bull. De La Soc. Imp. d'Acclimat.' tome 8 page 563. For analogous facts from Dahlbom on Hymenoptera see Westwood 'Modern Class. of Insects' volume 2 page 98. See also Dr. L. Moller 'Die Abhangigkeit der Insecten' 1867 s. 70.)

It would be travelling beyond my proper limits here to discuss how far organic beings in a state of nature are definitely modified by changed conditions. In my 'Origin of Species' I have given a brief abstract of the facts bearing on this point, and have shown the influence of light on the colours of birds, and of residence near the sea on the lurid tints of insects, and on the succulency of plants. Mr. Herbert Spencer (23/45. 'The Principles of Biology' volume 2 1866. The present chapters were written before I had read Mr. Herbert Spencer's work, so that I have not been able to make so much use of it as I should otherwise probably have done.) has recently discussed with much ability this whole subject on general grounds. He argues, for instance, that with all animals the external and internal tissues are differently acted on by the surrounding conditions, and they invariably differ in intimate structure. So again the upper and lower surfaces of true leaves, as well as of stems and petioles, when these assume the function and occupy the position of leaves, are differently circumstanced with respect to light, etc., and apparently in consequence differ in structure. But, as Mr. Herbert Spencer admits, it is most difficult in all such cases to distinguish between the effects of the definite action of physical conditions and the accumulation through natural selection of inherited variations which are serviceable to the organism, and which have arisen independently of the definite action of these conditions.]

Although we are not here concerned with the definite action of the conditions of life on organisms in a state of nature, I may state that much evidence has been gained during the last few years on this subject. In the United States, for instance, it has been clearly proved, more especially by Mr. J.A. Allen, that, with birds, many species differ in tint, size of body and of beak, and in length of tail, in proceeding from the North to the South; and it appears that these differences must be attributed to the direct action of temperature. (23/46. Professor Weismann comes to the same conclusion with respect to certain European butterflies in his valuable essay 'Ueber den Saison- Dimorphismus' 1875. I might also refer to the recent works of several other authors on the present subject; for instance to Kerner's 'Gute und schlechte Arten' 1866.) With respect to plants I will give a somewhat analogous case: Mr. Meehan (23/47. 'Proc. Acad. Nat. Soc. of Philadelphia' January 28, 1862.), has compared twenty-nine kinds of American trees with their nearest European allies, all grown in close proximity and under as nearly as possible the same conditions. In the American species he finds, with the rarest exceptions, that the leaves fall earlier in the season, and assume before their fall a brighter tint; that they are less deeply toothed or serrated; that the buds are smaller; that the trees are more diffuse in growth and have fewer branchlets; and, lastly, that the seeds are smaller—all in comparison with the corresponding European species. Now considering that these corresponding trees belong to several distinct orders, and that they are adapted to widely different stations, it can hardly be supposed that their differences are of any special service to them in the New and Old worlds; and if so such differences cannot have been gained through natural selection, and must be attributed to the long continued action of a different climate.

## GALLS.

Another class of facts, not relating to cultivated plants, deserves attention. I allude to the production of galls. Every one knows the curious, bright-red, hairy productions on the wild rose-tree, and the various different galls produced by the oak. Some of the latter resemble fruit, with one face as rosy as the rosiest apple. These bright colours can be of no service either to the gall-forming insect or to the tree, and probably are the direct result of the action of the light, in the same manner as the apples of Nova Scotia or Canada are brighter coloured than English apples. According to Osten Sacken's latest revision, no less than fifty-eight kinds of galls are produced on the several species of oak, by Cynips with its sub-genera; and Mr. B.D. Walsh (23/48. See Mr. B.D. Walsh's excellent papers in 'Proc. Entomolog. Soc. Philadelphia' December 1866 page 284. With respect to the willow see ibid 1864 page 546.) states that he can add many others to the list. One American species of willow, the Salix humilis, bears ten distinct kinds of galls. The leaves which spring from the galls of various English willows differ completely in shape from the natural leaves. The young shoots of junipers and firs, when punctured by certain insects, yield monstrous growths resembling flowers and fir-cones; and the flowers of some plants become from the same cause wholly changed in appearance. Galls are produced in every quarter of the world; of several sent to me by Mr. Thwaites from Ceylon, some were as symmetrical as a composite flower when in bud, others smooth and spherical like a berry; some protected by long spines, others clothed with yellow wool formed of long cellular hairs, others with regularly tufted hairs. In some galls the internal structure is simple, but in others it is highly complex; thus M. Lacaze-Duthiers (23/49. See his admirable 'Histoire des Galles' in 'Annal. des Sc. Nat. Bot.' 3rd series tome 19 1853 page 273.) has figured in the common ink-gall no less than seven concentric layers, composed of distinct tissue, namely, the epidermic, sub-epidermic, spongy, intermediate, and the hard protective layer formed of curiously thickened woody cells, and, lastly, the central mass, abounding with starch-granules on which the larvae feed.

Galls are produced by insects of various orders, but the greater number by species of Cynips. It is impossible to read M. Lacaze-Duthiers' discussion and doubt that the poisonous secretion of the insect causes the growth of the gall; and every one knows how virulent is the poison secreted by wasps and bees, which belong to the same group with Cynips. Galls grow with extraordinary rapidity, and it is said that they attain their full size in a few days

(23/50. Kirby and Spence 'Entomology' 1818 volume 1 page 450; Lacaze-Duthiers ibid page 284.); it is certain that they are almost completely developed before the larvae are hatched. Considering that many gall-insects are extremely small, the drop of secreted poison must be excessively minute; it probably acts on one or two cells alone, which, being abnormally stimulated, rapidly increase by a process of self-division. Galls, as Mr. Walsh (23/51. 'Proc. Entomolog. Soc. Philadelphia' 1864 page 558.) remarks, afford good, constant, and definite characters, each kind keeping as true to form as does any independent organic being. This fact becomes still more remarkable when we hear that, for instance, seven out of the ten different kinds of galls produced on Salix humilis are formed by gall-gnats (Cecidomyidae) which "though essentially distinct species, yet resemble one another so closely that in almost all cases it is difficult, and in most cases impossible, to distinguish the full-grown insects one from the other." (23/52. Mr. B.D. Walsh ibid page 633 and December 1866 page 275.) For in accordance with a wide-spread analogy we may safely infer that the poison secreted by insects so closely allied would not differ much in nature; yet this slight difference is sufficient to induce widely different results. In some few cases the same species of gall-gnat produces on distinct species of willows galls which cannot be distinguished; the Cynips fecundatrix, also, has been known to produce on the Turkish oak, to which it is not properly attached, exactly the same kind of gall as on the European oak. (23/53. Mr. B.D. Walsh ibid 1864 pages 545, 411, 495; and December 1866 page 278. See also Lacaze-Duthiers.) These latter facts apparently prove that the nature of the poison is a more powerful agent in determining the form of the gall than the specific character of the tree which is acted on.

As the poisonous secretion of insects belonging to various orders has the special power of affecting the growth of various plants; as a slight difference in the nature of the poison suffices to produce widely different results; and lastly, as we know that the chemical compounds secreted by plants are eminently liable to be modified by changed conditions of life, we may believe it possible that various parts of a plant might be modified through the agency of its own altered secretions. Compare, for instance, the mossy and viscid calyx of a moss-rose, which suddenly appears through bud-variation on a Provence-rose, with the gall of red moss growing from the inoculated leaf of a wild rose, with each filament symmetrically branched like a microscopical spruce-fir, bearing a glandular tip and secreting odoriferous gummy matter. (23/54. Lacaze-Duthiers ibid pages 325, 328.) Or compare, on the one hand, the fruit of the peach, with its hairy skin, fleshy covering, hard shell and kernel, and on the other hand one of the more complex galls with its epidermic, spongy, and woody layers, surrounding tissue loaded with starch granules. These normal and abnormal structures manifestly present a certain degree of resemblance. Or, again, reflect on the cases above given of parrots which have had their plumage brightly decorated through some change in their blood, caused by having been fed on certain fishes, or locally inoculated with the poison of a toad. I am far from wishing to maintain that the moss-rose or the hard shell of the peach-stone or the bright colours of birds are actually due to any chemical change in the sap or blood; but these cases of galls and of parrots are excellently adapted to show us how powerfully and singularly external agencies may affect structure. With such facts before us, we need feel no surprise at the appearance of any modification in any organic being.

[I may, also, here allude to the remarkable effects which parasitic fungi sometimes produce on plants. Reissek (23/55. 'Linnaea' volume 17 1843; quoted by Dr. M.T. Masters, Royal Institution, March 16, 1860.) has described a Thesium, affected by an Oecidium, which was greatly modified, and assumed some of the characteristic features of certain allied species, or even genera. Suppose, says Reissek, "the condition originally caused by the fungus to become constant in the course of time, the plant would, if found growing wild, be considered as a distinct species or even as belonging to a new genus." I quote this remark to show how profoundly, yet in how natural a manner, this plant must have been modified by the parasitic fungus. Mr. Meehan (23/56. 'Proc. Acad. Nat. Sc., Philadelphia' June 16, 1874 and July 23, 1875.) also states that three species of Euphorbia and Portulaca olereacea, which naturally grow prostrate, become erect when they are attacked by the Oecidium. Euphorbia maculata in this case also becomes nodose, with the branchlets comparatively smooth and the leaves modified in shape, approaching in these respects to a distinct species, namely, the E. hypericifolia.]

## FACTS AND CONSIDERATIONS OPPOSED TO THE BELIEF THAT THE CONDITIONS OF LIFE ACT IN A POTENT MANNER IN CAUSING DEFINITE MODIFICATIONS OF STRUCTURE.

I have alluded to the slight differences in species naturally living in distinct countries under different conditions; and such differences we feel at first inclined to attribute, probably often with justice, to the definite action of the surrounding conditions. But it must be borne in mind that there exist many animals and plants which range widely and have been exposed to great diversities of climate, yet remain uniform in character. Some authors, as previously remarked, account for the varieties of our culinary and agricultural plants by the definite action of the conditions to which they have been exposed in the different parts of Great Britain; but there are about 200 plants (23/57. Hewett C. Watson 'Cybele Britannica' volume 1 1847 page 11.) which are found in every single English county; and these plants must have been exposed for an immense period to considerable differences of climate and soil, yet do not differ. So, again,, some animals and plants range over a large portion of the world, yet retain the same character.

[Notwithstanding the facts previously given on the occurrence of highly peculiar local diseases and on the strange modifications of structure in plants caused by the inoculated poison of insects, and other analogous cases; still there are a multitude of variations—such as the modified skull of the niata ox and bulldog, the long horns of Caffre cattle, the conjoined toes of the solid-hoofed swine, the immense crest and protuberant skull of Polish fowls, the crop of the pouter-pigeon, and a host of other such cases—which we can hardly attribute to the definite action, in the sense before specified, of the external conditions of life. No doubt in every case there must have been some exciting cause; but as we see innumerable individuals exposed to nearly the same conditions, and one alone is affected, we may conclude that the constitution of the individual is of far higher importance than the conditions to which it has been exposed. It seems, indeed, to be a general rule that conspicuous variations occur rarely, and in one individual alone out of millions, though all may have been exposed, as far as we can judge, to nearly the same conditions. As the most strongly marked variations graduate insensibly into the most trifling, we are led by the same train of thought to attribute each slight variation much more to innate differences of constitution, however caused, than to the definite action of the surrounding conditions.

We are led to the same conclusion by considering the cases, formerly alluded to, of fowls and pigeons, which have varied and will no doubt go on varying in directly opposite ways, though kept during many generations under nearly the same conditions. Some, for instance, are born with their beaks, wings, tails, legs, etc., a little longer, and others with these same parts a little shorter. By the long-continued selection of such slight individual differences which occur in birds kept in the same aviary, widely different races could certainly be formed; and long-continued selection, important as is the result, does nothing but preserve the variations which arise, as it appears to us, spontaneously.

In these cases we see that domesticated animals vary in an indefinite number of particulars, though treated as uniformly as is possible. On the other hand, there are instances of animals and plants, which, though they have been exposed to very different conditions, both under nature and domestication, have varied in nearly the same manner. Mr. Layard informs me that he has observed amongst the Caffres of South Africa a dog singularly like an arctic Esquimaux dog. Pigeons in India present nearly the same wide diversities of colour as in Europe; and I have seen chequered and simply barred pigeons, and pigeons with blue and white loins, from Sierra Leone, Madeira, England, and India. New varieties of flowers are continually raised in different parts of

Great Britain, but many of these are found by the judges at our exhibitions to be almost identical with old varieties. A vast number of new fruit-trees and culinary vegetables have been produced in North America: these differ from European varieties in the same general manner as the several varieties raised in Europe differ from one another; and no one has ever pretended that the climate of America has given to the many American varieties any general character by which they can be recognised. Nevertheless, from the facts previously advanced on the authority of Mr. Meehan with respect to American and European forest-trees it would be rash to affirm that varieties raised in the two countries would not in the course of ages assume a distinctive character. Dr. M. Masters has recorded a striking fact (23/58. 'Gardener's Chronicle' 1857 page 629.) bearing on this subject: he raised numerous plants of Hybiscus syriacus from seed collected in South Carolina and the Holy Land, where the parent-plants must have been exposed to considerably different conditions; yet the seedlings from both localities broke into two similar strains, one with obtuse leaves and purple or crimson flowers, and the other with elongated leaves and more or less pink flowers.

We may, also, infer the prepotent influence of the constitution of the organism over the definite action of the conditions of life, from the several cases given in the earlier chapters of parallel series of varieties,—an important subject, hereafter to be more fully discussed. Sub-varieties of the several kinds of wheat, gourds, peaches, and other plants, and to a limited extent sub-varieties of the fowl, pigeon, and dog, have been shown either to resemble or to differ from one another in a closely corresponding or parallel manner. In other cases, a variety of one species resembles a distinct species; or the varieties of two distinct species resemble one another. Although these parallel resemblances no doubt often result from reversion to the former characters of a common progenitor; yet in other cases, when new characters first appear, the resemblance must be attributed to the inheritance of a similar constitution, and consequently to a tendency to vary in the same manner. We see something of a similar kind in the same monstrosity appearing and reappearing many times in the same species of animal, and, as Dr. Maxwell Masters has remarked to me, in the same species of plant.]

We may at least conclude, that the amount of modification which animals and plants have undergone under domestication does not correspond with the degree to which they have been subjected to changed circumstances. As we know the parentage of domesticated birds far better than of most quadrupeds, we will glance through the list. The pigeon has varied in Europe more than almost any other bird; yet it is a native species, and has not been exposed to any extraordinary change of conditions. The fowl has varied equally, or almost equally, with the pigeon, and is a native of the hot jungles of India. Neither the peacock, a native of the same country, nor the guinea-fowl, an inhabitant of the dry deserts of Africa, has varied at all, or only in colour. The turkey, from Mexico, has varied but little. The duck, on the other hand, a native of Europe, has yielded some well-marked races; and as this is an aquatic bird, it must have been subjected to a far more serious change in its habits than the pigeon or even the fowl, which nevertheless have varied in a much higher degree. The goose, a native of Europe and aquatic like the duck, has varied less than any other domesticated bird, except the peacock.

Bud-variation is, also, important under our present point of view, in some few cases, as when all the eyes on the same tuber of the potato, or all the fruit on the same plum-tree, or all the flowers on the same plant, have suddenly varied in the same manner, it might be argued that the variation had been definitely caused by some change in the conditions to which the plants had been exposed; yet, in other cases, such an admission is extremely difficult. As new characters sometimes appear by bud-variation, which do not occur in the parent-species or in any allied species, we may reject, at least in these cases, the idea that they are due to reversion. Now it is well worth while to reflect maturely on some striking case of bud-variation, for instance that of the peach. This tree has been cultivated by the million in various parts of the world, has been treated differently, grown on its own roots and grafted on various stocks, planted as a standard, trained against a wall, or under glass; yet each bud of each sub-variety keeps true to its kind. But occasionally, at long intervals of time, a tree in England, or under the widely different climate of Virginia, produces a single bud, and this yields a branch which ever afterwards bears nectarines. Nectarines differ, as every one knows, from peaches in their smoothness, size, and flavour; and the difference is so great that some botanists have maintained that they are specifically distinct. So permanent are the characters thus suddenly acquired, that a nectarine produced by bud-variation has propagated itself by seed. To guard against the supposition that there is some fundamental distinction between bud and seminal variation, it is well to bear in mind that nectarines have likewise been produced from the stone of the peach; and, reversely, peaches from the stone of the nectarine. Now is it possible to conceive external conditions more closely alike than those to which the buds on the same tree are exposed? Yet one bud alone, out of the many thousands borne by the same tree, has suddenly, without any apparent cause, produced a nectarine. But the case is even stronger than this, for the same flower-bud has yielded a fruit, one-half or one-quarter a nectarine, and the other half or three-quarters a peach. Again, seven or eight varieties of the peach have yielded by bud-variation nectarines: the nectarines thus produced, no doubt, differ a little from one another; but still they are nectarines. Of course there must be some cause, internal or external, to excite the peach-bud to change its nature; but I cannot imagine a class of facts better adapted to force on our minds the conviction that what we call the external conditions of life are in many cases quite insignificant in relation to any particular variation, in comparison with the organisation or constitution of the being which varies.

It is known from the labours of Geoffroy Saint-Hilaire, and recently from those of Dareste and others, that eggs of the fowl, if shaken, placed upright, perforated, covered in part with varnish, etc., produce monstrous chickens. Now these monstrosities may be said to be directly caused by such unnatural conditions, but the modifications thus induced are not of a definite nature. An excellent observer, M. Camille Dareste (23/59. 'Memoire sur la Production Artificielle des Monstruosites' 1862 pages 8-12; 'Recherches sur les Conditions, etc., chez les Monstres' 1863 page 6. An abstract is given of Geoffroy's Experiments by his son, in his 'Vie, Travaux' etc. 1847 page 290.), remarks "that the various species of monstrosities are not determined by specific causes; the external agencies which modify the development of the embryo act solely in causing a perturbation—a perversion in the normal course of development." He compares the result to what we see in illness: a sudden chill, for instance, affects one individual alone out of many, causing either a cold, or sore-throat, rheumatism, or inflammation of the lungs or pleura. Contagious matter acts in an analogous manner. (23/60. Paget 'Lectures on Surgical Pathology' 1853 volume 1 page 483.) We may take a still more specific instance: seven pigeons were struck by rattle-snakes (23/61. 'Researches upon the Venom of the Rattle-snake' January 1861 by Dr. Mitchell page 67.): some suffered from convulsions; some had their blood coagulated, in others it was perfectly fluid; some showed ecchymosed spots on the heart, others on the intestines, etc.; others again showed no visible lesion in any organ. It is well known that excess in drinking causes different diseases in different men; but in the tropics the effects of intemperance differ from those caused in a cold climate (23/62. Mr. Sedgwick 'British and Foreign Medico-Chirurg. Review' July 1863 page 175.); and in this case we see the definite influence of opposite conditions. The foregoing facts apparently give us as good an idea as we are likely for a long time to obtain, how in many cases external conditions act directly, though not definitely, in causing modifications of structure.

**SUMMARY.**

There can be no doubt, from the facts given in this chapter, that extremely slight changes in the conditions of life sometimes, probably often, act in a definite manner on our domesticated productions; and, as the action of changed conditions in causing indefinite variability is accumulative, so it may be with their definite action. Hence considerable and definite modifications of structure probably follow from altered conditions acting during a long series of generations. In some few instances a marked effect has been produced quickly on all, or nearly all, the individuals which have been exposed to a marked change of climate, food, or other circumstance. This has occurred with European men in the United States, with European dogs in India, with horses in the Falkland Islands, apparently with various animals at Angora, with foreign oysters in the Mediterranean, and with maize transported from one climate to another. We have seen that the chemical compounds of some plants and the state of their tissues are readily affected by changed conditions. A relation apparently exists between certain characters and certain conditions, so that if the latter be changed the character is lost—as with the colours of flowers, the state of some culinary plants, the fruit of the melon, the tail of fat-tailed sheep, and the peculiar fleeces of other sheep.

The production of galls, and the change of plumage in parrots when fed on peculiar food or when inoculated by the poison of a toad, prove to us what great and mysterious changes in structure and colour, may be the definite result of chemical changes in the nutrient fluids or tissues.

We now almost certainly know that organic beings in a state of nature may be modified in various definite ways by the conditions to which they have been long exposed, as in the case of the birds and other animals in the northern and southern United States, and of American trees in comparison with their representatives in Europe. But in many cases it is most difficult to distinguish between the definite result of changed conditions, and the accumulation through natural selection of indefinite variations which have proved serviceable. If it profited a plant to inhabit a humid instead of an arid station, a fitting change in its constitution might possibly result from the direct action of the environment, though we have no grounds for believing that variations of the right kind would occur more frequently with plants inhabiting a station a little more humid than usual, than with other plants. Whether the station was unusually dry or humid, variations adapting the plant in a slight degree for directly opposite habits of life would occasionally arise, as we have good reason to believe from what we actually see in other cases.

The organisation or constitution of the being which is acted on, is generally a much more important element than the nature of the changed conditions, in determining the nature of the variation. We have evidence of this in the appearance of nearly similar modifications under different conditions, and of different modifications under apparently nearly the same conditions. We have still better evidence of this in closely parallel varieties being frequently produced from distinct races, or even distinct species; and in the frequent recurrence of the same monstrosity in the same species. We have also seen that the degree to which domesticated birds have varied, does not stand in any close relation with the amount of change to which they have been subjected.

To recur once again to bud-variations. When we reflect on the millions of buds which many trees have produced, before some one bud has varied, we are lost in wonder as to what the precise cause of each variation can be. Let us recall the case given by Andrew Knight of the forty-year-old tree of the yellow magnum bonum plum, an old variety which has been propagated by grafts on various stocks for a very long period throughout Europe and North America, and on which a single bud suddenly produced the red magnum bonum. We should also bear in mind that distinct varieties, and even distinct species,—as in the case of peaches, nectarines, and apricots,—of certain roses and camellias,— although separated by a vast number of generations from any progenitor in common, and although cultivated under diversified conditions, have yielded by bud-variation closely analogous varieties. When we reflect on these facts we become deeply impressed with the conviction that in such cases the nature of the variation depends but little on the conditions to which the plant has been exposed, and not in any especial manner on its individual character, but much more on the inherited nature or constitution of the whole group of allied beings to which the plant in question belongs. We are thus driven to conclude that in most cases the conditions of life play a subordinate part in causing any particular modification; like that which a spark plays, when a mass of combustibles bursts into flame—the nature of the flame depending on the combustible matter, and not on the spark. (23/63. Professor Weismann argues strongly in favour of this view in his 'Saison-Dimorphismus der Schmetterlinge' 1875 pages 40-43.)

No doubt each slight variation must have its efficient cause; but it is as hopeless an attempt to discover the cause of each, as to say why a chill or a poison affects one man differently from another. Even with modifications resulting from the definite action of the conditions of life, when all or nearly all the individuals, which have been similarly exposed, are similarly affected, we can rarely see the precise relation between cause and effect. In the next chapter it will be shown that the increased use or disuse of various organs produces an inherited effect. It will further be seen that certain variations are bound together by correlation as well as by other laws. Beyond this we cannot at present explain either the causes or nature of the variability of organic beings.

# CHAPTER 2.XXIV.

## LAWS OF VARIATION—USE AND DISUSE, ETC.

NISUS FORMATIVUS, OR THE CO-ORDINATING POWER OF THE ORGANISATION. ON THE EFFECTS OF THE INCREASED USE AND DISUSE OF ORGANS. CHANGED HABITS OF LIFE. ACCLIMATISATION WITH ANIMALS AND PLANTS. VARIOUS METHODS BY WHICH THIS CAN BE EFFECTED. ARRESTS OF DEVELOPMENT. RUDIMENTARY ORGANS.

In this and the two following chapters I shall discuss, as well as the difficulty of the subject permits, the several laws which govern Variability. These may be grouped under the effects of use and disuse, including changed habits and acclimatisation—arrest of development—correlated variation—the cohesion of homologous parts-the variability of multiple parts—compensation of growth—the position of buds with respect to the axis of the plant—and lastly, analogous variation. These several subjects so graduate into one another that their distinction is often arbitrary.

It may be convenient first briefly to discuss that coordinating and reparative power which is common, in a higher or lower degree, to all organic beings, and which was formerly designated by physiologists as nisus formativus.

[Blumenbach and others (24/1. 'An Essay on Generation' English translation page 18; Paget 'Lectures on Surgical Pathology' 1853 volume 1 page 209.) have insisted that the principle which permits a Hydra, when cut into fragments, to develop itself into two or more perfect animals, is the same with that which causes a wound in the higher animals to heal by a cicatrice. Such cases as that of the Hydra are evidently analogous to the spontaneous division or fissiparous generation of the lowest animals, and likewise to the budding of plants. Between these extreme cases and that of a mere cicatrice we have every gradation. Spallanzani (24/2. 'An Essay on Animal Reproduction' English translation 1769 page 79.) by cutting off the legs and tail of a Salamander, got in the course of three months six crops of these members; so that 687 perfect bones were reproduced by one animal during one season. At whatever point the limb was cut off, the deficient part, and no more, was exactly reproduced. When a diseased bone has been removed, a new one sometimes "gradually assumes the regular form, and all the attachments of muscles, ligaments, etc., become as complete as before." (24/3. Carpenter 'Principles of Comp. Physiology' 1854 page 479.)

This power of regrowth does not, however, always act perfectly; the reproduced tail of a lizard differs in the form of the scales from the normal tail: with certain Orthopterous insects the large hind legs are reproduced of smaller size (24/4. Charlesworth 'Mag. of Nat. Hist.' volume 1 1837 page 145.): the white cicatrice which in the higher animals unites the edges of a deep wound is not formed of perfect skin, for elastic tissue is not produced till long afterwards. (24/5. Paget 'Lectures on Surgical Pathology' volume 1 page 239.) "The activity of the nisus formativus," says Blumenbach, "is in an inverse ratio to the age of the organised body." Its power is also greater with animals, the lower they stand in the scale of organisation; and animals low in the scale correspond with the embryos of higher animals belonging to the same class. Newport's observations (24/6. Quoted by Carpenter 'Comp. Phys.' page 479.) afford a good illustration of this fact, for he found that "myriapods, whose highest development scarcely carries them beyond the larva of perfect insects, can regenerate limbs and antennae up to the time of their last moult;" and so can the larvae of true insects, but, except in one order, not in the mature insect. Salamanders correspond in development with the tadpoles or larvae of the tailless Batrachians, and both possess to a large extent the power of regrowth; but not so the mature tailless Batrachians.

Absorption often plays an important part in the repair of injuries. When a bone is broken and does not unite, the ends are absorbed and rounded, so that a false joint is formed; or if the ends unite, but overlap, the projecting parts are removed. (24/7. Prof. Marey's discussion on the power of co-adaptation in all parts of the organisation is excellent. 'La Machine Animale' 1873 chapter 9. See also Paget 'Lectures' etc. page 257.) A dislocated bone will form for itself a new socket. Displaced tendons and varicose veins excavate new channels in the bones against which they press. But absorption comes into action, as Virchow remarks, during the normal growth of bones; parts which are solid during youth become hollowed out for the medullary tissue as the bone increases in size. In trying to understand the many well-adapted cases of regrowth when aided by absorption, we should remember that almost all parts of the organisation, even whilst retaining the same form, undergo constant renewal; so that a part which is not renewed would be liable to absorption.

Some cases, usually classed under the so-called nisus formativus, at first appear to come under a distinct head; for not only are old structures reproduced, but new structures are formed. Thus, after inflammation "false membranes," furnished with blood-vessels, lymphatics, and nerves, are developed; or a foetus escapes from the Fallopian tubes, and falls into the abdomen, "nature pours out a quantity of plastic lymph, which forms itself into organised membrane, richly supplied with blood-vessels," and the foetus is nourished for a time. In certain cases of hydrocephalus the open and dangerous spaces in the skull are filled up with new bones, which interlock by perfect serrated sutures. (24/8. These cases are given by Blumenbach in his 'Essay on Generation' pages 52, 54.) But most physiologists, especially on the Continent, have now given up the belief in plastic lymph or blastema, and Virchow (24/9. 'Cellular Pathology' translation by Dr. Chance 1860 pages 27, 441.) maintains that every structure, new or old, is formed by the proliferation of pre-existing cells. On this view false membranes, like cancerous or other tumours, are merely abnormal developments of normal growths; and we can thus understand how it is that they resemble adjoining structures; for instance, that a "false membrane in the serous cavities acquires a covering of epithelium exactly like that which covers the original serous membrane; adhesions of the iris may become black apparently from the production of pigment-cells like those of the uvea." (24/10. Paget 'Lectures on Pathology' volume 1 1853 page 357.)

No doubt the power of reparation, though not always perfect, is an admirable provision, ready for various emergencies, even for such as occur only at long intervals of time. (24/11. Paget ibid page 150.) Yet this power is not more wonderful than the growth and development of every single creature, more especially of those which are propagated by fissiparous generation. This subject has been here noticed, because we may infer that, when any part or organ is either greatly increased in size or wholly suppressed through variation and continued selection, the co-ordinating power of the organisation will continually tend to bring again all the parts into harmony with one another.]

## ON THE EFFECTS OF THE INCREASED USE AND DISUSE OF ORGANS.

It is notorious, and we shall immediately adduce proofs, that increased use or action strengthens muscles, glands, sense-organs, etc.; and that disuse, on the other hand, weakens them. It has been experimentally proved by Ranke (24/12. 'Die Blutvertheilung, etc. der Organe' 1871 as quoted by Jaeger 'In Sachen Darwin's' 1874 page 48. See also H. Spencer 'The Principles of Biology' volume 2 1866 chapters 3-5.) that the flow of blood is greatly increased towards any part which is performing work, and sinks again when the part is at rest. Consequently, if the work is frequent, the vessels increase in size and the part is better nourished. Paget (24/13. 'Lectures on Pathology' 1853 volume 1 page 71.) also accounts for the long, thick, dark-coloured hairs which occasionally grow, even in young children, near old-standing inflamed surfaces or fractured bones by an increased flow of blood to the part. When Hunter inserted the spur of a cock into the comb, which is well supplied with blood-vessels, it grew in one case spirally to a length of six inches, and in another case forward, like a horn, so that the bird could not touch the ground with its beak. According to the interesting observations of M. Sedillot (24/14. 'Comptes Rendus' September 26, 1864 page 539.), when a portion of one of the bones of the leg of an animal is removed, the associated bone enlarges till it attains a bulk equal to that of the two bones, of which it has to perform the functions. This is best exhibited in dogs in which the tibia has been removed; the companion bone, which is naturally almost filiform and not one-fifth the size of the other, soon acquires a size equal to or greater than that of the tibia. Now, it is at first difficult to believe that increased weight acting on a straight bone could, by alternately increasing and diminishing the pressure, cause the blood to flow more freely in the vessels which permeate the periosteum and thus supply more nutriment to the bone. Nevertheless the observations adduced by Mr. Spencer (24/15. H. Spencer 'The Principles of Biology' volume 2 page 243.), on the strengthening of the bowed bones of rickety children, along their concave sides, leads to the belief that this is possible.

The rocking of the stem of a tree increases in a marked manner the growth of the woody tissue in the parts which are strained. Prof. Sachs believes, from reasons which he assigns, that this is due to the pressure of the bark being relaxed in such parts, and not as Knight and H. Spencer maintain, to an increased flow of sap caused by the movement of the trunk. (24/16. Ibid volume 2 page 269. Sachs 'Text-book of Botany' 1875 page 734.) But hard woody tissue may be developed without the aid of any movement, as we see with ivy closely attached to an old wall. In all such cases, it is very difficult to distinguish between the effects of long-continued selection and those which follow from the increased action of the part, or directly from some other cause. Mr. H. Spencer (24/17. Ibid volume 2 page 273.) acknowledges this difficulty, and gives as an instance the thorns on trees and the shells of nuts. Here we have extremely hard woody tissue without the possibility of any movement, and without, as far as we can see, any other directly exciting cause; and as the hardness of these parts is of manifest service to the plant, we may look at the result as probably due to the selection of so-called spontaneous variations. Every one knows that hard work thickens the epidermis on the hands; and when we hear that with infants, long before birth, the epidermis is thicker on the palms and soles of the feet than on any other part of the body, as was observed with admiration by Albinus (24/18. Paget 'Lectures on Pathology' volume 2 page 209.), we are naturally inclined to attribute this to the inherited effects of long-continued use or pressure. We are tempted to extend the same view even to the hoofs of quadrupeds; but who will pretend to determine how far natural selection may have aided in the formation of structures of such obvious importance to the animal?

[That use strengthens the muscles may be seen in the limbs of artisans who follow different trades; and when a muscle is strengthened, the tendons, and the crests of bone to which they are attached, become enlarged; and this must likewise be the case with the blood-vessels and nerves. On the other hand, when a limb is not used, as by Eastern fanatics, or when the nerve supplying it with nervous power is effectually destroyed, the muscles wither. So again, when the eye is destroyed the optic nerve becomes atrophied, sometimes even in the course of a few months. (24/19. Muller 'Phys.' English translation pages 54, 791. Prof. Reed has given ('Physiological and Anat. Researches' page 10) a curious account of the atrophy of the limbs of rabbits after the destruction of the nerve.) The Proteus is furnished with branchiae as well as with lungs: and Schreibers (24/20. Quoted by Lecoq in 'Geograph. Bot.' tome 1 1854 page 182.) found that when the animal was compelled to live in deep water, the branchiae were developed to thrice their ordinary size, and the lungs were partially atrophied. When, on the other hand, the animal was compelled to live in shallow water, the lungs became larger and more vascular, whilst the branchiae disappeared in a more or less complete degree. Such modifications as these are, however, of comparatively little value for us, as we do not actually know that they tend to be inherited.

In many cases there is reason to believe that the lessened use of various organs has affected the corresponding parts in the offspring. But there is no good evidence that this ever follows in the course of a single generation. It appears, as in the case of general or indefinite variability, that several generations must be subjected to changed habits for any appreciable result. Our domestic fowls, ducks, and geese have almost lost, not only in the individual but in the race, their power of flight; for we do not see a young fowl, when frightened, take flight like a young pheasant. Hence I was led carefully to compare the limb-bones of fowls, ducks, pigeons, and rabbits, with the same bones in the wild parent-species. As the measurements and weights were fully given in the earlier chapters I need here only recapitulate the results. With domestic pigeons, the length of the sternum, the prominence of its crest, the length of the scapulae and furculum, the length of the wings as measured from tip to tip of the radii, are all reduced relatively to the same parts in the wild pigeon. The wing and tail feathers, however, are increased in length, but this may have as little connection with the use of the wings or tail, as the lengthened hair on a dog with the amount of exercise which it has habitually taken. The feet of pigeons, except in the long-beaked races, are reduced in size. With fowls the crest of the sternum is less prominent, and is often distorted or monstrous; the wing-bones have become lighter relatively to the leg-bones, and are apparently a little shorter in comparison with those of the parent-form, the Gallus bankiva. With ducks, the crest of the sternum is affected in the same manner as in the foregoing cases: the furculum, coracoids, and scapulae are all reduced in weight relatively to the whole skeleton: the bones of the wings are shorter and lighter, and the bones of the legs longer and heavier, relatively to each other, and relatively to the whole skeleton, in comparison with the same bones in the wild-duck. The decreased weight and size of the bones, in the foregoing cases, is probably the indirect result of the reaction of the weakened muscles on the bones. I failed to compare the feathers of the wings of the tame and wild duck; but Gloger (24/21. 'Das Abandern der Vogel' 1833 s. 74.) asserts that in the wild duck the tips of the wing-feathers reach almost to the end of the tail, whilst in the domestic duck they often hardly reach to its base. He remarks also on the greater thickness of the legs, and says that the swimming membrane between the toes is reduced; but I was not able to detect this latter difference.

With the domesticated rabbit the body, together with the whole skeleton, is generally larger and heavier than in the wild animal, and the leg-bones are heavier in due proportion; but whatever standard of comparison be taken, neither the leg-bones nor the scapulae have increased in length proportionally with the increased dimensions of the rest of the skeleton. The skull has become in a marked manner narrower, and, from the measurements of its capacity formerly given, we may conclude, that this narrowness results from the decreased size of the brain, consequent on the mentally inactive life led by these closely-confined animals.

We have seen in the eighth chapter that silk-moths, which have been kept during many centuries closely confined, emerge from their cocoons with their wings distorted, incapable of flight, often greatly reduced in size, or even, according to Quatrefages, quite rudimentary. This condition of the wings may be largely owing to the same kind of monstrosity which often affects wild Lepidoptera when artificially reared from the cocoon; or it may be in part due to an inherent tendency, which is common to the females of many Bombycidae, to have their wings in a more or less rudimentary state; but part of the effect may be attributed to long-continued disuse.]

From the foregoing facts there can be no doubt that with our anciently domesticated animals, certain bones have increased or decreased in size and weight owing to increased or decreased use; but they have not been modified, as shown in the earlier chapters, in shape or structure. With animals living a free life and occasionally exposed to severe competition the reduction would tend to be greater, as it would be an advantage to them to have the development of every superfluous part saved. With highly-fed domesticated animals, on the other hand, there seems to be no economy of growth, nor any tendency to the elimination of superfluous details. But to this subject I shall recur.

Turning now to more general observations, Nathusius has shown that with the improved races of the pig, the shortened legs and snout, the form of the articular condyles of the occiput, and the position of the jaws with the upper canine teeth projecting in a most anomalous manner in front of the lower canines, may be attributed to these parts not having been fully exercised. For the highly-cultivated races do not travel in search of food, nor root up the ground with their ringed muzzles. (24/22. Nathusius 'Die Racen des Schweines' 1860 s. 53, 57; 'Vorstudien...Schweineschadel' 1864 s. 103, 130, 133. Prof. Lucae supports and extends the conclusions of Von Nathusius: 'Der Schadel des Maskenschweines' 1870.) These modifications of structure, which are all strictly inherited, characterise several improved breeds, so that they cannot have been derived from any single domestic stock. With respect to cattle, Professor Tanner has remarked that the lungs and liver in the improved breeds "are found to be considerably reduced in size when compared with those possessed by animals having perfect liberty" (24/23. 'Journal of Agriculture of Highland Soc.' July 1860 page 321.); and the reduction of these organs affects the general shape of the body. The cause of the reduced lungs in highly-bred animals which take little exercise is obvious; and perhaps the liver may be affected by the nutritious and artificial food on which they largely subsist. Again, Dr. Wilckens asserts (24/24. 'Landwirth. Wochenblatt' No. 10.) that various parts of the body certainly differ in Alpine and lowland breeds of several domesticated animals, owing to their different habits of life; for instance, the neck and fore-legs in length, and the hoofs in shape.

[It is well known that, when an artery is tied, the anastomosing branches, from being forced to transmit more blood, increase in diameter; and this increase cannot be accounted for by mere extension, as their coats gain in strength. With respect to glands, Sir J. Paget observes that "when one kidney is destroyed the other often becomes much larger, and does double work." (24/25. 'Lectures on Surgical Pathology' 1853 volume 1 page 27.) If we compare the size of the udders and their power of secretion in cows which have been long domesticated, and in certain breeds of the goat in which the udders nearly touch the ground, with these organs in wild or half-domesticated animals, the difference is great. A good cow with us daily yields more than five gallons, or forty pints of milk, whilst a first-rate animal, kept, for instance, by the Damaras of South Africa (24/26. Andersson 'Travels in South Africa' page 318. For analogous cases in South America see Aug. St.-Hilaire 'Voyage dans la Province de Goyaz' tome 1 page 71.), "rarely gives more than two or three pints of milk daily, and, should her calf be taken from her, she absolutely refuses to give any." We may attribute the excellence of our cows and of certain goats, partly to the continued selection of the best milking animals, and partly to the inherited effects of the increased action, through man's art, of the secreting glands.

It is notorious that short-sight is inherited; and we have seen in the twelfth chapter from the statistical researches of M. Giraud-Teulon, that the habit of viewing near objects gives a tendency to short-sight. Veterinarians are unanimous that horses are affected with spavins, splints, ringbones, etc., from being shod and from travelling on hard roads, and they are almost equally unanimous that a tendency to these malformations is transmitted. Formerly horses were not shod in North Carolina, and it has been asserted that they did not then suffer from these diseases of the legs and feet. (24/27. Brickell 'Nat. Hist. of North Carolina' 1739 page 53.)]

Our domesticated quadrupeds are all descended, as far as is known, from species having erect ears; yet few kinds can be named, of which at least one race has not drooping ears. Cats in China, horses in parts of Russia, sheep in Italy and elsewhere, the guinea-pig formerly in Germany, goats and cattle in India, rabbits, pigs, and dogs in all long-civilised countries have dependent ears. With wild animals, which constantly use their ears like funnels to catch every passing sound, and especially to ascertain the direction whence it comes, there is not, as Mr. Blyth has remarked, any species with drooping ears except the elephant. Hence the incapacity to erect the ears is certainly in some manner the result of domestication; and this incapacity has been attributed by various authors (24/28. Livingstone quoted by Youatt on 'Sheep' page 142. Hodgson in 'Journal of Asiatic Soc. of Bengal' volume 16 1847 page 1006 etc. etc. On the other hand Dr. Wilckens argues strongly against the belief that the drooping of the ears is the result of disuse: 'Jahrbuch der deutschen Viehzucht' 1866.) to disuse, for animals protected by man are not compelled habitually to use their ears. Col. Hamilton Smith (24/29. 'Naturalist's Library' Dogs volume 2 1840 page 104.) states that in ancient effigies of the dog, "with the exception of one Egyptian instance, no sculpture of the earlier Grecian era produces representations of hounds with completely drooping ears; those with them half pendulous are missing in the most ancient; and this character increases, by degrees, in the works of the Roman period." Godron also has remarked "that the pigs of the ancient Egyptians had not their ears enlarged and pendent." (24/30. 'De l'Espece' tome 1 1859 page 367.) But it is remarkable that the drooping of the ear is not accompanied by any decrease in size; on the contrary, animals so different as fancy rabbits, certain Indian breeds of the goat, our petted spaniels, bloodhounds, and other dogs, have enormously elongated ears, so that it would appear as if their weight had caused them to droop, aided perhaps by disuse. With rabbits, the drooping of the much elongated ears has affected even the structure of the skull.

The tail of no wild animal, as remarked to me by Mr. Blyth, is curled; whereas pigs and some races of dogs have their tails much curled. This deformity, therefore, appears to be the result of domestication, but whether in any way connected with the lessened use of the tail is doubtful.

The epidermis on our hands is easily thickened, as every one knows, by hard work. In a district of Ceylon the sheep have "horny callosities that defend their knees, and which arise from their habit of kneeling down to crop the short herbage, and this distinguishes the Jaffna flocks from those of other portions of the island;" but it is not stated whether this peculiarity is inherited. (24/31. 'Ceylon' by Sir J.E. Tennent 1859 volume 2 page 531.)

The mucous membrane which lines the stomach is continuous with the external skin of the body; therefore it is not surprising that its texture should be affected by the nature of the food consumed, but other and more interesting changes likewise follow. Hunter long ago observed that the muscular coat of the stomach of a gull (Larus tridactylus) which had been fed for a year chiefly on grain was thickened; and, according to Dr. Edmondston, a similar change periodically occurs in the Shetland Islands in the stomach of the Larus argentatus, which in the spring frequents the cornfields and feeds on the seed. The same careful observer has noticed a great change in the stomach of a raven which had been long fed on vegetable food. In the case of an owl (Strix grallaria), similarly treated, Menetries states that the form of the stomach was changed, the inner coat became leathery, and the liver increased in size. Whether these modifications in the digestive organs would in the course of generations become inherited is not known. (24/32. For the foregoing statements see Hunter 'Essays and Observations' 1861 volume 2 page 329; Dr. Edmondston, as quoted in Macgillivray 'British Birds' volume 5 page 550: Menetries as quoted in Bronn 'Geschichte der Natur' b. 2 s. 110.)

The increased or diminished length of the intestines, which apparently results from changed diet, is a more remarkable case, because it is characteristic of certain animals in their domesticated condition, and therefore must be inherited. The complex absorbent system, the blood-vessels, nerves, and muscles, are necessarily all modified together with the intestines. According to Daubenton, the intestines of the domestic cat are one-third longer than those of the wild cat of Europe; and although this species is not the parent-stock of the domestic animal, yet, as Isidore Geoffroy has remarked, the several species of cats are so closely allied that the comparison is probably a fair one. The increased length appears to be due to the domestic cat being less strictly carnivorous in its diet than any wild feline species; for instance, I have seen a French kitten eating vegetables as readily as meat. According to Cuvier, the intestines of the domesticated pig exceed greatly in proportionate length those of the wild boar. In the tame and wild rabbit the change is of an opposite nature, and probably results from the nutritious food given to the tame rabbit. (24/33. These statements on the intestines are taken from Isidore Geoffroy Saint-Hilaire 'Hist. Nat. Gen.' tome 3 pages 427, 441.)

## CHANGED AND INHERITED HABITS OF LIFE.

This subject, as far as the mental powers of animals are concerned, so blends into instinct, that I will here only remind the reader of such cases as the tameness of our domesticated animals—the pointing or retrieving of dogs—their not attacking the smaller animals kept by man—and so forth. How much of these changes ought to be attributed to mere habit, and how much to the selection of individuals which have varied in the desired manner, irrespectively of the special circumstances under which they have been kept, can seldom be told.

We have already seen that animals may be habituated to a changed diet; but some additional instances may be given. In the Polynesian Islands and in China the dog is fed exclusively on vegetable matter, and the taste for this kind of food is to a certain extent inherited. (24/34. Gilbert White 'Nat. Hist. Selborne' 1825 volume 2 page 121.) Our sporting dogs will not touch the bones of game birds, whilst most other dogs devour them with greediness. In some parts of the world sheep have been largely fed on fish. The domestic hog is fond of barley, the wild boar is said to disdain it; and the disdain is partially inherited, for some young wild pigs bred in captivity showed an aversion for this grain, whilst others of the same brood relished it. (24/35. Burdach 'Traite de Phys.' tome 2 page 267 as quoted by Dr. P. Lucas 'L'Hered. Nat.' tome 1 page 388.) One of my relations bred some young pigs from a Chinese sow by a wild Alpine boar; they lived free in the park, and were so tame that they came to the house to be fed; but they would not touch swill, which was devoured by the other pigs. An animal when once accustomed to an unnatural diet, which can generally be effected only during youth, dislikes its proper food, as Spallanzani found to be the case with a pigeon which had been long fed on meat. Individuals of the same species take to new food with different degrees of readiness; one horse, it is stated, soon learned to eat meat, whilst another would have perished from hunger rather than have partaken of it. (24/36. This and several other cases are given by Colin 'Physiologie Comp. des Animaux Dom.' 1854 tome 1 page 426.) The caterpillars of the Bombyx hesperus feed in a state of nature on the leaves of the Cafe diable, but, after having been reared on the Ailanthus, they would not touch the Cafe diable, and actually died of hunger. (24/37. M. Michely de Cayenne in 'Bull. Soc. d'Acclimat.' tome 8 1861 page 563.)

It has been found possible to accustom marine fish to live in fresh water; but as such changes in fish and other marine animals have been chiefly observed in a state of nature, they do not properly belong to our present subject. The period of gestation and of maturity, as shown in the earlier chapters,—the season and the frequency of the act of breeding,—have all been greatly modified under domestication. With the Egyptian goose the rate of change with respect to the season has been recorded. (24/38. Quatrefages 'Unite de l'Espece Humaine' 1861 page 79.) The wild drake pairs with one female, the domestic drake is polygamous. Certain breeds of fowls have lost the habit of incubation. The paces of the horse, and the manner of flight of certain breeds of the pigeon, have been modified and are inherited. Cattle, horses, and pigs have learnt to browse under water in the St. John's River, East Florida, where the Vallisneria has been largely naturalised. The cows were observed by Prof. Wyman to keep their heads immersed for "a period varying from fifteen to thirty-five seconds." (24/39. 'The American Naturalist' April 1874 page 237.) The voice differs much in certain kinds of fowls and pigeons. Some varieties are clamorous and others silent, as the Call and common duck, or the Spitz and pointer dog. Every one knows how the breeds of the dog differ from one another in their manner of hunting, and in their ardour after different kinds of game or vermin.

With plants the period of vegetation is easily changed and is inherited, as in the case of summer and winter wheat, barley, and vetches; but to this subject we shall immediately return under acclimatisation. Annual plants sometimes become perennial under a new climate, as I hear from Dr. Hooker is the case with the stock and mignonette in Tasmania. On the other hand, perennials sometimes become annuals, as with the Ricinus in England, and as, according to Captain Mangles, with many varieties of the heartsease. Von Berg (24/40. 'Flora' 1835 b. 2 page 504.) raised from seed of Verbascum phoeniceum, which is usually a biennial, both annual and perennial varieties. Some deciduous bushes become evergreen in hot countries. (24/41. Alph. de Candolle 'Geograph. Bot.' tome 2 page 1078.) Rice requires much water, but there is one variety in India which can be grown without irrigation. (24/42. Royle 'Illustrations of the Botany of the Himalaya' page 19.) Certain varieties of the oat and of our other cereals are best fitted for certain soils. (24/43. 'Gardener's Chronicle' 1850 pages 204, 219.) Endless similar facts could be given in the animal and vegetable kingdoms. They are noticed here because they illustrate analogous differences in closely allied natural species, and because such changed habits of life, whether due to habit, or to the direct action of external conditions, or to so-called spontaneous variability, would be apt to lead to modifications of structure.

**ACCLIMATISATION.**

From the previous remarks we are naturally led to the much disputed subject of acclimatisation. There are two distinct questions: Do varieties descended from the same species differ in their power of living under different climates? And secondly, if they so differ, how have they become thus adapted? We have seen that European dogs do not succeed well in India, and it is asserted (24/44. Rev. R. Everest 'Journal As. Soc. of Bengal' volume 3 page 19.), that no one has there succeeded in keeping the Newfoundland long alive; but then it may be argued, and probably with truth, that these northern breeds are specifically distinct from the native dogs which flourish in India. The same remark may be made with respect to different breeds of sheep, of which, according to Youatt (24/45. Youatt on 'Sheep' 1838 page 491.), not one brought "from a torrid climate lasts out the second year," in the Zoological Gardens. But sheep are capable of some degree of acclimatisation, for Merino sheep bred at the Cape of Good Hope have been found far better adapted for India than those imported from England. (24/46. Royle 'Prod. Resources of India' page 153.) It is almost certain that all the breeds of the fowl are descended from one species; but the Spanish breed, which there is good reason to believe originated near the Mediterranean (24/47. Tegetmeier 'Poultry Book' 1866 page 102.), though so fine and vigorous in England, suffers more from frost than any other breed. The Arrindy silk moth introduced from Bengal, and the Ailanthus moth from the temperate province of Shan Tung, in China, belong to the same species, as we may infer from their identity in the caterpillar, cocoon, and mature states (24/48. Dr. R. Paterson in a paper communicated to Bot. Soc. of Canada quoted in the 'Reader' 1863 November 13.); yet they differ much in constitution: the Indian form "will flourish only in warm latitudes," the other is quite hardy and withstands cold and rain.

[Plants are more strictly adapted to climate than are animals. The latter when domesticated withstand such great diversities of climate, that we find nearly the same species in tropical and temperate countries; whilst the cultivated plants are widely dissimilar. Hence a larger field is open for inquiry in regard to the acclimatisation of plants than of animals. It is no exaggeration to say that with almost every plant which has long been cultivated, varieties exist which are endowed with constitutions fitted for very different climates; I will select only a few of the more striking cases, as it would be tedious to give all. In North America numerous fruit-trees have been raised, and in horticultural publications,—for instance, in that by Downing,—lists are given of the varieties which are best able to withstand the severe climate of the northern States and Canada. Many American varieties of the pear, plum, and peach are excellent in their own country, but until recently, hardly one was known that succeeded in England; and with apples (24/49. See remarks by Editor in 'Gardener's Chronicle' 1848 page 5.), not one succeeds. Though the American varieties can withstand a severer winter than ours, the summer here is not hot enough. Fruit-trees have also originated in Europe with different constitutions, but they are not much noticed, because nurserymen here do not supply wide areas. The Forelle pear flowers early, and when the flowers have just set, and this is the critical period, they have been observed, both in France and England, to withstand with complete impunity a frost of 18 deg and even 14 deg Fahr., which killed the flowers, whether fully expanded or in bud, of all other kinds of pears. (24/50. 'Gardener's Chronicle' 1860 page 938. Remarks by Editor and quotation from Decaisne.) This power in the flower of resisting cold and afterwards producing fruit does not invariably depend, as we know on good authority (24/51. J. de Jonghe of Brussels in 'Gardener's Chronicle' 1857 page 612.), on general constitutional vigour. In proceeding northward, the number of varieties which are found capable of resisting the climate rapidly decreases, as may be seen in the list of the varieties of the cherry, apple, and pear, which can be cultivated in the neighbourhood of Stockholm. (24/52. Ch. Martius 'Voyage Bot. Cotes Sept. de la Norvege' page 26.) Near Moscow, Prince Troubetzkoy planted for experiment in the open ground several varieties of the pear, but one alone, the Poire sans Pepins, withstood the cold of winter. (24/53. 'Journal de l'Acad. Hort. de Gand' quoted in 'Gardener's Chronicle' 1859 page 7.) We thus see that our fruit-trees, like distinct species of the same genus, certainly differ from each other in their constitutional adaptation to different climates.

With the varieties of many plants, the adaptation to climate is often very close. Thus it has been proved by repeated trials "that few if any of the English varieties of wheat are adapted for cultivation in Scotland" (24/54. 'Gardener's Chronicle' 1851 page 396.); but the failure in this case is at first only in the quantity, though ultimately in the quality, of the grain produced. The Rev. M.J. Berkeley sowed wheat-seed from India, and got "the most meagre ears," on land which would certainly have yielded a good crop from English wheat. (24/55. Ibid 1862 page 235.) In these cases varieties have been carried from a warmer to a cooler climate; in the reverse case, as "when wheat was imported directly from France into the West Indian Islands, it produced either wholly barren spikes or furnished with only two or three miserable seeds, while West Indian seed by its side yielded an enormous harvest." (24/56. On the authority of Labat quoted in 'Gardener's Chronicle' 1862 page 235.) Here is another case of close adaptation to a slightly cooler climate; a kind of wheat which in England may be used indifferently either as a winter or summer variety, when sown under the warmer climate of Grignan, in France, behaved exactly as if it had been a true winter wheat. (24/57. MM. Edwards and Colin 'Annal. des Sc. Nat.' 2nd series Bot. tome 5 page 22.)

Botanists believe that all the varieties of maize belong to the same species; and we have seen that in North America, in proceeding northward, the varieties cultivated in each zone produce their flowers and ripen their seed within shorter and shorter periods. So that the tall, slowly maturing southern varieties do not succeed in New England, and the New English varieties do not succeed in Canada. I have not met with any statement that the southern varieties are actually injured or killed by a degree of cold which the northern varieties can withstand with impunity, though this is probable; but the production of early flowering and early seeding varieties deserves to be considered as one form of acclimatisation. Hence it has been found possible, according to Kalm, to cultivate maize further and further northwards in America. In Europe, also, as we learn from the evidence given by Alph. De Candolle, the culture of maize has extended since the end of the last century thirty leagues north of its former boundary. (24/58. 'Geograph. Bot.' page 337.) On the authority of Linnaeus (24/59. 'Swedish Acts' English translation 1739-40 volume 1. Kalm in his 'Travels' volume 2 page 166 gives an analogous case with cotton-plants raised in New Jersey from Carolina seed.), I may quote an analogous case, namely, that in Sweden tobacco raised from home-grown seed ripens its seed a month sooner and is less liable to miscarry than plants raised from foreign seed.

With the Vine, differently from the maize, the line of practical culture has retreated a little southward since the middle ages (24/60. De Candolle 'Geograph. Bot.' page 339.); but this seems due to commerce being now easier, so that it is better to import wine from the south than to make it in northern districts. Nevertheless the fact of the vine not having spread northward shows that acclimatisation has made no progress during several centuries. There is, however, a marked difference in the constitution of the several varieties,— some being hardy, whilst others, like the muscat of Alexandria, require a very high temperature to come to perfection. According to Labat (24/61. 'Gardener's Chronicle' 1862 page 235.), vines taken from France to the West Indies succeed with extreme difficulty, whilst those imported from Madeira or the Canary Islands thrive admirably.

Gallesio gives a curious account of the naturalisation of the Orange in Italy. During many centuries the sweet orange was propagated exclusively by grafts, and so often suffered from frosts, that it required protection. After the severe frost of 1709, and more especially after that of 1763, so many trees were destroyed, that seedlings from the sweet orange were raised, and, to the surprise of the inhabitants, their fruit was found to be sweet. The trees thus raised were larger, more productive, and hardier than the old kinds; and seedlings are now continually raised. Hence Gallesio concludes that much more was effected for the naturalisation of the orange in Italy by the accidental production of new kinds during a period of about sixty years, than had been effected by grafting old varieties during many ages. (24/62. Gallesio 'Teoria della Riproduzione Veg.' 1816 page 125; and 'Traite du Citrus' 1811 page 359.) I may add that Risso (24/63. 'Essai sur l'Hist. des Orangers' 1813 page 20 etc.) describes some Portuguese varieties of the orange as extremely sensitive to cold, and as much tenderer than certain other varieties.

The peach was known to Theophrastus, 322 B.C. (24/64. Alph. de Candolle 'Geograph. Bot.' page 882.) According to the authorities quoted by Dr. F. Rolle (24/65. 'Ch. Darwin's Lehre von der Entstehung' etc. 1862 s. 87.), it was tender when first introduced into Greece, and even in the island of Rhodes only occasionally bore fruit. If this be correct, the peach, in spreading during the last two thousand years over the middle parts of Europe, must have become much hardier. At the present day different varieties differ much in hardiness: some French varieties will not succeed in England; and near Paris, the Pavie de Bonneuil does not ripen its fruit till very late in the season, even when grown on a wall; "it is, therefore, only fit for a very hot southern climate." (24/66. Decaisne quoted in 'Gardener's Chronicle' 1865 page 271.)

I will briefly give a few other cases. A variety of Magnolia grandiflora, raised by M. Roy, withstands a temperature several degrees lower than that which any other variety can resist. With camellias there is much difference in hardiness. One particular variety of the Noisette rose withstood the severe frost of 1860 "untouched and hale amidst a universal destruction of other Noisettes." In New York the "Irish yew is quite hardy, but the common yew is liable to be cut down." I may add that there are varieties of the sweet potato (Convolvulus batatas) which are suited for warmer, as well as for colder, climates. (24/67. For the magnolia see Loudon's 'Gardener's Mag.' volume 13 1837 page 21. For camellias and roses see 'Gardener's Chronicle' 1860 page 384. For the yew 'Journal of Hort.' March 3, 1863 p 174. For sweet potatoes see Col. von Siebold in 'Gardener's Chronicle' 1855 page 822.)]

The plants as yet mentioned have been found capable of resisting an unusual degree of cold or heat, when fully grown. The following cases refer to plants whilst young. In a large bed of young Araucarias of the same age, growing close together and equally exposed, it was observed (24/68. The Editor 'Gardener's Chronicle' 1861 page 239.), after the unusually severe winter of 1860-61, that, "in the midst of the dying, numerous individuals remained on which the frost had absolutely made no kind of impression." Dr. Lindley, after alluding to this and other similar cases, remarks, "Among the lessons which the late formidable winter has taught us, is that, even in their power of resisting cold, individuals of the same species of plants are remarkably different." Near Salisbury, there was a sharp frost on the night of May 24, 1836, and all the French beans (Phaseolus vulgaris) in a bed were killed except about one in thirty, which completely escaped. (24/69. Loudon's 'Gardener's Mag.' volume 12 1836 page 378.) On the same day of the month, but in the year 1864, there was a severe frost in Kent, and two rows of scarlet-runners (P. multiflorus) in my garden, containing 390 plants of the same age and equally exposed, were all blackened and killed except about a dozen plants. In an adjoining row of "Fulmer's dwarf bean" (P. vulgaris), one single plant escaped. A still more severe frost occurred four days afterwards, and of the dozen plants which had previously escaped only three survived; these were not taller or more vigorous than the other young plants, but they escaped completely, with not even the tips of their leaves browned. It was impossible to behold these three plants, with their blackened, withered, and dead brethren all around, and not see at a glance that they differed widely in constitutional power of resisting frost.

This work is not the proper place to show that wild plants of the same species, naturally growing at different altitudes or under different latitudes, become to a certain extent acclimatised, as is proved by the different behaviour of their seedlings when raised in another country. In my 'Origin of Species' I have alluded to some cases, and I could add many others. One instance must suffice: Mr. Grigor, of Forres (24/70. 'Gardener's Chronicle' 1865 page 699. Mr. G. Maw gives ('Gardener's Chronicle' 1870 page 895) a number of striking cases; he brought home from southern Spain and northern Africa several plants, which he cultivated in England alongside specimens from northern districts; and he found a great difference not only in their hardiness during the winter, but in the behaviour of some of them during the summer.), states that seedlings of the Scotch fir (Pinus sylvestris), raised from seed from the Continent and from the forests of Scotland, differ much. "The difference is perceptible in one-year-old, and more so in two-year- old seedlings; but the effects of the winter on the second year's growth almost uniformly make those from the Continent quite brown, and so damaged, that by the month of March they are quite unsaleable, while the plants from the native Scotch pine, under the same treatment, and standing alongside, although considerably shorter, are rather stouter and quite green, so that the beds of the one can be known from the other when seen from the distance of a mile." Closely similar facts have been observed with seedling larches.

[Hardy varieties would alone be valued or noticed in Europe; whilst tender varieties, requiring more warmth, would generally be neglected; but such occasionally arise. Thus Loudon (24/71. 'Arboretum et Fruticetum' volume 3 page 1376.) describes a Cornish variety of the elm which is almost an evergreen, and of which the shoots are often killed by the autumnal frosts, so that its timber is of little value. Horticulturists know that some varieties are much more tender than others: thus all the varieties of the broccoli are more tender than cabbages; but there is much difference in this respect in the sub-varieties of the broccoli; the pink and purple kinds are a little hardier than the white Cape broccoli, "but they are not to be depended on after the thermometer falls below 24 deg Fahr.;" the Walcheren broccoli is less tender than the Cape, and there are several varieties which will stand much severer cold than the Walcheren. (24/72. Mr. Robson in 'Journal of Horticulture' 1861 page 23.) Cauliflowers seed more freely in India than cabbages. (24/73. Dr. Bonavia 'Report of the Agri.-Hort. Soc. of Oudh' 1866.) To give one instance with flowers: eleven plants raised from a hollyhock, called the Queen of the Whites (24/74. 'Cottage Gardener' 1860 April 24 page 57.) were found to be much more tender than various other seedlings. It may be presumed that all tender varieties would succeed better under a climate warmer than ours. With fruit-trees, it is well known that certain varieties, for instance of the peach, stand forcing in a hot-house better than others; and this shows either pliability of organisation or some constitutional difference. The same individual cherry-tree, when forced, has been observed during successive years gradually to change its period of vegetation. (24/75. 'Gardener's Chronicle' 1841 page 291.) Few pelargoniums can resist the heat of a stove, but Alba Multiflora will, as a most skilful gardener asserts, "stand pine-apple top and bottom heat the whole winter; without looking any more drawn than if it had stood in a common greenhouse; and Blanche Fleur seems as if it had been made on purpose for growing in winter, like many bulbs, and to rest all summer." (24/76. Mr. Beaton in 'Cottage Gardener' March 20, 1860 page 377. Queen Mab will also stand stove heat. See 'Gardener's Chronicle' 1845 page 226.) There can hardly be a doubt that the Alba Multiflora pelargonium must have a widely different constitution from that of most other varieties of this plant; it would probably withstand even an equatorial climate.

We have seen that according to Labat the vine and wheat require acclimatisation in order to succeed in the West Indies. Similar facts have been observed at Madras: "two parcels of mignonette-seed, one direct from Europe, the other saved at Bangalore (of which the mean temperature is much below that of Madras), were sown at the same time: they both vegetated equally favourably, but the former all died off a few days after they appeared above ground; the latter still survive, and are vigorous, healthy plants." "So again, turnip and carrot seed saved at Hyderabad are found to answer better at Madras than seed from Europe or from the Cape of Good Hope." (24/77. 'Gardener's Chronicle' 1841 page 439.) Mr. J. Scott of the Calcutta Botanic Gardens, informs me that seeds of the sweet-pea (Lathyrus odoratus) imported from England produce plants, with thick, rigid stems and small leaves, which rarely blossom and never yield seed; plants raised from French seed blossom sparingly, but all the flowers are sterile; on the other hand, plants raised from sweet-peas grown near Darjeeling in Upper India, but originally derived from England, can be successfully cultivated on the plains of India; for they flower and seed profusely, and their stems are lax and scandent. In some of the foregoing cases, as Dr. Hooker has remarked to me, the greater success may perhaps be attributed to the seeds having been more fully ripened under a more favourable climate; but this view can hardly be extended to so many cases, including plants, which, from being cultivated under a climate hotter than their native one, become fitted for a still hotter climate. We may therefore safely conclude that plants can to a certain extent become accustomed to a climate either hotter or colder than their own; although the latter cases have been more frequently observed.]

We will now consider the means by which acclimatisation may be effected, namely, through the appearance of varieties having a different constitution, and through the effects of habit. In regard to new varieties, there is no evidence that a change in the constitution of the offspring necessarily stands in any direct relation with the nature of the climate inhabited by the parents. On the contrary, it is certain that hardy and tender varieties of the same species appear in the same country. New varieties thus spontaneously arising become fitted to slightly different climates in two different ways; firstly, they may have the power, either as seedlings or when full-grown, of resisting intense cold, as with the Moscow pear, or of resisting intense heat, as with some kinds of Pelargonium, or the flowers may withstand severe frost, as with the Forelle pear. Secondly, plants may become adapted to climates widely different from their own, from flowering and fruiting either earlier or later in the season. In both these cases the power of acclimatisation by man consists simply in the selection and preservation of new varieties. But without any direct intention on his part of securing a hardier variety, acclimatisation may be unconsciously effected by merely raising tender plants from seed, and by occasionally attempting their cultivation further and further northwards, as in the case of maize, the orange and the peach.

How much influence ought to be attributed to inherited habit or custom in the acclimatisation of animals and plants is a much more difficult question. In many cases natural selection can hardly have failed to have come into play and complicated the result. It is notorious that mountain sheep resist severe weather and storms of snow which would destroy lowland breeds; but then mountain sheep have been thus exposed from time immemorial, and all delicate individuals will have been destroyed, and the hardiest preserved. So with the Arrindy silk-moths of China and India; who can tell how far natural selection may have taken a share in the formation of the two races, which are now fitted for such widely different climates? It seems at first probable that the many fruit-trees which are so well fitted for the hot summers and cold winters of North America, in contrast with their poor success under our climate, have become adapted through habit; but when we reflect on the multitude of seedlings annually raised in that country, and that none would succeed unless born with a fitting constitution, it is possible that mere habit may have done nothing towards their acclimatisation. On the other hand, when we hear that Merino sheep, bred during no great number of generations at the Cape of Good Hope—that some European plants raised during only a few generations in the cooler parts of India, withstand the hotter parts of that country much better than the sheep or seeds imported directly from England, we must attribute some influence to habit. We are led to the same conclusion when we hear from Naudin (24/78. Quoted by Asa Gray in 'Am. Journ. of Sc.' 2nd series January 1865 page 106.) that the races of melons, squashes, and gourds, which have long been cultivated in Northern Europe, are comparatively more precocious, and need much less heat for maturing their fruit, than the varieties of the same species recently brought from tropical regions. In the reciprocal conversion of summer and winter wheat, barley, and vetches into each other, habit produces a marked effect in the course of a very few generations. The same thing apparently occurs with the varieties of maize, which, when carried from the Southern States of America, or into Germany, soon became accustomed to their new homes. With vine-plants taken to the West Indies from Madeira, which are said to succeed better than plants brought directly from France, we have some degree of acclimatisation in the individual, independently of the production of new varieties by seed.

The common experience of agriculturists is of some value, and they often advise persons to be cautious in trying the productions of one country in another. The ancient agricultural writers of China recommend the preservation and cultivation of the varieties peculiar to each country. During the classical period, Columella wrote, "Vernaculum pecus peregrino longe praestantius est." (24/79. For China see 'Memoire sur les Chinois' tome 11 1786 page 60. Columella is quoted by Carlier in 'Journal de Physique' tome 24 1784.)

I am aware that the attempt to acclimatise either animals or plants has been called a vain chimera. No doubt the attempt in most cases deserves to be thus called, if made independently of the production of new varieties endowed with a different constitution. With plants propagated by buds, habit rarely produces any effect; it apparently acts only through successive seminal generations. The laurel, bay, laurestinus, etc., and the Jerusalem artichoke, which are propagated by cuttings or tubers, are probably now as tender in England as when first introduced; and this appears to be the case with the potato, which until recently was seldom multiplied by seed. With plants propagated by seed, and with animals, there will be little or no acclimatisation unless the hardier individuals are either intentionally or unconsciously preserved. The kidney-bean has often been advanced as an instance of a plant which has not become hardier since its first introduction into Britain. We hear, however, on excellent authority (24/80. Messrs. Hardy and Son in 'Gardener's Chronicle' 1856 page 589.) that some very fine seed, imported from abroad, produced plants "which blossomed most profusely, but were nearly all but abortive, whilst plants grown alongside from English seed podded abundantly;" and this apparently shows some degree of acclimatisation in our English plants. We have also seen that seedlings of the kidney-bean occasionally appear with a marked power of resisting frost; but no one, as far as I can hear, has ever separated such hardy seedlings, so as to prevent accidental crossing, and then gathered their seed, and repeated the process year after year. It may, however, be objected with truth that natural selection ought to have had a decided effect on the hardiness of our kidney-beans; for the tenderest individuals must have been killed during every severe spring, and the hardier preserved. But it should be borne in mind that the result of increased hardiness would simply be that gardeners, who are always anxious for as early a crop as possible, would sow their seed a few days earlier than formerly. Now, as the period of sowing depends much on the soil and elevation of each district, and varies with the season; and as new varieties have often been imported from abroad, can we feel sure that our kidney-beans are not somewhat hardier? I have not been able, by searching old horticultural works, to answer this question satisfactorily.

On the whole the facts now given show that, though habit does something towards acclimatisation, yet that the appearance of constitutionally different individuals is a far more effective agent. As no single instance has been recorded either with animals or plants of hardier individuals having been long and steadily selected, though such selection is admitted to be indispensable for the improvement of any other character, it is not surprising that man has done little in the acclimatisation of domesticated animals and cultivated plants. We need not, however, doubt that under nature new races and new species would become adapted to widely different climates, by variation, aided by habit, and regulated by natural selection.

## [ARRESTS OF DEVELOPMENT: RUDIMENTARY AND ABORTED ORGANS.

Modifications of structure from arrested development, so great or so serious as to deserve to be called monstrosities, are not infrequent with domesticated animals, but, as they differ much from any normal structure, they require only a passing notice. Thus the whole head may be represented by a soft nipple-like projection, and the limbs by mere papillae. These rudiments of limbs are sometimes inherited, as has been observed in a dog. (24/81. Isid. Geoffroy Saint-Hilaire 'Hist. Nat. des Anomalies' 1836 tome 2 pages 210, 223, 224, 395; 'Philosoph. Transact.' 1775 page 313.)

Many lesser anomalies appear to be due to arrested development. What the cause of the arrest may be, we seldom know, except in the case of direct injury to the embryo. That the cause does not generally act at an extremely early embryonic period we may infer from the affected organ seldom being wholly aborted,—a rudiment being generally preserved. The external ears are represented by mere vestiges in a Chinese breed of sheep; and in another breed, the tail is reduced "to a little button, suffocated in a manner, by fat." (24/82. Pallas quoted by Youatt on 'Sheep' page 25.) In tailless dogs and cats a stump is left. In certain breeds of fowls the comb and wattles are reduced to rudiments; in the Cochin-China breed scarcely more than rudiments of spurs exist. With polled Suffolk cattle, "rudiments of horns can often be felt at an early age" (24/83. Youatt on 'Cattle' 1834 page 174.); and with species in a state of nature, the relatively great development of rudimentary organs at an early period of life is highly characteristic of such organs. With hornless breeds of cattle and sheep, another and singular kind of rudiment has been observed, namely, minute dangling horns attached to the skin alone, and which are often shed and grow again. With hornless goats, according to Desmarest (24/84. 'Encyclop. Method.' 1820 page 483: see page 500, on the Indian zebu casting its horns. Similar cases in European cattle were given in the third chapter.), the bony protuberance which properly supports the horn exists as a mere rudiment.

With cultivated plants it is far from rare to find the petals, stamens, and pistils represented by rudiments, like those observed in natural species. So it is with the whole seed in many fruits; thus, near Astrakhan there is a grape with mere traces of seeds, "so small and lying so near the stalk that they are not perceived in eating the grape." (24/85. Pallas 'Travels' English Translat. volume 1 page 243.) In certain varieties of the gourd, the tendrils, according to Naudin, are represented by rudiments or by various monstrous growths. In the broccoli and cauliflower the greater number of the flowers are incapable of expansion, and include rudimentary organs. In the Feather hyacinth (Muscari comosum) in its natural state the upper and central flowers are brightly coloured but rudimentary; under cultivation the tendency to abortion travels downwards and outwards, and all the flowers become rudimentary; but the abortive stamens and pistils are not so small in the lower as in the upper flowers. In the Viburnum opulus, on the other hand, the outer flowers naturally have their organs of fructification in a rudimentary state, and the corolla is of large size; under cultivation, the change spreads to the centre, and all the flowers become affected. In the compositae, the so-called doubling of the flowers consists in the greater development of the corolla of the central florets, generally accompanied with some degree of sterility; and it has been observed (24/86. Mr. Beaton in 'Journal of Horticulture' May 21, 1861 page 133.) that the progressive doubling invariably spreads from the circumference to the centre,—that is, from the ray florets, which so often include rudimentary organs, to those of the disc. I may add, as bearing on this subject, that with Asters, seeds taken from the florets of the circumference have been found to yield the greatest number of double flowers. (24/87. Lecoq 'De la Fecondation' 1862 page 233.) In the above cases we have a natural tendency in certain parts to be rudimentary, and this under culture spreads either to, or from, the axis of the plant. It deserves notice, as showing how the same laws govern the changes which natural species and artificial varieties undergo, that in the species of Carthamus, one of the Compositae, a tendency to the abortion of the pappus may be traced extending from the circumference to the centre of the disc as in the so-called doubling of the flowers in the members of the same family. Thus, according to A. de Jussieu (24/88. 'Annales du Museum' tome 6 page 319.), the abortion is only partial in Carthamus creticus, but more extended in C. lanatus; for in this species only two or three of the central seeds are furnished with a pappus, the surrounding seeds being either quite naked or furnished with a few hairs; and lastly in C. tinctorius, even the central seeds are destitute of pappus, and the abortion is complete.

With animals and plants under domestication, when an organ disappears, leaving only a rudiment, the loss has generally been sudden, as with hornless and tailless breeds; and such cases may be ranked as inherited monstrosities. But in some few cases the loss has been gradual, and has been effected partly by selection, as with the rudimentary combs and wattles of certain fowls. We have also seen that the wings of some domesticated birds have been slightly reduced by disuse, and the great reduction of the wings in certain silk-moths, with mere rudiments left, has probably been aided by disuse.]

With species in a state of nature, rudimentary organs are extremely common. Such organs are generally variable, as several naturalists have observed; for, being useless, they are not regulated by natural selection, and they are more or less liable to reversion. The same rule certainly holds good with parts which have become rudimentary under domestication. We do not know through what steps under nature rudimentary organs have passed in being reduced to their present condition; but we so incessantly see in species of the same group the finest gradations between an organ in a rudimentary and perfect state, that we are led to believe that the passage must have been extremely gradual. It may be doubted whether a change of structure so abrupt as the sudden loss of an organ would ever be of service to a species in a state of nature; for the conditions to which all organisms are closely adapted usually change very slowly. Even if an organ did suddenly disappear in some one individual by an arrest of development, intercrossing with the other individuals of the same species would tend to cause its partial reappearance; so that its final reduction could only be effected by some other means. The most probable view is, that a part which is now rudimentary, was formerly, owing to changed habits of life, used less and less, being at the same time reduced in size by disuse, until at last it became quite useless and superfluous. But as most parts or organs are not brought into action during an early period of life, disuse or decreased action will not lead to their reduction until the organism arrives at a somewhat advanced age; and from the principle of inheritance at corresponding ages the reduction will be transmitted to the offspring at the same advanced stage of growth. The part or organ will thus retain its full size in the embryo, as we know to be the case with most rudiments. As soon as a part becomes useless, another principle, that of economy of growth, will come into play, as it would be an advantage to an organism exposed to severe competition to save the development of any useless part; and individuals having the part less developed will have a slight advantage over others. But, as Mr. Mivart has justly remarked, as soon as a part is much reduced, the saving from its further reduction will be utterly insignificant; so that this cannot be effected by natural selection. This manifestly holds good if the part be formed of mere cellular tissue, entailing little expenditure of nutriment. How then can the further reduction of an already somewhat reduced part be effected? That this has occurred repeatedly under Nature is shown by the many gradations which exist between organs in a perfect state and the merest vestiges of them. Mr. Romanes (24/89. I suggested in 'Nature' (volume 8 pages 432, 505) that with organisms subjected to unfavourable conditions all the parts would tend towards reduction, and that under such circumstances any part which was not kept up to its standard size by natural selection would, owing to intercrossing, slowly but steadily decrease. In three subsequent communications to 'Nature' (March 12, April 9, and July 2, 1874), Mr. Romanes gives his improved view.) has, I think, thrown much light on this difficult problem. His view, as far as it can be given in a few words, is as follows: all parts are somewhat variable and fluctuate in size round an average point. Now, when a part has already begun from any cause to decrease, it is very improbable that the variations should be as great in the direction of increase as of diminution; for the previous reduction shows that circumstances have not been favourable for its development; whilst there is nothing to check variations in the opposite direction. If this be so, the long continued crossing of many individuals furnished with an organ which fluctuates in a greater degree towards decrease than towards increase, will slowly but steadily lead to its diminution. With respect to the complete and absolute abortion of a part, a distinct principle, which will be discussed in the chapter on pangenesis, probably comes into action.

With animals and plants reared by man there is no severe or recurrent struggle for existence, and the principle of economy will not come into action, so that the reduction of an organ will not thus be aided. So far, indeed, is this from being the case, that in some few instances organs, which are naturally rudimentary in the parent-species, become partially redeveloped in the domesticated descendants. Thus cows, like most other ruminants, properly have four active and two rudimentary mamma; but in our domesticated animals, the latter occasionally become considerably developed and yield milk. The atrophied mammae, which, in male domesticated animals, including man, have in some rare cases grown to full size and secreted milk, perhaps offer an analogous case. The hind feet of dogs naturally include rudiments of a fifth toe, and in certain large breeds these toes, though still rudimentary, become considerably developed and are furnished with claws. In the common Hen, the spurs and comb are rudimentary, but in certain breeds these become, independently of age or disease of the ovaria, well developed. The stallion has canine teeth, but the mare has only traces of the alveoli, which, as I am informed by the eminent veterinarian Mr. G.T. Brown, frequently contain minute irregular nodules of bone. These nodules, however, sometimes become developed into imperfect teeth, protruding through the gums and coated with enamel; and occasionally they grow to a fourth or even a third of the length of the canines in the stallion. With plants I do not know whether the redevelopment of rudimentary organs occurs more frequently under culture than under nature. Perhaps the pear-tree may be a case in point, for when wild it bears thorns, which consist of branches in a rudimentary condition and serve as a protection, but, when the tree is cultivated, they are reconverted into branches.

# CHAPTER 2.XXV.

LAWS OF VARIATION, continued.—CORRELATED VARIABILITY.

**EXPLANATION OF TERM CORRELATION. CONNECTED WITH DEVELOPMENT. MODIFICATIONS CORRELATED WITH THE INCREASED OR DECREASED SIZE OF PARTS. CORRELATED VARIATION OF HOMOLOGOUS PARTS. FEATHERED FEET IN BIRDS ASSUMING THE STRUCTURE OF THE WINGS. CORRELATION BETWEEN THE HEAD AND THE EXTREMITIES. BETWEEN THE SKIN AND DERMAL APPENDAGES. BETWEEN THE ORGANS OF SIGHT AND HEARING. CORRELATED MODIFICATIONS IN THE ORGANS OF PLANTS. CORRELATED MONSTROSITIES. CORRELATION BETWEEN THE SKULL AND EARS. SKULL AND CREST OF FEATHERS. SKULL AND HORNS. CORRELATION OF GROWTH COMPLICATED BY THE ACCUMULATED EFFECTS OF NATURAL SELECTION. COLOUR AS CORRELATED WITH CONSTITUTIONAL PECULIARITIES.**

All parts of the organisation are to a certain extent connected together; but the connection may be so slight that it hardly exists, as with compound animals or the buds on the same tree. Even in the higher animals various parts are not at all closely related; for one part may be wholly suppressed or rendered monstrous without any other part of the body being affected. But in some cases, when one part varies, certain other parts always, or nearly always, simultaneously vary; they are then subject to the law of correlated variation. The whole body is admirably co-ordinated for the peculiar habits of life of each organic being, and may be said, as the Duke of Argyll insists in his 'Reign of Law' to be correlated for this purpose. Again, in large groups of animals certain structures always co-exist: for instance, a peculiar form of stomach with teeth of peculiar form, and such structures may in one sense be said to be correlated. But these cases have no necessary connection with the law to be discussed in the present chapter; for we do not know that the initial or primary variations of the several parts were in any way related: slight modifications or individual differences may have been preserved, first in one and then in another part, until the final and perfectly co-adapted structure was acquired; but to this subject I shall presently recur. Again, in many groups of animals the males alone are furnished with weapons, or are ornamented with gay colours; and these characters manifestly stand in some sort of correlation with the male reproductive organs, for when the latter are destroyed these characters disappear. But it was shown in the twelfth chapter that the very same peculiarity may become attached at any age to either sex, and afterwards be exclusively transmitted to the same sex at a corresponding age. In these cases we have inheritance limited by both sex and age; but we have no reason for supposing that the original cause of the variation was necessarily connected with the reproductive organs, or with the age of the affected being.

In cases of true correlated variation, we are sometimes able to see the nature of the connection; but in most cases it is hidden from us, and certainly differs in different cases. We can seldom say which of two correlated parts first varies, and induces a change in the other; or whether the two are the effects of some common cause. Correlated variation is an important subject for us; for when one part is modified through continued selection, either by man or under nature, other parts of the organisation will be unavoidably modified. From this correlation it apparently follows that with our domesticated animals and plants, varieties rarely or never differ from one another by a single character alone.

One of the simplest cases of correlation is that a modification which arises during an early stage of growth tends to influence the subsequent development of the same part, as well as of other and intimately connected parts. Isidore Geoffroy Saint-Hilaire states (25/1. 'Hist. des Anomalies' tome 3 page 392. Prof. Huxley applies the same principle in accounting for the remarkable, though normal, differences in the arrangement of the nervous system in the Mollusca, in his paper on the Morphology of the Cephalous Mollusca in 'Phil. Transact.' 1853 page 56.) that this may constantly be observed with monstrosities in the animal kingdom; and Moquin-Tandon (25/2. 'Elements de Teratologie Veg.' 1841 page 13.) remarks, that, as with plants the axis cannot become monstrous without in some way affecting the organs subsequently produced from it, so axial anomalies are almost always accompanied by deviations of structure in the appended parts. We shall presently see that with short-muzzled races of the dog certain histological changes in the basal elements of the bones arrest their development and shorten them, and this affects the position of the subsequently developed molar teeth. It is probable that certain modifications in the larvae of insects would affect the structure of the mature insects. But we must be careful not to extend this view too far, for during the normal course of development, certain species pass through an extraordinary course of change, whilst other and closely allied species arrive at maturity with little change of structure.

Another simple case of correlation is that with the increased or decreased dimensions of the whole body, or of any particular part, certain organs are increased or diminished in number, or are otherwise modified. Thus pigeon fanciers have gone on selecting pouters for length of body, and we have seen that their vertebrae are generally increased not only in size but in number, and their ribs in breadth. Tumblers have been selected for their small bodies, and their ribs and primary wing-feathers are generally lessened in number. Fantails have been selected for their large widely-expanded tails, with numerous tail-feathers, and the caudal vertebrae are increased in size and number. Carriers have been selected for length of beak, and their tongues have become longer, but not in strict accordance with the length of beak. In this latter breed and in others having large feet, the number of the scutellae on the toes is greater than in the breeds with small feet. Many similar cases could be given. In Germany it has been observed that the period of gestation is longer in large than in small breeds of cattle. With our highly-improved breeds of all kinds, the periods of maturity and of reproduction have advanced with respect to the age of the animal; and, in correspondence with this, the teeth are now developed earlier than formerly, so that, to the surprise of agriculturists, the ancient rules for judging of the age of an animal by the state of its teeth are no longer trustworthy. (25/3. Prof. J.B. Simonds on the Age of the Ox, Sheep, etc. quoted in 'Gardener's Chronicle' 1854 page 588.)

## CORRELATED VARIATION OF HOMOLOGOUS PARTS.

Parts which are homologous tend to vary in the same manner; and this is what might have been expected, for such parts are identical in form and structure during an early period of embryonic development, and are exposed in the egg or womb to similar conditions. The symmetry, in most kinds of animals, of the corresponding or homologous organs on the right and left sides of the body, is the simplest case in point; but this symmetry sometimes fails, as with rabbits having only one ear, or stags with one horn, or with many-horned sheep which sometimes carry an additional horn on one side of their heads. With flowers which have regular corollas, all the petals generally vary in the same manner, as we see in the complicated and symmetrical pattern, on the flowers, for instance, of the Chinese pink; but with irregular flowers, though the petals are of course homologous, this symmetry often fails, as with the varieties of the Antirrhinum or snapdragon, or that variety of the kidney-bean (Phaseolus) which has a white standard-petal.

In the Vertebrata the front and hind limbs are homologous, and they tend to vary in the same manner, as we see in long and short legged, or in thick and thin legged races of the horse and dog. Isidore Geoffroy (25/4. 'Hist. des Anomalies' tome 1 page 674.) has remarked on the tendency of supernumerary digits in man to appear, not only on the right and left sides, but on the upper and lower extremities. Meckel has insisted (25/5. Quoted by Isid. Geoffroy ibid tome 1 page 635.) that, when the muscles of the arm depart in number or arrangement from their proper type, they almost always imitate those of the leg; and so conversely the varying muscles of the leg imitate the normal muscles of the arm.

In several distinct breeds of the pigeon and fowl, the legs and the two outer toes are heavily feathered, so that in the trumpeter pigeon they appear like little wings. In the feather-legged bantam the "boots" or feathers, which grow from the outside of the leg and generally from the two outer toes, have, according to the excellent authority of Mr. Hewitt (25/6. 'The Poultry Book' by W.B. Tegetmeier 1866 page 250.), been seen to exceed the wing-feathers in length, and in one case were actually nine and a half inches long! As Mr. Blyth has remarked to me, these leg-feathers resemble the primary wing-feathers, and are totally unlike the fine down which naturally grows on the legs of some birds, such as grouse and owls. Hence it may be suspected that excess of food has first given redundancy to the plumage, and then that the law of homologous variation has led to the development of feathers on the legs, in a position corresponding with those on the wing, namely, on the outside of the tarsi and toes. I am strengthened in this belief by the following curious case of correlation, which for a long time seemed to me utterly inexplicable, namely, that in pigeons of any breed, if the legs are feathered, the two outer toes are partially connected by skin. These two outer toes correspond with our third and fourth toes. (25/7. Naturalists differ with respect to the homologies of the digits of birds; but several uphold the view above advanced. See on this subject Dr. E.S. Morse in 'Annals of the Lyceum of Nat. Hist. of New York' volume 10 1872 page 16.) Now, in the wing of the pigeon or of any other bird, the first and fifth digits are aborted; the second is rudimentary and carries the so-called "bastard-wing;" whilst the third and fourth digits are completely united and enclosed by skin, together forming the extremity of the wing. So that in feather-footed pigeons, not only does the exterior surface support a row of long feathers, like wing-feathers, but the very same digits which in the wing are completely united by skin become partially united by skin in the feet; and thus by the law of the correlated variation of homologous parts we can understand the curious connection of feathered legs and membrane between the two outer toes.

Andrew Knight (24/8. A. Walker on 'Intermarriage' 1838 page 160.) has remarked that the face or head and the limbs usually vary together in general proportions. Compare, for instance, the limbs of a dray and race horse, or of a greyhound and mastiff. What a monster a greyhound would appear with the head of a mastiff! The modern bulldog, however, has fine limbs, but this is a recently-selected character. From the measurements given in the sixth chapter, we see that in several breeds of the pigeon the length of the beak and the size of the feet are correlated. The view which, as before explained, seems the most probable is, that disuse in all cases tends to diminish the feet, the beak becoming at the same time shorter through correlation; but that in some few breeds in which length of beak has been a selected point, the feet, notwithstanding disuse, have increased in size through correlation. In the following case some kind of correlation is seen to exist between the feet and beak: several specimens have been sent to Mr. Bartlett at different times, as hybrids between ducks and fowls, and I have seen one; these were, as might be expected, ordinary ducks in a semi-monstrous condition, and in all of them the swimming-web between the toes was quite deficient or much reduced, and in all the beak was narrow and ill-shaped.

With the increased length of the beak in pigeons, not only the tongue increases in length, but likewise the orifice of the nostrils. But the increased length of the orifice of the nostrils perhaps stands in closer correlation with the development of the corrugated skin or wattle at the base of the beak, for when there is much wattle round the eyes, the eyelids are greatly increased or even doubled in length.

There is apparently some correlation even in colour between the head and the extremities. Thus with horses a large white star or blaze on the forehead is generally accompanied by white feet. (25/9. 'The Farrier and Naturalist' volume 1 1828 page 456. A gentleman who has attended to this point, tells me that about three-fourths of white-faced horses have white legs.) With white rabbits and cattle, dark marks often co-exist on the tips of the ears and on the feet. In black and tan dogs of different breeds, tan-coloured spots over the eyes and tan-coloured feet almost invariably go together. These latter cases of connected colouring may be due either to reversion or to analogous variation,—subjects to which I shall hereafter return,—but this does not necessarily determine the question of their original correlation. Mr. H.W. Jackson informs me that he has observed many hundred white-footed cats, and he finds that all are more or less conspicuously marked with white on the front of the neck or chest.

The lopping forwards and downwards of the immense ears of fancy rabbits seems partly due to the disuse of the muscles, and partly to the weight and length of the ears, which have been increased by selection during many generations. Now, with the increased size and changed direction of the ears not only has the bony auditory meatus become changed in outline, direction, and greatly in size, but the whole skull has been slightly modified. This could be clearly seen in "half-lops"—that is, in rabbits with only one ear lopping forward— for the opposite sides of their skulls were not strictly symmetrical. This seems to me a curious instance of correlation, between hard bones and organs so soft and flexible, as well as so unimportant under a physiological point of view, as the external ears. The result no doubt is largely due to mere mechanical action, that is, to the weight of the ears, on the same principle that the skull of a human infant is easily modified by pressure.

The skin and the appendages of hair, feathers, hoofs, horns, and teeth, are homologous over the whole body. Every one knows that the colour of the skin and that of the hair usually vary together; so that Virgil advises the shepherd to look whether the mouth and tongue of the ram are black, lest the lambs should not be purely white. The colour of the skin and hair, and the odour emitted by the glands of the skin, are said (25/10. Godron 'Sur l'Espèce' tome 2 page 217.) to be connected, even in the same race of men. Generally the hair varies in the same way all over the body in length, fineness, and curliness. The same rule holds good with feathers, as we see with the laced and frizzled breeds both of fowls and pigeons. In the common cock the feathers on the neck and loins are always of a particular shape, called hackles: now in the Polish breed, both sexes are characterised by a tuft of feathers on the head, and through correlation these feathers in the male always assume the form of hackles. The wing and tail-feathers, though arising from parts not homologous, vary in length together; so that long or short winged pigeons generally have long or short tails. The case of the Jacobin-pigeon is more curious, for the wing and tail feathers are remarkably long; and this apparently has arisen in correlation with the elongated and reversed feathers on the back of the neck, which form the hood.

The hoofs and hair are homologous appendages; and a careful observer, namely Azara (25/11. 'Quadrupèdes du Paraguay' tome 2 page 333.), states that in Paraguay horses of various colours are often born with their hair curled and twisted like that on the head of a negro. This peculiarity is strongly inherited. But what is remarkable is that the hoofs of these horses "are absolutely like those of a mule." The hair also of their manes and tails is invariably much shorter than usual, being only from four to twelve inches in length; so that curliness and shortness of the hair are here, as with the negro, apparently correlated.

With respect to the horns of sheep, Youatt (25/12. 'On Sheep' page 142.) remarks that "multiplicity of horns is not found in any breed of much value; it is generally accompanied by great length and coarseness of the fleece." Several tropical breeds of sheep which are clothed with hair instead of wool, have horns almost like those of a goat. Sturm (25/13. 'Ueber Racen, Kreuzungen' etc. 1825 s. 24.) expressly declares that in different races the more the wool is curled the more the horns are spirally twisted. We have seen in the third chapter, where other analogous facts have been given, that the parent of the Mauchamp breed, so famous for its fleece, had peculiarly shaped horns. The inhabitants of Angora assert (25/14. Quoted from Conolly in 'The Indian Field' February 1859 volume 2 page 266.) that "only the white goats which have horns wear the fleece in the long curly locks that are so much admired; those which are not horned having a comparatively close coat." From these cases we may infer that the hair or wool and the horns tend to vary in a correlated manner. (25/15. In the third chapter I have said that "the hair and horns are so closely related to each other, that they are apt to vary together." Dr. Wilckens ("Darwin's Theorie" 'Jahrbuch der

Deutschen Viehzucht' 1866 1. Heft) translates my words into "lang- und grobhaarige Thiere sollen geneigter sein, lange und viele Horner zu bekommen" and he then justly disputes this proposition; but what I have really said, in accordance with the authorities just quoted, may, I think, be trusted.) Those who have tried hydropathy are aware that the frequent application of cold water stimulates the skin; and whatever stimulates the skin tends to increase the growth of the hair, as is well shown in the abnormal growth of hair near old inflamed surfaces. Now, Professor Low (25/16. 'Domesticated Animals of the British Islands' pages 307, 368. Dr. Wilckens argues ('Landwirth. Wochenblatt' Nr. 10 1869) to the same effect with respect to domestic animals in Germany.) is convinced that with the different races of British cattle thick skin and long hair depend on the humidity of the climate which they inhabit. We can thus see how a humid climate might act on the horns—in the first place directly on the skin and hair, and secondly by correlation on the horns. The presence or absence of horns, moreover, both in the case of sheep and cattle, acts, as will presently be shown, by some sort of correlation on the skull.

With respect to hair and teeth, Mr. Yarrell (25/17. 'Proceedings Zoolog. Soc.' 1833 page 113.) found many of the teeth deficient in three hairless "Egyptian dogs," and in a hairless terrier. The incisors, canines, and the premolars suffered most, but in one case all the teeth, except the large tubercular molar on each side, were deficient. With man several striking cases have been recorded (25/18. Sedgwick 'Brit. and Foreign Medico-Chirurg. Review' April 1863 page 453.) of inherited baldness with inherited deficiency, either complete or partial, of the teeth. I may give an analogous case, communicated to me by Mr. W. Wedderburn, of a Hindoo family in Scinde, in which ten men, in the course of four generations, were furnished, in both jaws taken together, with only four small and weak incisor teeth and with eight posterior molars. The men thus affected have very little hair on the body, and become bald early in life. They also suffer much during hot weather from excessive dryness of the skin. It is remarkable that no instance has occurred of a daughter being thus affected; and this fact reminds us how much more liable men are in England to become bald than women. Though the daughters in the above family are never affected, they transmit the tendency to their sons; and no case has occurred of a son transmitting it to his sons. The affection thus appears only in alternate generations, or after longer intervals. There is a similar connection between hair and teeth, according to Mr. Sedgwick, in those rare cases in which the hair has been renewed in old age, for this has "usually been accompanied by a renewal of the teeth." I have remarked in a former part of this volume that the great reduction in the size of the tusks in domestic boars probably stands in close relation with their diminished bristles, due to a certain amount of protection; and that the reappearance of the tusks in boars, which have become feral and are fully exposed to the weather, probably depends on the reappearance of the bristles. I may add, though not strictly connected with our present point, that an agriculturist (25/19. 'Gardener's Chronicle' 1849 page 205.) asserts that "pigs with little hair on their bodies are most liable to lose their tails, showing a weakness of the tegumental structure. It may be prevented by crossing with a more hairy breed."

In the previous cases deficient hair, and teeth deficient in number or size, are apparently connected. In the following cases abnormally redundant hair, and teeth either deficient or redundant, are likewise connected. Mr. Crawfurd (25/20. 'Embassy to the Court of Ava' volume 1 page 320.) saw at the Burmese Court a man, thirty years old, with his whole body, except the hands and feet, covered with straight silky hair, which on the shoulders and spine was five inches in length. At birth the ears alone were covered. He did not arrive at puberty, or shed his milk teeth, until twenty years old; and at this period he acquired five teeth in the upper jaw, namely, four incisors and one canine, and four incisor teeth in the lower jaw; all the teeth were small. This man had a daughter who was born with hair within her ears; and the hair soon extended over her body. When Captain Yule (25/21. 'Narrative of a Mission to the Court of Ava in 1855' page 94.) visited the Court, he found this girl grown up; and she presented a strange appearance with even her nose densely covered with soft hair. Like her father, she was furnished with incisor teeth alone. The King had with difficulty bribed a man to marry her, and of her two children, one, a boy fourteen months old, had hair growing out of his ears, with a beard and moustache. This strange peculiarity has, therefore, been inherited for three generations, with the molar teeth deficient in the grandfather and mother; whether these teeth would likewise fail in the infant could not then be told.

A parallel case of a man fifty-five years old, and of his son, with their faces covered with hair, has recently occurred in Russia. Dr. Alex. Brandt has sent me an account of this case, together with specimens of the extremely fine hair from the cheeks. The man is deficient in teeth, possessing only four incisors in the lower and two in the upper jaw. His son, about three years old, has no teeth except four lower incisors. The case, as Dr. Brandt remarks in his letter, no doubt is due to an arrest of development in the hair and teeth. We here see how independent of the ordinary conditions of existence such arrests must be, for the lives of a Russian peasant and of a native of Burmah are as different as possible. (25/22. I owe to the kindness of M. Chauman, of St. Petersburg, excellent photographs of this man and his son, both of whom have since been exhibited in Paris and London.)

Here is another and somewhat different case communicated to me by Mr. Wallace on the authority of Dr. Purland, a dentist: Julia Pastrana, a Spanish dancer, was a remarkably fine woman, but she had a thick masculine beard and a hairy forehead; she was photographed, and her stuffed skin was exhibited as a show; but what concerns us is, that she had in both the upper and lower jaw an irregular double set of teeth, one row being placed within the other, of which Dr. Purland took a cast. From the redundancy of teeth her mouth projected, and her face had a gorilla-like appearance. These cases and those of the hairless dogs forcibly call to mind the fact, that the two orders of mammals—namely, the Edentata and Cetacea—which are the most abnormal in their dermal covering, are likewise the most abnormal either by deficiency or redundancy of teeth.

The organs of sight and hearing are generally admitted to be homologous with one another and with various dermal appendages; hence these parts are liable to be abnormally affected in conjunction. Mr. White Cowper says "that in all cases of double microphthalmia brought under his notice he has at the same time met with defective development of the dental system." Certain forms of blindness seem to be associated with the colour of the hair; a man with black hair and a woman with light-coloured hair, both of sound constitution, married and had nine children, all of whom were born blind; of these children, five "with dark hair and brown iris were afflicted with amaurosis; the four others, with light-coloured hair and blue iris, had amaurosis and cataract conjoined." Several cases could be given, showing that some relation exists between various affections of the eyes and ears; thus Liebreich states that out of 241 deaf-mutes in Berlin, no less than fourteen suffered from the rare disease called pigmentary retinitis. Mr. White Cowper and Dr. Earle have remarked that inability to distinguish different colours, or colour-blindness, "is often associated with a corresponding inability to distinguish musical sounds." (25/23. These statements are taken from Mr. Sedgwick in the 'Medico-Chirurg. Review' July 1861 page 198; April 1863 pages 455 and 458. Liebreich is quoted by Professor Devay in his 'Mariages Consanguins' 1862 page 116.)

Here is a more curious case: white cats, if they have blue eyes, are almost always deaf. I formerly thought that the rule was invariable, but I have heard of a few authentic exceptions. The first two notices were published in 1829 and relate to English and Persian cats: of the latter, the Rev. W.T. Bree possessed a female, and he states, "that of the offspring produced at one and the same birth, such as, like the mother, were entirely white (with blue eyes) were, like her, invariably deaf; while those that had the least speck of colour on their fur, as invariably possessed the usual faculty of hearing." (25/24. Loudon's 'Mag. of Nat. Hist.' volume 1 1829 pages 66, 178. See also Dr. P. Lucas 'L'Hered. Nat.' tome 1 page 428 on the inheritance of deafness in cats. Mr. Lawson Tait states ('Nature' 1873 page 323) that only male cats are thus affected; but this must be a hasty generalisation. The first case recorded in England by Mr. Bree related to a female, and Mr. Fox informs me that he has bred kittens from a white female with blue eyes, which was completely deaf; he has also observed other females in the same condition.) The Rev. W. Darwin Fox informs me that he has seen more than a dozen instances of this correlation in English, Persian, and Danish cats; but he adds "that, if one eye, as I have several times observed, be not blue, the cat hears. On the other hand, I have never seen a white cat with eyes of the common colour that was deaf." In France Dr. Sichel (25/25. 'Annales des Sc. Nat.' Zoolog. 3rd series 1847 tome 8 page 239.) has observed during twenty years similar facts; he adds the remarkable case of the iris beginning, at the end of four months, to grow dark-coloured, and then the cat first began to hear.

This case of correlation in cats has struck many persons as marvellous. There is nothing unusual in the relation between blue eyes and white fur; and we have already seen that the organs of sight and hearing are often simultaneously affected. In the present instance the cause probably lies in a slight arrest of development in the nervous system in connection with the sense-organs. Kittens during the first nine days, whilst their eyes are closed, appear to be completely deaf; I have made a great clanging noise with a poker and shovel close to their heads, both when they were asleep and awake, without producing any effect. The trial must not be made by shouting close to their ears, for they are, even when asleep, extremely sensitive to a breath of air. Now, as long as the eyes continue closed, the iris is no doubt blue, for in all the kittens which I have seen this colour remains for some time after the eyelids open. Hence, if we suppose the development of the organs of sight and hearing to be arrested at the stage of the closed eyelids, the eyes would remain permanently blue and the ears would be incapable of perceiving sound; and we should thus understand this curious case. As, however, the colour of the fur is determined long before birth, and as the blueness of the eyes and the whiteness of the fur are obviously connected, we must believe that some primary cause acts at a much earlier period.

The instances of correlated variability hitherto given have been chiefly drawn from the animal kingdom, and we will now turn to plants. Leaves, sepals, petals, stamens, and pistils are all homologous. In double flowers we see that the stamens and pistils vary in the same manner, and assume the form and colour of the petals. In the double columbine (Aquilegia vulgaris), the successive whorls of stamens are converted into cornucopias, which are enclosed within one another and resemble the true petals. In hose-in-hose flowers the sepals mock the petals. In some cases the flowers and leaves vary together in tint: in all the varieties of the common pea, which have purple flowers, a purple mark may be seen on the stipules.

M. Faivre states that with the varieties of Primula sinensis the colour of the flower is evidently correlated with the colour of the under side of the leaves; and he adds that the varieties with fimbriated flowers almost always have voluminous, balloon-like calyces. (25/26. 'Revue des Cours Scientifiques' June 5, 1869 page 430.) With other plants the leaves and fruit or seeds vary together in colour, as in a curious pale-leaved variety of the sycamore, which has recently been described in France (25/27. 'Gardener's Chronicle' 1864 page 1202.), and as in the purple-leaved hazel, in which the leaves, the husk of the nut, and the pellicle round the kernel are all coloured purple. (25/28. Verlot gives several other instances 'Des Varietes' 1865 page 72.) Pomologists can predict to a certain extent, from the size and appearance of the leaves of their seedlings, the probable nature of the fruit; for, as Van Mons remarks (25/29. 'Arbres Fruitiers' 1836 tome 2 pages 204, 226.) variations in the leaves are generally accompanied by some modification in the flower, and consequently in the fruit. In the Serpent melon, which has a narrow tortuous fruit above a yard in length, the stem of the plant, the peduncle of the female flower, and the middle lobe of the leaf, are all elongated in a remarkable manner. On the other hand, several varieties of Cucurbita, which have dwarfed stems, all produce, as Naudin remarks, leaves of the same peculiar shape. Mr. G. Maw informs me that all the varieties of the scarlet Pelargoniums which have contracted or imperfect leaves have contracted flowers: the difference between "Brilliant" and its parent "Tom Thumb" is a good instance of this. It may be suspected that the curious case described by Risso (25/30. 'Annales du Museum' tome 20 page 188.), of a variety of the Orange which produces on the young shoots rounded leaves with winged petioles, and afterwards elongated leaves on long but wingless petioles, is connected with the remarkable change in form and nature which the fruit undergoes during its development.

In the following instance we have the colour and the form of the petals apparently correlated, and both dependent on the nature of the season. An observer, skilled in the subject, writes (25/31. 'Gardener's Chronicle' 1843 page 877.), "I noticed, during the year 1842, that every Dahlia of which the colour had any tendency to scarlet, was deeply notched—indeed, to so great an extent as to give the petals the appearance of a saw; the indentures were, in some instances, more than a quarter of an inch deep." Again, Dahlias which have their petals tipped with a different colour from the rest of the flower are very inconstant, and during certain years some, or even all the flowers, become uniformly coloured; and it has been observed with several varieties (25/32. Ibid 1845 page 102.) that when this happens the petals grow much elongated and lose their proper shape. This, however, may be due to reversion, both in colour and form, to the aboriginal species.

In this discussion on correlation, we have hitherto treated of cases in which we can partly understand the bond of connection; but I will now give cases in which we cannot even conjecture, or can only very obscurely see, the nature of the bond. Isidore Geoffroy Saint-Hilaire, in his work on

Monstrosities, insists (25/33. 'Hist. des Anomalies' tome 3 page 402. See also M. Camille Dareste 'Recherches sur les Conditions' etc. 1863 pages 16, 48.), "que certaines anomalies coexistent rarement entr'elles, d'autres frequemment, d'autres enfin presque constamment, malgre la difference tres-grande de leur nature, et quoiqu'elles puissent paraitre COMPLETEMENT INDEPENDANTES les unes des autres." We see something analogous in certain diseases: thus in a rare affection of the renal capsules (of which the functions are unknown), the skin becomes bronzed; and in hereditary syphilis, as I hear from Sir J. Paget, both the milk and the second teeth assume a peculiar and characteristic form. Professor Rolleston, also, informs me that the incisor teeth are sometimes furnished with a vascular rim in correlation with intra-pulmonary deposition of tubercles. In other cases of phthisis and of cyanosis the nails and finger- ends become clubbed like acorns. I believe that no explanation has been offered of these and of many other cases of correlated disease.

What can be more curious and less intelligible than the fact previously given, on the authority of Mr. Tegetmeier, that young pigeons of all breeds, which when mature have white, yellow, silver-blue, or dun-coloured plumage, come out of the egg almost naked; whereas pigeons of other colours when first born are clothed with plenty of down? White Pea-fowls, as has been observed both in England and France (25/34. Rev. E.S. Dixon 'Ornamental Poultry' 1848 page 111; Isidore Geoffroy 'Hist. Anomalies' tome 1 page 211.), and as I have myself seen, are inferior in size to the common coloured kind; and this cannot be accounted for by the belief that albinism is always accompanied by constitutional weakness; for white or albino moles are generally larger than the common kind.

To turn to more important characters: the niata cattle of the Pampas are remarkable from their short foreheads, upturned muzzles, and curved lower jaws. In the skull the nasal and premaxillary bones are much shortened, the maxillaries are excluded from any junction with the nasals, and all the bones are slightly modified, even to the plane of the occiput. From the analogous case of the dog, hereafter to be given, it is probable that the shortening of the nasal and adjoining bones is the proximate cause of the other modifications in the skull, including the upward curvature of the lower jaw, though we cannot follow out the steps by which these changes have been effected.

Polish fowls have a large tuft of feathers on their heads; and their skulls are perforated by numerous holes, so that a pin can be driven into the brain without touching any bone. That this deficiency of bone is in some way connected with the tuft of feathers is clear from tufted ducks and geese likewise having perforated skulls. The case would probably be considered by some authors as one of balancement or compensation. In the chapter on

Fowls, I have shown that with Polish fowls the tuft of feathers was probably at first small; by continued selection it became larger, and then rested on a fibrous mass; and finally, as it became still larger, the skull itself became more and more protuberant until it acquired its present extraordinary structure. Through correlation with the protuberance of the skull, the shape and even the relative connection of the premaxillary and nasal bones, the shape of the orifice of the nostrils, the breadth of the frontal bone, the shape of the post-lateral processes of the frontal and squamosal bones, and the direction of the bony cavity of the ear, have all been modified. The internal configuration of the skull and the whole shape of the brain have likewise been altered in a truly marvellous manner.

After this case of the Polish fowl it would be superfluous to do more than refer to the details previously given on the manner in which the changed form of the comb has affected the skull, in various breeds of the fowl, causing by correlation crests, protuberances, and depressions on its surface.

With our cattle and sheep the horns stand in close connection with the size of the skull, and with the shape of the frontal bones; thus Cline (25/35. 'On the Breeding of Domestic Animals' 1829 page 6.) found that the skull of a horned ram weighed five times as much as that of a hornless ram of the same age. When cattle become hornless, the frontal bones are "materially diminished in breadth towards the poll;" and the cavities between the bony plates "are not so deep, nor do they extend beyond the frontals." (25/36. Youatt on 'Cattle' 1834 page 283.)

It may be well here to pause and observe how the effects of correlated variability, of the increased use of parts, and of the accumulation of so-called spontaneous variations through natural selection, are in many cases inextricably commingled. We may borrow an illustration from Mr. Herbert Spencer, who remarks that, when the Irish elk acquired its gigantic horns, weighing above one hundred pounds, numerous co-ordinated changes of structure would have been indispensable,—namely, a thickened skull to carry the horns; strengthened cervical vertebrae, with strengthened ligaments; enlarged dorsal vertebrae to support the neck, with powerful fore-legs and feet; all these parts being supplied with proper muscles, blood-vessels, and nerves. How then could these admirably co-ordinated modifications of structure have been acquired? According to the doctrine which I maintain, the horns of the male elk were slowly gained through sexual selection,—that is, by the best-armed males conquering the worse-armed, and leaving a greater number of descendants. But it is not at all necessary that the several parts of the body should have simultaneously varied. Each stag presents individual characteristics, and in the same district those which had slightly heavier horns, or stronger necks, or stronger bodies, or were the most courageous, would secure the greater number of does, and consequently have a greater number of offspring. The offspring would inherit, in a greater or less degree, these same qualities, would occasionally intercross with one another, or with other individuals varying in some favourable manner; and of their offspring, those which were the best endowed in any respect would continue multiplying; and so onwards, always progressing, sometimes in one direction, and sometimes in another, towards the excellently co-ordinated structure of the male elk. To make this clear, let us reflect on the probable steps, as shown in the twentieth chapter, by which our race and dray horses have arrived at their present state of excellence; if we could view the whole series of intermediate forms between one of these animals and an early unimproved progenitor, we should behold a vast number of animals, not equally improved in each generation throughout their entire structure, but sometimes a little more in one point, and sometimes in another, yet on the whole gradually approaching in character to our present race or dray horses, which are so admirably fitted in the one case for fleetness and in the other for draught.

Although natural selection would thus (25/37. Mr. Herbert Spencer 'Principles of Biology' 1864 volume 1 pages 452, 468 takes a different view; and in one place remarks: "We have seen reason to think that, as fast as essential faculties multiply, and as fast as the number of organs that co-operate in any given function increases, indirect equilibration through natural selection becomes less and less capable of producing specific adaptations; and remains fully capable only of maintaining the general fitness of constitution to conditions." This view that natural selection can do little in modifying the higher animals surprises me, seeing that man's selection has undoubtedly effected much with our domesticated quadrupeds and birds.) tend to give to the male elk its present structure, yet it is probable that the inherited effects of use, and of the mutual action of part on part, have been equally or more important. As the horns gradually increased in weight the muscles of the neck, with the bones to which they are attached, would increase in size and strength; and these parts would react on the body and legs. Nor must we overlook the fact that certain parts of the skull and the extremities would, judging by analogy, tend from the first to vary in a correlated manner. The increased weight of the horns would also act directly on the skull, in the same manner as when one bone is removed in the leg of a dog, the other bone, which has to carry the whole weight of the body, increases in thickness. But from the fact given with respect to horned and hornless cattle, it is probable that the horns and skull would immediately act on each other through the principle of correlation. Lastly, the growth and subsequent wear and tear of the augmented muscles and bones would require an increased supply of blood, and consequently increased supply of food; and this again would require increased powers of mastication, digestion, respiration, and excretion.

# COLOUR AS CORRELATED WITH CONSTITUTIONAL PECULIARITIES.

It is an old belief that with man there is a connection between complexions and constitution; and I find that some of the best authorities believe in this to the present day. (25/38. Dr. Prosper Lucas apparently disbelieves in any such connection; 'L'Hered. Nat.' tome 2 pages 88-94.) Thus Dr. Beddoe by his tables shows (25/39. 'British Medical Journal' 1862 page 433.) that a relation exists between liability to consumption and the colour of the hair, eyes, and skin. It has been affirmed (25/40. Boudin 'Geograph. Medicale' tome 1 page 406.) that, in the French army which invaded Russia, soldiers having a dark complexion from the southern parts of Europe, withstood the intense cold better than those with lighter complexions from the north; but no doubt such statements are liable to error.

In the second chapter on Selection I have given several cases proving that with animals and plants differences in colour are correlated with constitutional differences, as shown by greater or less immunity from certain diseases, from the attacks of parasitic plants and animals, from scorching by the sun, and from the action of certain poisons. When all the individuals of any one variety possess an immunity of this nature, we do not know that it stands in any sort of correlation with their colour; but when several similarly coloured varieties of the same species are thus characterised, whilst other coloured varieties are not thus favoured, we must believe in the existence of a correlation of this kind. Thus, in the United States purple-fruited plums of many kinds are far more affected by a certain disease than green or yellow-fruited varieties. On the other hand, yellow-fleshed peaches of various kinds suffer from another disease much more than the white-fleshed varieties. In the Mauritius red sugar-canes are much less affected by a particular disease than the white canes. White onions and verbenas are the most liable to mildew; and in Spain the green-fruited grapes suffered from the vine-disease more than other coloured varieties. Dark-coloured pelargoniums and verbenas are more scorched by the sun than varieties of other colours. Red wheats are believed to be hardier than white; and red-flowered hyacinths were more injured during one particular winter in Holland than other coloured varieties. With animals, white terriers suffer most from the distemper, white chickens from a parasitic worm in their tracheae, white pigs from scorching by the sun, and white cattle from flies; but the caterpillars of the silk-moth which yield white cocoons suffered in France less from the deadly parasitic fungus than those producing yellow silk.

The cases of immunity from the action of certain vegetable poisons, in connexion with colour, are more interesting, and are at present wholly inexplicable. I have already given a remarkable instance, on the authority of Professor Wyman, of all the hogs, excepting those of a black colour, suffering severely in Virginia from eating the root of the Lachnanthes tinctoria. According to Spinola and others (25/41. This fact and the following cases, when not stated to the contrary, are taken from a very curious paper by Prof. Heusinger in 'Wochenschrift fur Heilkunde' May 1846 s. 277. Settegast 'Die Thierzucht' 1868 page 39 says that white or white-spotted sheep suffer like pigs, or even die from eating buckwheat; whilst black or dark-woolled individuals are not in the least affected.), buckwheat (Polygonum fagopyrum), when in flower, is highly injurious to white or white-spotted pigs, if they are exposed to the heat of the sun, but is quite innocuous to black pigs. According to two accounts, the Hypericum crispum in Sicily is poisonous to white sheep alone; their heads swell, their wool falls off, and they often die; but this plant, according to Lecce, is poisonous only when it grows in swamps; nor is this improbable, as we know how readily the poisonous principle in plants is influenced by the conditions under which they grow.

Three accounts have been published in Eastern Prussia, of white and white-spotted horses being greatly injured by eating mildewed and honeydewed vetches; every spot of skin bearing white hairs becoming inflamed and gangrenous. The Rev. J. Rodwell informs me that his father turned out about fifteen cart-horses into a field of tares which in parts swarmed with black aphides, and which no doubt were honeydewed, and probably mildewed; the horses, with two exceptions, were chestnuts and bays with white marks on their faces and pasterns, and the white parts alone swelled and became angry scabs. The two bay horses with no white marks entirely escaped all injury. In Guernsey, when horses eat fool's parsley (Aethusa cynapium) they are sometimes violently purged; and this plant "has a peculiar effect on the nose and lips, causing deep cracks and ulcers, particularly on horses with white muzzles." (25/42. Mr. Mogford in the 'Veterinarian' quoted in 'The Field' January 22, 1861 page 545.) With cattle, independently of the action of any poison, cases have been published by Youatt and Erdt of cutaneous diseases with much constitutional disturbance (in one instance after exposure to a hot sun) affecting every single point which bore a white hair, but completely passing over other parts of the body. Similar cases have been observed with horses. (25/43. 'Edinburgh Veterinary Journal' October 1860 page 347.)

We thus see that not only do those parts of the skin which bear white hair differ in a remarkable manner from those bearing hair of any other colour, but that some great constitutional difference must be correlated with the colour of the hair; for in the above-mentioned cases, vegetable poisons caused fever, swelling of the head, as well as other symptoms, and even death, to all the white, or white-spotted animals.

# CHAPTER 2.XXVI.

LAWS OF VARIATION, continued.—SUMMARY.

**THE FUSION OF HOMOLOGOUS PARTS. THE VARIABILITY OF MULTIPLE AND HOMOLOGOUS PARTS. COMPENSATION OF GROWTH. MECHANICAL PRESSURE. RELATIVE POSITION OF FLOWERS WITH RESPECT TO THE AXIS, AND OF SEEDS IN THE OVARY, AS INDUCING VARIATION. ANALOGOUS OR PARALLEL VARIETIES. SUMMARY OF THE THREE LAST CHAPTERS.**

**THE FUSION OF HOMOLOGOUS PARTS.**

Geoffroy Saint-Hilaire formerly propounded what he called la loi de l'affinite de soi pour soi, which has been discussed and illustrated by his son, Isidore, with respect to monsters in the animal kingdom (26/1. 'Hist. des Anomalies' 1832 tome 1 pages 22, 537-556; tome 3 page 462.), and by Moquin-Tandon, with respect to monstrous plants.

This law seems to imply that homologous parts actually attract one another and then unite. No doubt there are many wonderful cases, in which such parts become intimately fused together. This is perhaps best seen in monsters with two heads, which are united, summit to summit, or face to face, or Janus-like, back to back, or obliquely side to side. In one instance of two heads united almost face to face, but a little obliquely, four ears were developed, and on one side a perfect face, which was manifestly formed by the fusion of two half-faces. Whenever two bodies or two heads are united, each bone, muscle, vessel, and nerve on the line of junction appears as if it had sought out its fellow, and had become completely fused with it. Lereboullet (26/2. 'Comptes Rendus' 1855 pages 855, 1039.), who carefully studied the development of double monsters in fishes, observed in fifteen instances the steps by which two heads gradually became united into one. In all such cases it is now thought by the greater number of capable judges that the homologous parts do not attract each other, but that in the words of Mr. Lowne (26/3. 'Catalogue of the Teratological Series in the Museum of the R. Coll. of Surgeons' 1872 page 16.): "As union takes place before the differentiation of distinct organs occurs, these are formed in continuity with each other." He adds that organs already differentiated probably in no case become united to homologous ones. M. Dareste does not speak (26/4. 'Archives de Zoolog. Exper.' January 1874 page 78.) quite decisively against the law of soi pour soi, but concludes by saying, "On se rend parfaitement compte de la formation des monstres, si l'on admet que les embryons qui se soudent appartiennent a un meme oeuf; qu'ils s'unissent en meme temps qu'ils se forment, et que la soudure ne se produit que pendant la premiere periode de la vie embryonnaire, celle ou les organes ne sont encore constitues que par des blastemes homogenes."

By whatever means the abnormal fusion of homologous parts is effected, such cases throw light on the frequent presence of organs which are double during an embryonic period (and throughout life in other and lower members of the same class) but which afterwards unite by a normal process into a single medial organ. In the vegetable kingdom Moquin-Tandon (26/5. 'Teratologie Veg.' 1841 livre 3.) gives a long list of cases, showing how frequently homologous parts, such as leaves, petals, stamens, and pistils, flowers, and aggregates of homologous parts, such as buds, as well as fruit, become blended, both normally and abnormally, with perfect symmetry into one another.

## THE VARIABILITY OF MULTIPLE AND HOMOLOGOUS PARTS.

Isidore Geoffroy (26/6. 'Hist. des Anomalies' tome 3 pages 4, 5, 6.) insists that, when any part or organ is repeated many times in the same animal, it is particularly liable to vary both in number and structure. With respect to number, the proposition may, I think, be considered as fully established; but the evidence is chiefly derived from organic beings living under their natural conditions, with which we are not here concerned. Whenever such parts as the vertebrae or teeth, the rays in the fins of fishes, or the feathers in the tails of birds, or petals, stamens, pistils, or seeds, are very numerous, the number is generally variable. With respect to the structure of multiple parts, the evidence of variability is not so decisive; but the fact, as far as it may be trusted, probably depends on multiple parts being of less physiological importance than single parts; consequently their structure has been less rigorously guarded by natural selection.

## COMPENSATION OF GROWTH, OR BALANCEMENT.

This law, as applied to natural species, was propounded by Goethe and Geoffroy Saint-Hilaire at nearly the same time. It implies that, when much organised matter is used in building up some one part, other parts are starved and become reduced. Several authors, especially botanists, believe in this law; others reject it. As far as I can judge, it occasionally holds good; but its importance has probably been exaggerated. It is scarcely possible to distinguish between the supposed effects of such compensation, and the effects of long-continued selection which may lead to the augmentation of one part, and simultaneously to the diminution of another. Anyhow, there can be no doubt that an organ may be greatly increased without any corresponding diminution of an adjoining part. To recur to our former illustration of the Irish elk, it may be asked what part has suffered in consequence of the immense development of the horns?

It has already been observed that the struggle for existence does not bear hard on our domesticated productions, and consequently the principle of economy of growth will seldom come into play, so that we ought not to expect to find with them frequent evidence of compensation. We have, however, some such cases. Moquin-Tandon describes a monstrous bean (26/7. 'Teratologie Veg.' page 156. See also my book on 'The Movements and Habits of Climbing Plants' 2nd edition 1875 page 202.), in which the stipules were enormously developed, and the leaflets apparently in consequence completely aborted; this case is interesting, as it represents the natural condition of Lathyrus aphaca, with its stipules of great size, and its leaves reduced to mere threads, which act as tendrils. De Candolle (26/8. 'Memoires du Museum' etc. tome 8 page 178.) has remarked that the varieties of Raphanus sativus which have small roots yield numerous seed containing much oil, whilst those with large roots are not productive in oil; and so it is with Brassica asperifolia. The varieties of Cucurbita pepo which bear large fruit yield a small crop, according to Naudin; whilst those producing small fruit yield a vast number. Lastly, I have endeavoured to show in the eighteenth chapter that with many cultivated plants unnatural treatment checks the full and proper action of the reproductive organs, and they are thus rendered more or less sterile; consequently, in the way of compensation, the fruit becomes greatly enlarged, and, in double flowers, the petals are greatly increased in number.

With animals, it has been found difficult to produce cows which yield much milk, and are afterwards capable of fattening well. With fowls which have large top-knots and beards the comb and wattles are generally much reduced in size; though there are exceptions to this rule. Perhaps the entire absence of the oil-gland in fantail pigeons may be connected with the great development of their tails.

**MECHANICAL PRESSURE AS A CAUSE OF MODIFICATIONS.**

In some few cases there is reason to believe that mere mechanical pressure has affected certain structures. Vrolik and Weber (26/9. Prichard 'Phys. Hist. of Mankind' 1851 volume 1 page 324.) maintain that the shape of the human head is influenced by the shape of the mother's pelvis. The kidneys in different birds differ much in form, and St. Ange (26/10. 'Annales des Sc. Nat.' 1st series tome 19 page 327.) believes that this is determined by the form of the pelvis, which again, no doubt, stands in close relation with their power of locomotion. In snakes, the viscera are curiously displaced, in comparison with their position in other vertebrates; and this has been attributed by some authors to the elongation of their bodies; but here, as in so many previous cases, it is impossible to disentangle a direct result of this kind from that consequent on natural selection. Godron has argued (26/11. 'Comptes Rendus' December 1864 page 1039.) that the abortion of the spur on the inner side of the flowers in Corydalis, is caused by the buds at a very early period of growth whilst underground being closely pressed against one another and against the stem. Some botanists believe that the singular difference in the shape both of the seed and corolla, in the interior and exterior florets in certain Compositous and Umbelliferous plants, is due to the pressure to which the inner florets are subjected; but this conclusion is doubtful.

The facts just given do not relate to domesticated productions, and therefore do not strictly concern us. But here is a more appropriate case: H. Muller (26/12. "Ueber fotale Rachites" 'Wurzburger Medicin. Zeitschrift' 1860 b. 1 s. 265.) has shown that in shortfaced races of the dog some of the molar teeth are placed in a slightly different position to that which they occupy in other dogs, especially in those having elongated muzzles; and as he remarks, any inherited change in the arrangement of the teeth deserves notice, considering their classificatory importance. This difference in position is due to the shortening of certain facial bones and the consequent want of space; and the shortening results from a peculiar and abnormal state of the embryonal cartilages of the bones.

## [RELATIVE POSITION OF FLOWERS WITH RESPECT TO THE AXIS, AND OF SEEDS IN THE OVARY, AS INDUCING VARIATION.

In the thirteenth chapter various peloric flowers were described, and their production was shown to be due either to arrested development, or to reversion to a primordial condition. Moquin-Tandon has remarked that the flowers which stand on the summit of the main stem or of a lateral branch are more liable to become peloric than those on the sides (26/13. 'Teratologie Veg.' page 192.); and he adduces, amongst other instances, that of Teucrium campanulatum. In another Labiate plant grown by me, viz., the Galeobdolon luteum, the peloric flowers were always produced on the summit of the stem, where flowers are not usually borne. In Pelargonium, a SINGLE flower in the truss is frequently peloric, and when this occurs I have during several years invariably observed it to be the central flower. This is of such frequent occurrence that one observer (26/14. 'Journal of Horticulture' July 2, 1861 page 253.) gives the names of ten varieties flowering at the same time, in every one of which the central flower was peloric. Occasionally more than one flower in the truss is peloric, and then of course the additional ones must be lateral. These flowers are interesting as showing how the whole structure is correlated. In the common Pelargonium the upper sepal is produced into a nectary which coheres with the flower-peduncle; the two upper petals differ a little in shape from the three lower ones, and are marked with dark shades of colour; the stamens are graduated in length and upturned. In the peloric flowers, the nectary aborts; all the petals become alike both in shape and colour; the stamens are generally reduced in number and become straight, so that the whole flower resembles that of the allied genus Erodium. The correlation between these changes is well shown when one of the two upper petals alone loses its dark mark, for in this case the nectary does not entirely abort, but is usually much reduced in length. (26/15. It would be worth trial to fertilise with the same pollen the central and lateral flowers of the pelargonium, or of other highly cultivated plants, protecting them of course from insects: then to sow the seed separately, and observe whether the one or the other lot of seedlings varied the most.)

Morren has described (26/16. Quoted in 'Journal of Horticulture' February 24, 1863 page 152.) a marvellous flask-shaped flower of the Calceolaria, nearly four inches in length, which was almost completely peloric; it grew on the summit of the plant, with a normal flower on each side; Prof. Westwood also has described (26/17. 'Gardener's Chronicle' 1866 page 612. For the Phalaenopsis see ibid 1867 page 211.) three similar peloric flowers, which all occupied a central position on the flower-branches. In the Orchideous genus, Phalaenopsis, the terminal flower has been seen to become peloric.

In a Laburnum-tree I observed that about a fourth part of the racemes produced terminal flowers which had lost their papilionaceous structure. These were produced after almost all the other flowers on the same racemes had withered. The most perfectly pelorised examples had six petals, each marked with black striae like those on the standard-petal. The keel seemed to resist the change more than the other petals. Dutrochet has described (26/18. 'Memoires...des Vegetaux' 1837 tome 2 page 170.) an exactly similar case in France, and I believe these are the only two instances of pelorism in the laburnum which have been recorded. Dutrochet remarks that the racemes on this tree do not properly produce a terminal flower, so that (as in the case of the Galeobdolon) their position as well as structure are both anomalies, which no doubt are in some manner related. Dr. Masters has briefly described another leguminous plant (26/19. 'Journal of Horticulture' July 23, 1861 page 311.), namely, a species of clover, in which the uppermost and central flowers were regular or had lost their papilionaceous structure. In some of these plants the flower-heads were also proliferous.

Lastly, Linaria produces two kinds of peloric flowers, one having simple petals, and the other having them all spurred. The two forms, as Naudin remarks (26/20. 'Nouvelles Archives du Museum' tome 1 page 137.), not rarely occur on the same plant, but in this case the spurred form almost invariably stands on the summit of the spike.

The tendency in the terminal or central flower to become peloric more frequently than the other flowers, probably results from "the bud which stands on the end of a shoot receiving the most sap; it grows out into a stronger shoot than those situated lower down." (26/21. Hugo von Mohl 'The Vegetable Cell' English translation 1852 page 76.) I have discussed the connection between pelorism and a central position, partly because some few plants are known normally to produce a terminal flower different in structure from the lateral ones; but chiefly on account of the following case, in which we see a tendency to variability or to reversion connected with the same position. A great judge of Auriculas (26/22. The Rev. H.H. Dombrain in 'Journal of Horticulture' 1861 June 4 page 174; and June 25 page 234; 1862 April 29 page 83.) states that when one throws up a side bloom it is pretty sure to keep its character; but that if it grows from the centre or heart of the plant, whatever the colour of the edging ought to be, "it is just as likely to come in any other class as in the one to which it properly belongs." This is so notorious a fact, that some florists regularly pinch off the central trusses of flowers. Whether in the highly improved varieties the departure of the central trusses from their proper type is due to reversion, I do not know. Mr. Dombrain insists that, whatever may be the commonest kind of imperfection in each variety, this is generally exaggerated in the central truss. Thus one variety "sometimes has the fault of producing a little green floret in the centre of the flower," and in central blooms these become excessive in size. In some central blooms, sent to me by Mr. Dombrain, all the organs of the flower were rudimentary in structure, of minute size, and of a green colour, so that by a little further change all would have been converted into small leaves. In this case we clearly see a tendency to prolification—a term which I may explain, for those who have never attended to botany, to mean the production of a branch or flower, or head of flowers, out of another flower. Now Dr. Masters (26/23. 'Transact. Linn. Soc.' volume 23 1861 page 360.) states that the central or uppermost flower on a plant is generally the most liable to prolification. Thus, in the varieties of the Auricula, the loss of their proper character and a tendency to prolification, also a tendency to prolification with pelorism, are all connected together, and are due either to arrested development, or to reversion to a former condition.

The following is a more interesting case; Metzger (26/24. 'Die Getreidearten' 1845 s. 208, 209.) cultivated in Germany several kinds of maize brought from the hotter parts of America, and he found, as previously described, that in two or three generations the grains became greatly changed in form, size, and colour; and with respect to two races he expressly states that in the first generation, whilst the lower grains on each head retained their proper character, the uppermost grains already began to assume that character which in the third generation all the grains acquired. As we do not know the aboriginal parent of the maize, we cannot tell whether these changes are in any way connected with reversion.

In the two following cases, reversion comes into play and is determined by the position of the seed in the capsule. The Blue Imperial pea is the offspring of the Blue Prussian, and has larger seed and broader pods than its parent. Now Mr. Masters, of Canterbury, a careful observer and a raiser of new varieties of the pea, states (26/25. 'Gardener's Chronicle' 1850 page 198.) that the Blue Imperial always has a strong tendency to revert to its parent-stock, and the reversion "occurs in this manner: the last (or uppermost) pea in the pod is frequently much smaller than the rest; and if these small peas are carefully collected and sown separately, very many more, in proportion, will revert to their origin, than those taken from the other parts of the pod." Again, M. Chate (26/26. Quoted in 'Gardener's Chronicle' 1866 page 74.) says that in raising seedling stocks he succeeds in getting eighty per cent to bear double flowers, by leaving only a few of the secondary branches to seed; but in addition to this, "at the time of extracting the seeds, the upper portion of the pod is separated and placed aside, because it has been ascertained that the plants coming from the seeds situated in this portion of the pod, give eighty per cent of single flowers." Now the production of single-flowering plants from the seed of double-flowering plants is clearly a case of reversion. These latter facts, as well as the connection between a central position and pelorism and prolification, show in an interesting manner how small a difference—namely, a little greater or less freedom in the flow of sap towards one part of the plant—determines important changes of structure.]

## ANALOGOUS OR PARALLEL VARIATION.

By this term I mean that similar characters occasionally make their appearance in the several varieties or races descended from the same species, and more rarely in the offspring of widely distinct species. We are here concerned, not as hitherto with the causes of variation, but with the results; but this discussion could not have been more conveniently introduced elsewhere. The cases of analogous variation, as far as their origin is concerned, may be grouped, disregarding minor subdivisions, under two main heads; firstly, those due to unknown causes acting on similarly constituted organisms, and which consequently have varied in a similar manner; and secondly, those due to the reappearance of characters which were possessed by a more or less remote progenitor. But these two main divisions can often be separated only conjecturally, and graduate, as we shall presently see, into each other.

[Under the first head of analogous variations, not due to reversion, we have the many cases of trees belonging to quite different orders which have produced pendulous and fastigiate varieties. The beech, hazel, and barberry have given rise to purple-leaved varieties; and, as Bernhardi remarks (26/27. 'Ueber den Begriff der Pflanzenart' 1834 s. 14.), a multitude of plants, as distinct as possible, have yielded varieties with deeply-cut or laciniated leaves.

Varieties descended from three distinct species of Brassica have their stems, or so-called roots, enlarged into globular masses. The nectarine is the offspring of the peach; and the varieties of peaches and nectarines offer a remarkable parallelism in the fruit being white, red, or yellow fleshed—in being clingstones or freestones—in the flowers being large or small—in the leaves being serrated or crenated, furnished with globose or reniform glands, or quite destitute of glands. It should be remarked that each variety of the nectarine has not derived its character from a corresponding variety of the peach. The several varieties also of a closely allied genus, namely the apricot, differ from one another in nearly the same parallel manner. There is no reason to believe that any of these varieties have merely reacquired long-lost characters; and in most of them this certainly is not the case.

Three species of Cucurbita have yielded a multitude of races which correspond so closely in character that, as Naudin insists, they may be arranged in almost strictly parallel series. Several varieties of the melon are interesting from resembling, in important characters, other species, either of the same genus or of allied genera; thus, one variety has fruit so like, both externally and internally, the fruit of a perfectly distinct species, namely, the cucumber, as hardly to be distinguished from it; another has long cylindrical fruit twisting about like a serpent; in another the seeds adhere to portions of the pulp; in another the fruit, when ripe, suddenly cracks and falls into pieces; and all these highly remarkable peculiarities are characteristic of species belonging to allied genera. We can hardly account for the appearance of so many unusual characters by reversion to a single ancient form; but we must believe that all the members of the family have inherited a nearly similar constitution from an early progenitor. Our cereal and many other plants offer similar cases.

With animals we have fewer cases of analogous variation, independently of direct reversion. We see something of the kind in the resemblance between the short-muzzled races of the dog, such as the pug and bull-dog; in feather-footed races of the fowl, pigeon, and canary-bird; in horses of the most different races presenting the same range of colour; in all black-and-tan dogs having tan-coloured eye-spots and feet, but in this latter case reversion may possibly have played a part. Low has remarked (26/28. 'Domesticated Animals' 1845 page 351.) that several breeds of cattle are "sheeted,"—that is, have a broad band of white passing round their bodies like a sheet; this character is strongly inherited, and sometimes originates from a cross; it may be the first step in reversion to an early type, for, as was shown in the third chapter, white cattle with dark ears, dark feet and tip of tail, formerly existed, and now exist in feral or semi-feral condition in several quarters of the world.

Under our second main division, namely, of analogous variations due to reversion, the best cases are afforded by pigeons. In all the most distinct breeds, sub-varieties occasionally appear coloured exactly like the parent rock-pigeon, with black wing-bars, white loins, banded tail, etc.; and no one can doubt that these characters are due to reversion. So with minor details; turbits properly have white tails, but occasionally a bird is born with a dark-coloured and banded tail; pouters properly have their primary wing-feathers white, but not rarely a "sword-flighted" bird appears, that is, one with the few first primaries dark-coloured; and in these cases we have characters proper to the rock-pigeon, but new to the breed, evidently appearing from reversion. In some domestic varieties the wing-bars, instead of being simply black, as in the rock-pigeon, are beautifully edged with different zones of colour, and they then present a striking analogy with the wing-bars in certain natural species of the same family, such as Phaps chalcoptera; and this may probably be accounted for by all the species of the family being descended from the same remote progenitor and having a tendency to vary in the same manner. Thus, also, we can perhaps understand the fact of some Laugher-pigeons cooing almost like turtle-doves, and for several races having peculiarities in their flight, since certain natural species (viz., C. torquatrix and palumbus), display singular vagaries in this respect. In other cases a race, instead of imitating a distinct species, resembles some other race; thus, certain runts tremble and slightly elevate their tails, like fantails; and turbits inflate the upper part of their oesophagus, like pouter-pigeons.

It is a common circumstance to find certain coloured marks persistently characterising all the species of a genus, but differing much in tint; and the same thing occurs with the varieties of the pigeon: thus, instead of the general plumage being blue, with the wing-bars black, there are snow-white varieties with red bars, and black varieties with white bars; in other varieties the wing-bars, as we have seen, are elegantly zoned with different tints. The Spot pigeon is characterised by the whole plumage being white, excepting a spot on the forehead and the tail; but these parts may be red, yellow, or black. In the rock-pigeon and in many varieties the tail is blue, with the outer edges of the outer feathers white; but in the sub-variety of the monk-pigeon we have a reversed style of coloration, for the tail is white, except the outer edges of the outer feathers, which are black. (26/29. Bechstein 'Naturgeschichte Deutschlands' b. 4 1795 s. 31.)

With some species of birds, for instance with gulls, certain coloured parts appear as if almost washed out, and I have observed exactly the same appearance in the terminal dark tail-bar in certain pigeons, and in the whole plumage of certain varieties of the duck. Analogous facts in the vegetable kingdom could be given.

Many sub-varieties of the pigeon have reversed and somewhat lengthened feathers on the back part of their heads, and this is certainly not due to reversion to the parent-species, which shows no trace of such structure: but when we remember that sub-varieties of the fowl, turkey, canary-bird, duck, and goose, all have either topknots or reversed feathers on their heads; and when we remember that scarcely a single large natural group of birds can be named, in which some members have not a tuft of feathers on their heads, we may suspect that reversion to some extremely remote form has come into action.

Several breeds of the fowl have either spangled or pencilled feathers; and these cannot be derived from the parent-species, the Gallus bankiva; though of course it is possible that one early progenitor of this species may have been spangled, and another pencilled. But, as many gallinaceous birds are either spangled or pencilled, it is a more probable view that the several domestic breeds of the fowl have acquired this kind of plumage from all the members of the family inheriting a tendency to vary in a like manner. The same principle may account for the ewes in certain breeds of sheep being hornless, like the females of some other hollow-horned ruminants; it may account for certain domestic cats having slightly-tufted ears, like those of the lynx; and for the skulls of domestic rabbits often differing from one another in the same characters by which the skulls of the various species of the genus Lepus differ.

I will only allude to one other case, already discussed. Now that we know that the wild parent of the ass commonly has striped legs, we may feel confident that the occasional appearance of stripes on the legs of the domestic ass is due to reversion; but this will not account for the lower end of the shoulder-stripe being sometimes angularly bent or slightly forked. So, again, when we see dun and other coloured horses with stripes on the spine, shoulders, and legs, we are led, from reasons formerly given, to believe that they reappear through reversion to the wild parent-horse. But when horses have two or three shoulder-stripes, with one of them occasionally forked at the lower end, or when they have stripes on their faces, or are faintly striped as foals over nearly their whole bodies, with the stripes angularly bent one under the other on the forehead, or irregularly branched in other parts, it would be rash to attribute such diversified characters to the reappearance of those proper to the aboriginal wild horse. As three African species of the genus are much striped, and as we have seen that the crossing of the unstriped species often leads to the hybrid offspring being conspicuously striped—bearing also in mind that the act of crossing certainly causes the reappearance of long-lost characters—it is a more probable view that the above-specified stripes are due to reversion, not to the immediate wild parent-horse, but to the striped progenitor of the whole genus.]

I have discussed this subject of analogous variation at considerable length, because it is well known that the varieties of one species frequently resemble distinct species—a fact in perfect harmony with the foregoing cases, and explicable on the theory of descent. Secondly, because these facts are important from showing, as remarked in a former chapter, that each trifling variation is governed by law, and is determined in a much higher degree by the nature of the organisation, than by the nature of the conditions to which the varying being has been exposed. Thirdly, because these facts are to a certain extent related to a more general law, namely, that which Mr. B.D. Walsh (26/30. 'Proc. Entomolog. Soc. of Philadelphia' October 1863 page 213.) has called the "Law of EQUABLE VARIABILITY," or, as he explains it, "if any given character is very variable in one species of a group, it will tend to be variable in allied species; and if any given character is perfectly constant in one species of a group, it will tend to be constant in allied species."

This leads me to recall a discussion in the chapter on Selection, in which it was shown that with domestic races, which are now undergoing rapid improvement, those parts or characters vary the most, which are the most valued. This naturally follows from recently selected characters continually tending to revert to their former less improved standard, and from their being still acted on by the same agencies, whatever these may be, which first caused the characters in question to vary. The same principle is applicable to natural species, for, as stated in my 'Origin of Species' generic characters are less variable than specific characters; and the latter are those which have been modified by variation and natural selection, since the period when all the species belonging to the genus branched off from a common progenitor, whilst generic characters are those which have remained unaltered from a much more remote epoch, and accordingly are now less variable. This statement makes a near approach to Mr. Walsh's law of Equable Variability. Secondary sexual characters, it may be added, rarely serve to characterise distinct genera, for they usually differ much in the species of the same genus, and they are highly variable in the individuals of the same species; we have also seen in the earlier chapters of this work how variable secondary sexual characters become under domestication.

## SUMMARY OF THE THREE PREVIOUS CHAPTERS ON THE LAWS OF VARIATION.

In the twenty-third chapter we saw that changed conditions occasionally, or even often, act in a definite manner on the organisation, so that all, or nearly all, the individuals thus exposed become modified in the same manner. But a far more frequent result of changed conditions, whether acting directly on the organisation or indirectly through the reproductive system, is indefinite and fluctuating variability. In the three last chapters, some of the laws by which such variability is regulated have been discussed.

Increased use adds to the size of muscles, together with the blood-vessels, nerves, ligaments, the crests of bone and the whole bones, to which they are attached. Increased functional activity increases the size of various glands, and strengthens the sense-organs. Increased and intermittent pressure thickens the epidermis. A change in the nature of the food sometimes modifies the coats of the stomach, and augments or decreases the length of the intestines. Continued disuse, on the other hand, weakens and diminishes all parts of the organisation. Animals which during many generations have taken but little exercise, have their lungs reduced in size, and as a consequence the bony fabric of the chest and the whole form of the body become modified. With our anciently domesticated birds, the wings have been little used, and they are slightly reduced; with their decrease, the crest of the sternum, the scapulae, coracoids, and furculum, have all been reduced.

With domesticated animals, the reduction of a part from disuse is never carried so far that a mere rudiment is left; whereas we have reason to believe that this has often occurred under nature; the effects of disuse in this latter case being aided by economy of growth, together with the intercrossing of many varying individuals. The cause of this difference between organisms in a state of nature, and under domestication, probably is that in the latter case there has not been time sufficient for any very great change, and that the principle of economy of growth does not come into action. On the contrary, structures which are rudimentary in the parent-species, sometimes become partially redeveloped in our domesticated productions. Such rudiments as occasionally make their appearance under domestication, seem always to be the result of a sudden arrest of development; nevertheless they are of interest, as showing that rudiments are the relics of organs once perfectly developed.

Corporeal, periodical, and mental habits, though the latter have been almost passed over in this work, become changed under domestication, and the changes are often inherited. Such changed habits in an organic being, especially when living a free life, would often lead to the augmented or diminished use of various organs, and consequently to their modification. From long-continued habit, and more especially from the occasional birth of individuals with a slightly different constitution, domestic animals and cultivated plants become to a certain extent acclimatised or adapted to a climate different from that proper to the parent-species.

Through the principle of correlated variability, taken in its widest sense, when one part varies other parts vary, either simultaneously, or one after the other. Thus, an organ modified during an early embryonic period affects other parts subsequently developed. When an organ, such as the beak, increases or decreases in length, adjoining or correlated parts, as the tongue and the orifice of the nostrils, tend to vary in the same manner. When the whole body increases or decreases in size, various parts become modified; thus, with pigeons the ribs increase or decrease in number and breadth. Homologous parts which are identical during their early development and are exposed to similar conditions, tend to vary in the same or in some connected manner,—as in the case of the right and left sides of the body, and of the front and hind limbs. So it is with the organs of sight and hearing; for instance, white cats with blue eyes are almost always deaf. There is a manifest relation throughout the body between the skin and various dermal appendages, such as hair, feathers, hoofs, horns, and teeth. In Paraguay, horses with curly hair have hoofs like those of a mule; the wool and the horns of sheep often vary together; hairless dogs are deficient in their teeth; men with redundant hair have abnormal teeth, either by deficiency or excess. Birds with long wing-feathers usually have long tail-feathers. When long feathers grow from the outside of the legs and toes of pigeons, the two outer toes are connected by membrane; for the whole leg tends to assume the structure of the wing. There is a manifest relation between a crest of feathers on the head and a marvellous amount of change in the skull of various fowls; and in a lesser degree, between the greatly elongated, lopping ears of rabbits and the structure of their skulls. With plants, the leaves, various parts of the flower, and the fruit, often vary together to a correlated manner.

In some cases we find correlation without being able even to conjecture what is the nature of the connection, as with various monstrosities and diseases. This is likewise the case with the colour of the adult pigeon, in connection with the presence of down on the young bird. Numerous curious instances have been given of peculiarities of constitution, in correlation with colour, as shown by the immunity of individuals of one colour from certain diseases, from the attacks of parasites and from the action of certain vegetable poisons.

Correlation is an important subject; for with species, and in a lesser degree with domestic races, we continually find that certain parts have been greatly modified to serve some useful purpose; but we almost invariably find that other parts have likewise been more or less modified, without our being able to discover any advantage in the change. No doubt great caution is necessary with respect to this latter point, for it is difficult to overrate our ignorance on the use of various parts of the organisation; but from what we have seen, we may believe that many modifications are of no direct service, having arisen in correlation with other and useful changes.

Homologous parts during their early development often become fused together. Multiple and homologous organs are especially liable to vary in number and probably in form. As the supply of organised matter is not unlimited, the principle of compensation sometimes comes into action; so that, when one part is greatly developed, adjoining parts are apt to be reduced; but this principle is probably of much less importance than the more general one of the economy of growth. Through mere mechanical pressure hard parts occasionally affect adjoining parts. With plants the position of the flowers on the axis, and of the seeds in the ovary, sometimes leads, through a more or less free flow of sap, to changes of structure; but such changes are often due to reversion. Modifications, in whatever manner caused, will be to a certain extent regulated by that co-ordinating power, or so-called nisus formativus, which is in fact a remnant of that simple form of reproduction, displayed by many lowly organised beings in their power of fissiparous generation and budding. Finally, the effects of the laws which directly or indirectly govern variability, may be largely regulated by man's selection, and will so far be determined by natural selection that changes advantageous to any race will be favoured, and disadvantageous changes will be checked.

Domestic races descended from the same species, or from two or more allied species, are liable to revert to characters derived from their common progenitor; and, as they inherit a somewhat similar constitution, they are liable to vary in the same manner. From these two causes analogous varieties often arise. When we reflect on the several foregoing laws, imperfectly as we understand them, and when we bear in mind how much remains to be discovered, we need not be surprised at the intricate and to us unintelligible manner in which our domestic productions have varied, and still go on varying.

# CHAPTER 2.XXVII.

PROVISIONAL HYPOTHESIS OF PANGENESIS.

PRELIMINARY REMARKS. FIRST PART: THE FACTS TO BE CONNECTED UNDER A SINGLE POINT OF VIEW, NAMELY, THE VARIOUS KINDS OF REPRODUCTION. REGROWTH OF AMPUTATED PARTS. GRAFT-HYBRIDS. THE DIRECT ACTION OF THE MALE ELEMENT ON THE FEMALE. DEVELOPMENT. THE FUNCTIONAL INDEPENDENCE OF THE UNITS OF THE BODY. VARIABILITY. INHERITANCE. REVERSION.

SECOND PART: STATEMENT OF THE HYPOTHESIS. HOW FAR THE NECESSARY ASSUMPTIONS ARE IMPROBABLE. EXPLANATION BY AID OF THE HYPOTHESIS OF THE SEVERAL CLASSES OF FACTS SPECIFIED IN THE FIRST PART. CONCLUSION.

In the previous chapters large classes of facts, such as those bearing on bud-variation, the various forms of inheritance, the causes and laws of variation, have been discussed; and it is obvious that these subjects, as well as the several modes of reproduction, stand in some sort of relation to one another. I have been led, or rather forced, to form a view which to a certain extent connects these facts by a tangible method. Every one would wish to explain to himself, even in an imperfect manner, how it is possible for a character possessed by some remote ancestor suddenly to reappear in the offspring; how the effects of increased or decreased use of a limb can be transmitted to the child; how the male sexual element can act not solely on the ovules, but occasionally on the mother-form; how a hybrid can be produced by the union of the cellular tissue of two plants independently of the organs of generation; how a limb can be reproduced on the exact line of amputation, with neither too much nor too little added; how the same organism may be produced by such widely different processes, as budding and true seminal generation; and, lastly, how of two allied forms, one passes in the course of its development through the most complex metamorphoses, and the other does not do so, though when mature both are alike in every detail of structure. I am aware that my view is merely a provisional hypothesis or speculation; but until a better one be advanced, it will serve to bring together a multitude of facts which are at present left disconnected by any efficient cause. As Whewell, the historian of the inductive sciences, remarks:—"Hypotheses may often be of service to science, when they involve a certain portion of incompleteness, and even of error." Under this point of view I venture to advance the hypothesis of Pangenesis, which implies that every separate part of the whole organisation reproduces itself. So that ovules, spermatozoa, and pollen-grains,—the fertilised egg or seed, as well as buds,—include and consist of a multitude of germs thrown off from each separate part or unit. (27/1. This hypothesis has been severely criticised by many writers, and it will be fair to give references to the more important articles. The best essay which I have seen is by Prof. Delpino, entitled 'Sulla Darwiniana Teoria della Pangenesi, 1869' of which a translation appeared in 'Scientific Opinion' September 29, 1869 and the succeeding numbers. He rejects the hypothesis, but criticises it fairly, and I have found his criticisms very useful. Mr. Mivart ('Genesis of Species' 1871 chapter 10.) follows Delpino, but adds no new objections of any weight. Dr. Bastian ('The Beginnings of Life' 1872 volume 2 page 98) says that the hypothesis "looks like a relic of the old rather than a fitting appanage of the new evolution philosophy." He shows that I ought not to have used the term "pangenesis," as it had been previously used by Dr. Gros in another sense. Dr. Lionel Beale ('Nature' May 11, 1871 page 26) sneers at the whole doctrine with much acerbity and some justice. Prof. Wigand ('Schriften der Gesell. der gesammt. Naturwissen. zu Marburg' b. 9 1870) considers the hypothesis as unscientific and worthless. Mr. G.H. Lewes ('Fortnightly Review' November 1, 1868 page 503) seems to consider that it may be useful: he makes many good criticisms in a perfectly fair spirit. Mr. F. Galton, after describing his valuable experiments ('Proc. Royal Soc.' volume 19 page 393) on the intertransfusion of the blood of distinct varieties of the rabbit, concludes by saying that in his opinion the results negative beyond all doubt the doctrine of Pangenesis. He informs me that subsequently to the publication of his paper he continued his experiments on a still larger scale for two more generations, without any sign of mongrelism showing itself in the very numerous offspring. I certainly should have expected that gemmules would have been present in the blood, but this is no necessary part of the hypothesis, which manifestly applies to plants and the lowest animals. Mr. Galton, in a letter to 'Nature' (April 27, 1871 page 502), also criticises various incorrect expressions used by me. On the other hand, several writers have spoken favourably of the hypothesis, but there would be no use in giving references to their articles. I may, however, refer to Dr. Ross' work, 'The Graft Theory of Disease; being an application of Mr. Darwin's hypothesis of Pangenesis' 1872 as he gives several original and ingenious discussions.)

In the First Part I will enumerate as briefly as I can the groups of facts which seem to demand connection; but certain subjects, not hitherto discussed, must be treated at disproportionate length. In the Second Part the hypothesis will be given; and after considering how far the necessary assumptions are in themselves improbable, we shall see whether it serves to bring under a single point of view the various facts.

**PART I.**

Reproduction may be divided into two main classes, namely, sexual and asexual. The latter is effected in many ways—by the formation of buds of various kinds, and by fissiparous generation, that is by spontaneous or artificial division. It is notorious that some of the lower animals, when cut into many pieces, reproduce so many perfect individuals: Lyonnet cut a Nais or freshwater worm into nearly forty pieces, and these all reproduced perfect animals. (27/2. Quoted by Paget 'Lectures on Pathology' 1853 page 159.) It is probable that segmentation could be carried much further in some of the protozoa; and with some of the lowest plants each cell will reproduce the parent-form. Johannes Muller thought that there was an important distinction between gemmation and fission; for in the latter case the divided portion, however small, is more fully developed than a bud, which also is a younger formation; but most physiologists are now convinced that the two processes are essentially alike. (27/3. Dr. Lachmann also observes ('Annals and Mag. of Nat. History' 2nd series volume 19 1857 page 231) with respect to infusoria, that "fissation and gemmation pass into each other almost imperceptibly." Again, Mr. W.C. Minor ('Annals and Mag. of Nat. Hist.' 3rd series volume 11 page 328) shows that with Annelids the distinction that has been made between fission and budding is not a fundamental one. See also Professor Clark's work 'Mind in Nature' New York 1865 pages 62, 94.) Prof. Huxley remarks, "fission is little more than a peculiar mode of budding," and Prof. H.J. Clark shows in detail that there is sometimes "a compromise between self-division and budding." When a limb is amputated, or when the whole body is bisected, the cut extremities are said to bud forth (27/4. See Bonnet 'Oeuvres d'Hist. Nat.' tome 5 1781 page 339 for remarks on the budding-out of the amputated limbs of Salamanders.); and as the papilla, which is first formed, consists of undeveloped cellular tissue like that forming an ordinary bud, the expression is apparently correct. We see the connection of the two processes in another way; for Trembley observed with the hydra, that the reproduction of the head after amputation was checked as soon as the animal put forth reproductive gemmae. (27/5. Paget 'Lectures on Pathology' 1853 page 158.)

Between the production, by fissiparous generation, of two or more complete individuals, and the repair of even a very slight injury, there is so perfect a gradation, that it is impossible to doubt that the two processes are connected. As at each stage of growth an amputated part is replaced by one in the same state of development, we must also follow Sir J. Paget in admitting, "that the powers of development from the embryo, are identical with those exercised for the restoration from injuries: in other words, that the powers are the same by which perfection is first achieved, and by which, when lost, it is recovered." (27/6. Ibid pages 152, 164.) Finally, we may conclude that the several forms of budding, fissiparous generation, the repair of injuries, and development, are all essentially the results of one and the same power.

# SEXUAL GENERATION.

The union of the two sexual elements seems at first sight to make a broad distinction between sexual and asexual generation. But the conjugation of algae, by which process the contents of two cells unite into a single mass capable of development, apparently gives us the first step towards sexual union: and Pringsheim, in his memoir on the pairing of Zoospores (27/7. Translated in 'Annals and Mag. of Nat. Hist.' April 1870 page 272.), shows that conjugation graduates into true sexual reproduction. Moreover, the now well-ascertained cases of Parthenogenesis prove that the distinction between sexual and asexual generation is not nearly so great as was formerly thought; for ova occasionally, and even in some cases frequently, become developed into perfect beings, without the concourse of the male. With most of the lower animals and even with mammals, the ova show a trace of parthenogenetic power, for without being fertilised they pass through the first stages of segmentation. (27/8. Bischoff as quoted by von Siebold "Ueber Parthenogenesis" 'Sitzung der math. phys. Classe.' Munich November 4, 1871 page 240. See also Quatrefages 'Annales des Sc. Nat. Zoolog.' 3rd series 1850 page 138.) Nor can pseudova which do not need fertilisation, be distinguished from true ova, as was first shown by Sir J. Lubbock, and is now admitted by Siebold. So, again, the germ-balls in the larvae of Cecidomyia are said by Leuckart (27/9. 'On the Asexual Reproduction of Cecidomyide Larvae' translated in 'Annals and Mag. of Nat. Hist.' March 1866 pages 167, 171.) to be formed within the ovarium, but they do not require to be fertilised. It should also be observed that in sexual generation, the ovules and the male element have equal power of transmitting every single character possessed by either parent to their offspring. We see this clearly when hybrids are paired inter se, for the characters of both grandparents often appear in the progeny, either perfectly or by segments. It is an error to suppose that the male transmits certain characters and the female other characters; although no doubt, from unknown causes, one sex sometimes has a much stronger power of transmission than the other.

It has, however, been maintained by some authors that a bud differs essentially from a fertilised germ, in always reproducing the perfect character of the parent-stock; whilst fertilised germs give birth to variable beings. But there is no such broad distinction as this. In the eleventh chapter numerous cases were advanced showing that buds occasionally grow into plants having quite new characters; and the varieties thus produced can be propagated for a length of time by buds, and occasionally by seed. Nevertheless, it must be admitted that beings produced sexually are much more liable to vary than those produced asexually; and of this fact a partial explanation will hereafter be attempted. The variability in both cases is determined by the same general causes, and is governed by the same laws. Hence new varieties arising from buds cannot be distinguished from those arising from seed. Although bud-varieties usually retain their character during successive bud-generations, yet they occasionally revert, even after a long series of bud-generations, to their former character. This tendency to reversion in buds, is one of the most remarkable of the several points of agreement between the offspring from bud and seminal reproduction.

But there is one difference between organisms produced sexually and asexually, which is very general. The former pass in the course of their development from a very low stage to their highest stage, as we see in the metamorphoses of insects and of many other animals, and in the concealed metamorphoses of the vertebrata. Animals propagated asexually by buds or fission, on the other hand, commence their development at that stage at which the budding or self- dividing animal may happen to be, and therefore do not pass through some of the lower developmental stages. (27/10. Prof. Allman speaks ('Transact. R. Soc. of Edinburgh' volume 26 1870 page 102) decisively on this head with respect to the Hydroida: he says, "It is a universal law in the succession of zooids, that no retrogression ever takes place in the series.") Afterwards, they often advance in organisation, as we see in the many cases of "alternate generation." In thus speaking of alternate generation, I follow those naturalists who look at this process as essentially one of internal budding or of fissiparous generation. Some of the lower plants, however, such as mosses and certain algae, according to Dr. L. Radlkofer (27/11. 'Annals and Mag. of Nat. Hist.' 2nd series volume 20 1857 pages 153-455), when propagated asexually, do undergo a retrogressive metamorphosis. As far as the final cause is concerned, we can to a certain extent understand why beings propagated by buds should not pass through all the early stages of development; for with each organism the structure acquired at each stage must be adapted to its peculiar habits; and if there are places for the support of many individuals at some one stage, the simplest plan will be that they should be multiplied at this stage, and not that they should first retrograde in their development to an earlier or simpler structure, which might not be fitted for the then surrounding conditions.

From the several foregoing considerations we may conclude that the difference between sexual and asexual generation is not nearly so great as at first appears; the chief difference being that an ovule cannot continue to live and to be fully developed unless it unites with the male element; but even this difference is far from invariable, as shown by the many cases of parthenogenesis. We are therefore naturally led to inquire what the final cause can be of the necessity in ordinary generation for the concourse of the two sexual elements.

Seeds and ova are often highly serviceable as the means of disseminating plants and animals, and of preserving them during one or more seasons in a dormant state; but unimpregnated seeds or ova, and detached buds, would be equally serviceable for both purposes. We can, however, indicate two important advantages gained by the concourse of the two sexes, or rather of two individuals belonging to opposite sexes; for, as I have shown in a former chapter, the structure of every organism appears to be especially adapted for the concurrence, at least occasionally, of two individuals. When species are rendered highly variable by changed conditions of life, the free intercrossing of the varying individuals tends to keep each form fitted for its proper place in nature; and crossing can be effected only by sexual generation; but whether the end thus gained is of sufficient importance to account for the first origin of sexual intercourse is extremely doubtful. Secondly, I have shown from a large body of facts, that, as a slight change in the conditions of life is beneficial to each creature, so, in an analogous manner, is the change effected in the germ by sexual union with a distinct individual; and I have been led, from observing the many widely-extended provisions throughout nature for this purpose, and from the greater vigour of crossed organisms of all kinds, as proved by direct experiments, as well as from the evil effects of close interbreeding when long continued, to believe that the advantage thus gained is very great.

Why the germ, which before impregnation undergoes a certain amount of development, ceases to progress and perishes, unless it be acted on by the male element; and why conversely the male element, which in the case of some insects is enabled to keep alive for four or five years, and in the case of some plants for several years, likewise perishes, unless it acts on or unites with the germ, are questions which cannot be answered with certainty. It is, however, probable that both sexual elements perish, unless brought into union, simply from including too little formative matter for independent development. Quatrefages has shown in the case of the Teredo (27/12. 'Annales des Sc. Nat.' 3rd series 1850 tome 13.), as did formerly Prevost and Dumas with other animals, that more than one spermatozoon is requisite to fertilise an ovum. This has likewise been shown by Newport (27/13. 'Transact. Phil. Soc.' 1851 pages 196, 208, 210; 1853 pages 245, 247.), who proved by numerous experiments, that, when a very small number of spermatozoa are applied to the ova of Batrachians, they are only partially impregnated, and an embryo is never fully developed. The rate also of the segmentation of the ovum is determined by the number of the spermatozoa. With respect to plants, nearly the same results were obtained by Kolreuter and Gartner. This last careful observer, after making successive trials on a Malva with more and more pollen- grains, found (27/14. 'Beitrage zur Kenntniss' etc. 1844 s. 345.), that even thirty grains did not fertilise a single seed; but when forty grains were applied to the stigma, a few seeds of small size were formed. In the case of Mirabilis the pollen grains are extraordinarily large, and the ovarium contains only a single ovule; and these circumstances led Naudin (27/15. 'Nouvelles Archives du Museum' tome 1 page 27.) to make the following experiments: a flower was fertilised by three grains and succeeded perfectly; twelve flowers were fertilised by two grains, and seventeen flowers by a single grain, and of these one flower alone in each lot perfected its seed: and it deserves especial notice that the plants produced by these two seeds never attained their proper dimensions, and bore flowers of remarkably small size. From these facts we clearly see that the quantity of the peculiar formative matter which is contained within the spermatozoa and pollen-grains is an all-important element in the act of fertilisation, not only for the full development of the seed, but for the vigour of the plant produced from such seed. We see something of the same kind in certain cases of parthenogenesis, that is, when the male element is wholly excluded; for M. Jourdan (27/16. As quoted by Sir J. Lubbock in 'Nat. Hist. Review' 1862 page 345. Weijenbergh also raised ('Nature' December 21, 1871 page 149) two successive generations from unimpregnated females of another lepidopterous insect, Liparis dispar. These females did not produce at most one-twentieth of their full complement of eggs, and many of the eggs were worthless. Moreover the caterpillars raised from these unfertilised eggs "possessed far less vitality" than those from fertilised eggs. In the third parthenogenetic generation not a single egg yielded a caterpillar.) found that, out of about 58,000 eggs laid by unimpregnated silk-moths, many passed through their early embryonic stages, showing that they were capable of self-development, but only twenty-nine out of the whole number produced caterpillars. The same principle of quantity seems to hold good even in artificial fissiparous reproduction, for Hackel (27/17. 'Entwickelungsgeschichte der Siphonophora' 1869 page 73.) found that by cutting the segmented and fertilised ova or larva of Siphonophorae (jelly-fishes) into pieces, the smaller the pieces were, the slower was the rate of development, and the larvae thus produced were by so much the more imperfect and inclined to monstrosity. It seems, therefore, probable that with the separate sexual elements deficient quantity of formative matter is the main cause of their not having the capacity for prolonged existence and development, unless they combine and thus increase each other's bulk. The belief that it is the function of the spermatozoa to communicate life to the ovule seems a strange one, seeing that the unimpregnated ovule is already alive and generally undergoes a certain amount of independent development. Sexual and asexual reproduction are thus seen not to differ essentially; and we have already shown that asexual reproduction, the power of regrowth and development are all parts of one and the same great law.

**REGROWTH OF AMPUTATED PARTS.**

This subject deserves a little further discussion. A multitude of the lower animals and some vertebrates possess this wonderful power. For instance,

Spallanzani cut off the legs and tail of the same salamander six times successively, and Bonnet (27/18. Spallanzani 'An Essay on Animal Reproduction' translated by Dr. Maty 1769 page 79. Bonnet 'Oeuvres d'Hist. Nat.' tome 5 part 1 4to. edition 1781 pages 343, 350.) did so eight times; and on each occasion the limbs were reproduced on the exact line of amputation, with no part deficient or in excess. An allied animal, the axolotl, had a limb bitten off, which was reproduced in an abnormal condition, but when this was amputated it was replaced by a perfect limb. (27/19. Vulpian as quoted by Prof. Faivre 'La Variabilite des Especes' 1868 page 112.) The new limbs in these cases bud forth, and are developed in the same manner as during the regular development of a young animal. For instance, with the Amblystoma lurida, three toes are first developed, then the fourth, and on the hind-feet the fifth, and so it is with a reproduced limb. (27/20. Dr. P. Hoy 'The American Naturalist' September 1871 page 579.)

The power of regrowth is generally much greater during the youth of an animal or during the earlier stages of its development than during maturity. The larvae or tadpoles of the Batrachians are capable of reproducing lost members, but not so the adults. (27/21. Dr. Gunther in Owen 'Anatomy of Vertebrates' volume 1 1866 page 567. Spallanzani has made similar observations.) Mature insects have no power of regrowth, excepting in one order, whilst the larvae of many kinds have this power. Animals low in the scale are able, as a general rule, to reproduce lost parts far more easily than those which are more highly organised. The myriapods offer a good illustration of this rule; but there are some strange exceptions to it—thus Nemerteans, though lowly organised, are said to exhibit little power of regrowth. With the higher vertebrata, such as birds and mammals, the power is extremely limited. (27/22. A thrush was exhibited before the British Association at Hull in 1853 which had lost its tarsus, and this member, it was asserted, had been thrice reproduced; having been lost, I presume, each time by disease. Sir J. Paget informs me that he feels some doubt about the facts recorded by Sir J. Simpson ('Monthly Journal of Medical Science' Edinburgh 1848 new series volume 2 page 890) of the regrowth of limbs in the womb in the case of man.)

In the case of those animals which may be bisected or chopped into pieces, and of which every fragment will reproduce the whole, the power of regrowth must be diffused throughout the whole body. Nevertheless there seems to be much truth in the view maintained by Prof. Lessona (27/23. 'Atti della Soc. Ital. di Sc. Nat.' volume 11 1869 page 493.), that this capacity is generally a localised and special one, serving to replace parts which are eminently liable to be lost in each particular animal. The most striking case in favour of this view, is that the terrestrial salamander, according to Lessona, cannot reproduce lost parts, whilst another species of the same genus, the aquatic salamander, has extraordinary powers of regrowth, as we have just seen; and this animal is eminently liable to have its limbs, tail, eyes and jaws bitten off by other tritons. (27/24. Lessona states that this is so in the paper just referred to. See also 'The American Naturalist' September 1871 page 579.) Even with the aquatic salamander the capacity is to a certain extent localised, for when M. Philipeaux (27/25. 'Comptes Rendus' October 1, 1866 and June 1867.) extirpated the entire fore limb together with the scapula, the power of regrowth was completely lost. It is also a remarkable fact, standing in opposition to a very general rule, that the young of the aquatic salamander do not possess the power of repairing their limbs in an equal degree with the adults (27/26. Bonnet 'Oeuvres Hist. Nat.' volume 5 page 294, as quoted by Prof. Rolleston in his remarkable address to the 36th annual meeting of the British Medical Association.) but I do not know that they are more active, or can otherwise better escape the loss of their limbs, than the adults. The walking-stick insect, Diapheromera femorata, like other insects of the same order, can reproduce its legs in the mature state, and these from their great length must be liable to be lost: but the capacity is localised (as in the case of the salamander), for Dr. Scudder found (27/27. 'Proc. Boston Soc. of Nat. Hist.' volume 12 1868-69 page 1.), that if the limb was removed within the trochanto-femoral articulation, it was never renewed. When a crab is seized by one of its legs, this is thrown off at the basal joint, being afterwards replaced by a new leg; and it is generally admitted that this is a special provision for the safety of the animal. Lastly, with gasteropod molluscs, which are well known to have the power of reproducing their heads, Lessona shows that they are very liable to have their heads bitten off by fishes; the rest of the body being protected by the shell. Even with plants we see something of the same kind, for non-deciduous leaves and young stems have no power of regrowth, these parts being easily replaced by growth from new buds; whilst the bark and subjacent tissues of the trunks of trees have great power of regrowth, probably on account of their increase in diameter, and of their liability to injury from being gnawed by animals.

**GRAFT-HYBRIDS.**

It is well known from innumerable trials made in all parts of the world, that buds may be inserted into a stock, and that the plants thus raised are not affected in a greater degree than can be accounted for by changed nutrition. Nor do the seedlings raised from such inserted buds partake of the character of the stock, though they are more liable to vary than are seedlings from the same variety growing on its own roots. A bud, also, may sport into a new and strongly-marked variety without any other bud on the same plant being in the least degree affected. We may therefore infer, in accordance with the common view, that each bud is a distinct individual, and that its formative elements do not spread beyond the parts subsequently developed from it.

Nevertheless, we have seen in the abstract on graft-hybridisation in the eleventh chapter that buds certainly include formative matter, which can occasionally combine with that included in the tissues of a distinct variety or species; a plant intermediate between the two parent-forms being thus produced. In the case of the potato we have seen that the tubers produced from a bud of one kind inserted into another are intermediate in colour, size, shape and state of surface; that the stems, foliage, and even certain constitutional peculiarities, such as precocity, are likewise intermediate. With these well- established cases, the evidence that graft-hybrids have also been produced with the laburnum, orange, vine, rose, etc., seems sufficient. But we do not know under what conditions this rare form of reproduction is possible. From these several cases we learn the important fact that formative elements capable of blending with those of a distinct individual (and this is the chief characteristic of sexual generation), are not confined to the reproductive organs, but are present in the buds and cellular tissue of plants; and this is a fact of the highest physiological importance.

## DIRECT ACTION OF THE MALE ELEMENT ON THE FEMALE.

In the eleventh chapter, abundant proofs were given that foreign pollen occasionally affects in a direct manner the mother-plant. Thus, when Gallesio fertilised an orange-flower with pollen from the lemon, the fruit bore stripes of perfectly characterised lemon-peel. With peas, several observers have seen the colour of the seed-coats and even of the pod directly affected by the pollen of a distinct variety. So it has been with the fruit of the apple, which consists of the modified calyx and upper part of the flower-stalk. In ordinary cases these parts are wholly formed by the mother-plant. We here see that the formative elements included within the male element or pollen of one variety can affect and hybridise, not the part which they are properly adapted to affect, namely, the ovules, but the partially-developed tissues of a distinct variety or species. We are thus brought half-way towards a graft- hybrid, in which the formative elements included within the tissues of one individual combine with those included in the tissues of a distinct variety or species, thus giving rise to a new and intermediate form, independently of the male or female sexual organs.

With animals which do not breed until nearly mature, and of which all the parts are then fully developed, it is hardly possible that the male element should directly affect the female. But we have the analogous and perfectly well-ascertained case of the male element affecting (as with the quagga and Lord Morton's mare) the female or her ova, in such a manner that when she is impregnated by another male her offspring are affected and hybridised by the first male. The explanation would be simple if the spermatozoa could keep alive within the body of the female during the long interval which has sometimes elapsed between the two acts of impregnation; but no one will suppose that this is possible with the higher animals.

# DEVELOPMENT.

The fertilised germ reaches maturity by a vast number of changes: these are either slight and slowly effected, as when the child grows into the man, or are great and sudden, as with the metamorphoses of most insects. Between these extremes we have every gradation, even within the same class; thus, as Sir J. Lubbock has shown (27/28. 'Transact. Linn. Soc.' volume 24 1863 page 62.) there is an Ephemerous insect which moults above twenty times, undergoing each time a slight but decided change of structure; and these changes, as he further remarks, probably reveal to us the normal stages of development, which are concealed and hurried through or suppressed in most other insects. In ordinary metamorphoses, the parts and organs appear to become changed into the corresponding parts in the next stage of development; but there is another form of development, which has been called by Professor Owen metagenesis. In this case "the new parts are not moulded upon the inner surface of the old ones. The plastic force has changed its course of operation. The outer case, and all that gave form and character to the precedent individual, perish and are cast off; they are not changed into the corresponding parts of the new individual. These are due to a new and distinct developmental process," etc. (27/29. 'Parthenogenesis' 1849 pages 25, 26. Prof. Huxley has some excellent remarks ('Medical Times' 1856 page 637) on this subject in reference to the development of star-fishes, and shows how curiously metamorphosis graduates into gemmation or zoid-formation, which is in fact the same as metagenesis.) Metamorphosis, however, graduates so insensibly, into metagenesis, that the two processes cannot be distinctly separated. For instance, in the last change which Cirripedes undergo, the alimentary canal and some other organs are moulded on pre-existing parts; but the eyes of the old and the young animal are developed in entirely different parts of the body; the tips of the mature limbs are formed within the larval limbs, and may be said to be metamorphosed from them; but their basal portions and the whole thorax are developed in a plane at right angles to the larval limbs and thorax; and this may be called metagenesis. The metagenetic process is carried to an extreme point in the development of some Echinoderms, for the animal in the second stage of development is formed almost like a bud within the animal of the first stage, the latter being then cast off like an old vestment, yet sometimes maintaining for a short period an independent vitality. (27/30. Prof. J. Reay Greene in Gunther's 'Record of Zoolog. Lit.' 1865 page 625.) If, instead of a single individual, several were to be thus developed metagenetically within a pre-existing form, the process would be called one of alternate generation. The young thus developed may either closely resemble the encasing parent-form, as with the larvae of Cecidomyia, or may differ to an astonishing degree, as with many parasitic worms and jelly-fishes; but this does not make any essential difference in the process, any more than the greatness or abruptness of the change in the metamorphoses of insects.

The whole question of development is of great importance for our present subject. When an organ, the eye, for instance, is metagenetically formed in a part of the body where during the previous stage of development no eye existed, we must look at it as a new and independent growth. The absolute independence of new and old structures, although corresponding in structure and function, is still more obvious when several individuals are formed within a previous form, as in the cases of alternate generation. The same important principle probably comes largely into play even in the case of apparently continuous growth, as we shall see when we consider the inheritance of modifications at corresponding ages.

We are led to the same conclusion, namely, the independence of parts successively developed, by another and quite distinct group of facts. It is well known that many animals belonging to the same order, and therefore not differing widely from each other, pass through an extremely different course of development. Thus certain beetles, not in any way remarkably different from others of the same order, undergo what has been called a hyper-metamorphosis— that is, they pass through an early stage wholly different from the ordinary grub-like larva. In the same sub-order of crabs, namely, the Macroura, as Fritz Muller remarks, the river cray-fish is hatched under the same form which it ever afterwards retains; the young lobster has divided legs, like a Mysis; the Palaemon appears under the form of a Zoea, and Peneus under the Nauplius- form; and how wonderfully these larval forms differ from one another, is known to every naturalist. (27/31. Fritz Muller 'Fur Darwin' 1864 s. 65, 71. The highest authority on crustaceans, Prof. Milne-Edwards, insists ('Annal. des Sci. Nat.' 2nd series Zoolog. tome 3 page 322) on the difference in the metamorphosis of closely-allied genera.) Some other crustaceans, as the same author observes, start from the same point and arrive at nearly the same end, but in the middle of their development are widely different from one another. Still more striking cases could be given with respect to the Echinodermata. With the Medusae or jelly-fishes Professor Allman observes, "The classification of the Hydroida would be a comparatively simple task if, as has been erroneously asserted, generically-identical medusoids always arose from generically-identical polypoids; and, on the other hand, that generically- identical polypoids always gave origin to generically-identical medusoids." So again, Dr. Strethill Wright remarks, "In the life-history of the Hydroidae any phase, planuloid, polypoid, or medusoid, may be absent." (27/32. Prof. Allman 'Annals and Mag. of Nat.

Hist.' 3rd series volume 13 1864 page 348; Dr. S. Wright ibid volume 8 1861 page 127. See also page 358 for analogous statements by Sars.)

According to the belief now generally accepted by our best naturalists, all the members of the same order or class, for instance, the Medusae or the Macrourous crustaceans, are descended from a common progenitor. During their descent they have diverged much in structure, but have retained much in common; and this has occurred, though they have passed through and still pass through marvellously different metamorphoses. This fact well illustrates how independent each structure is from that which precedes and that which follows it in the course of development.

## THE FUNCTIONAL INDEPENDENCE OF THE ELEMENTS OR UNITS OF THE BODY.

Physiologists agree that the whole organism consists of a multitude of elemental parts, which are to a great extent independent of one another. Each organ, says Claude Bernard (27/33. 'Tissus Vivants' 1866 page 22.), has its proper life, its autonomy; it can develop and reproduce itself independently of the adjoining tissues. A great German authority, Virchow (27/34. 'Cellular Pathology' translated by Dr. Chance 1860 pages 14, 18, 83, 460.), asserts still more emphatically that each system consists of an "enormous mass of minute centres of action...Every element has its own special action, and even though it derive its stimulus to activity from other parts, yet alone effects the actual performance of duties...Every single epithelial and muscular fibre- cell leads a sort of parasitical existence in relation to the rest of the body...Every single bone-corpuscle really possesses conditions of nutrition peculiar to itself." Each element, as Sir J. Paget remarks, lives its appointed time and then dies, and is replaced after being cast off or absorbed. (27/35. Paget 'Surgical Pathology' volume 1 1853 pages 12-14.) I presume that no physiologist doubts that, for instance, each bone-corpuscle of the finger differs from the corresponding corpuscle in the corresponding joint of the toe; and there can hardly be a doubt that even those on the corresponding sides of the body differ, though almost identical in nature. This near approach to identity is curiously shown in many diseases in which the same exact points on the right and left sides of the body are similarly affected; thus Sir J. Paget (27/36. Ibid page 19.) gives a drawing of a diseased pelvis, in which the bone has grown into a most complicated pattern, but "there is not one spot or line on one side which is not represented, as exactly as it would be in a mirror, on the other."

Many facts support this view of the independent life of each minute element of the body. Virchow insists that a single bone-corpuscle or a single cell in the skin may become diseased. The spur of a cock, after being inserted into the ear of an ox, lived for eight years, and acquired a weight of 396 grammes (nearly fourteen ounces), and the astonishing length of twenty-four centimetres, or about nine inches; so that the head of the ox appeared to bear three horns. (27/37. See Prof. Mantegazza's interesting work 'Degli innesti Animali' etc. Milano 1865 page 51 tab. 3.) The tail of a pig has been grafted into the middle of its back, and reacquired sensibility. Dr. Ollier (27/38. 'De la Production Artificielle des Os' page 8.) inserted a piece of periosteum from the bone of a young dog under the skin of a rabbit, and true bone was developed. A multitude of similar facts could be given. The frequent presence of hairs and of perfectly developed teeth, even teeth of the second dentition, in ovarian tumours (27/39. Isidore Geoffroy Saint-Hilaire 'Hist. des Anomalies' tome 2 pages 549, 560, 562; Virchow ibid page 484. Lawson Tait 'The Pathology of Diseases of the Ovaries' 1874 pages 61, 62.), are facts leading to the same conclusion. Mr. Lawson Tait refers to a tumour in which "over 300 teeth were found, resembling in many respects milk-teeth;" and to another tumour, "full of hair which had grown and been shed from one little spot of skin not bigger than the tip of my little finger. The amount of hair in the sac, had it grown from a similarly sized area of the scalp, would have taken almost a lifetime to grow and be shed."

Whether each of the innumerable autonomous elements of the body is a cell or the modified product of a cell, is a more doubtful question, even if so wide a definition be given to the term, as to include cell-like bodies without walls and without nuclei. (27/40. For the most recent classification of cells, see Ernst Hackel 'Generelle Morpholog.' b. 2 1866 s. 275.) The doctrine of omnis cellula e cellula is admitted for plants, and widely prevails with respect to animals. (27/41. Dr. W. Turner 'The Present Aspect of Cellular Pathology' 'Edinburgh Medical Journal' April 1863.) Thus Virchow, the great supporter of the cellular theory, whilst allowing that difficulties exist, maintains that every atom of tissue is derived from cells, and these from pre-existing cells, and these primarily from the egg, which he regards as a great cell. That cells, still retaining the same nature, increase by self-division or proliferation, is admitted by every one. But when an organism undergoes great changes of structure during development, the cells, which at each stage are supposed to be directly derived from previously existing cells, must likewise be greatly changed in nature; this change is attributed by the supporters of the cellular doctrine to some inherent power which the cells possess, and not to any external agency. Others maintain that cells and tissues of all kinds may be formed, independently of pre-existing cells, from plastic lymph or blastema. Whichever view may be correct, every one admits that the body consists of a multitude of organic units, all of which possess their own proper attributes, and are to a certain extent independent of all others. Hence it will be convenient to use indifferently the terms cells or organic units, or simply units.

# VARIABILITY AND INHERITANCE.

We have seen in the twenty-second chapter that variability is not a principle co-ordinate with life or reproduction, but results from special causes, generally from changed conditions acting during successive generations. The fluctuating variability thus induced is apparently due in part to the sexual system being easily affected, so that it is often rendered impotent; and when not so seriously affected, it often fails in its proper function of transmitting truly the characters of the parents to the offspring. But variability is not necessarily connected with the sexual system, as we see in the cases of bud-variation. Although we are seldom able to trace the nature of the connection, many deviations of structure no doubt result from changed conditions acting directly on the organisation, independently of the reproductive system. In some instances we may feel sure of this, when all, or nearly all the individuals which have been similarly exposed are similarly and definitely affected, of which several instances have been given. But it is by no means clear why the offspring should be affected by the exposure of the parents to new conditions, and why it is necessary in most cases that several generations should have been thus exposed.

How, again, can we explain the inherited effects of the use or disuse of particular organs? The domesticated duck flies less and walks more than the wild duck, and its limb-bones have become diminished and increased in a corresponding manner in comparison with those of the wild duck. A horse is trained to certain paces, and the colt inherits similar consensual movements. The domesticated rabbit becomes tame from close confinement; the dog, intelligent from associating with man; the retriever is taught to fetch and carry; and these mental endowments and bodily powers are all inherited. Nothing in the whole circuit of physiology is more wonderful. How can the use or disuse of a particular limb or of the brain affect a small aggregate of reproductive cells, seated in a distant part of the body, in such a manner that the being developed from these cells inherits the characters of either one or both parents? Even an imperfect answer to this question would be satisfactory.

In the chapters devoted to inheritance it was shown that a multitude of newly acquired characters, whether injurious or beneficial, whether of the lowest or highest vital importance, are often faithfully transmitted—frequently even when one parent alone possesses some new peculiarity; and we may on the whole conclude that inheritance is the rule, and non-inheritance the anomaly. In some instances a character is not inherited, from the conditions of life being directly opposed to its development; in many instances, from the conditions incessantly inducing fresh variability, as with grafted fruit-trees and highly-cultivated flowers. In the remaining cases the failure may be attributed to reversion, by which the child resembles its grandparents or more remote progenitors, instead of its parents.

Inheritance is governed by various laws. Characters which first appear at any particular age tend to reappear at a corresponding age. They often become associated with certain seasons of the year, and reappear in the offspring at a corresponding season. If they appear rather late in life in one sex, they tend to reappear exclusively in the same sex at the same period of life.

The principle of reversion, recently alluded to, is one of the most wonderful of the attributes of Inheritance. It proves to us that the transmission of a character and its development, which ordinarily go together and thus escape discrimination, are distinct powers; and these powers in some cases are even antagonistic, for each acts alternately in successive generations. Reversion is not a rare event, depending on some unusual or favourable combination of circumstances, but occurs so regularly with crossed animals and plants, and so frequently with uncrossed breeds, that it is evidently an essential part of the principle of inheritance. We know that changed conditions have the power of evoking long-lost characters, as in the case of animals becoming feral. The act of crossing in itself possesses this power in a high degree. What can be more wonderful than that characters, which have disappeared during scores, or hundreds, or even thousands of generations, should suddenly reappear perfectly developed, as in the case of pigeons and fowls, both when purely bred and especially when crossed; or as with the zebrine stripes on dun-coloured horses, and other such cases? Many monstrosities come under this same head, as when rudimentary organs are redeveloped, or when an organ which we must believe was possessed by an early progenitor of the species, but of which not even a rudiment is left, suddenly reappears, as with the fifth stamen in some Scrophulariaceae. We have already seen that reversion acts in bud- reproduction; and we know that it occasionally acts during the growth of the same individual animal, especially, but not exclusively, if of crossed parentage,—as in the rare cases described of fowls, pigeons, cattle, and rabbits, which have reverted to the colours of one of their parents or ancestors as they advanced in years.

We are led to believe, as formerly explained, that every character which occasionally reappears is present in a latent form in each generation, in nearly the same manner as in male and female animals the secondary characters of the opposite sex lie latent and ready to be evolved when the reproductive organs are injured. This comparison of the secondary sexual characters which lie latent in both sexes, with other latent characters, is the more appropriate from the case recorded of a Hen, which assumed some of the masculine characters, not of her own race, but of an early progenitor; she thus exhibited at the same time the redevelopment of latent characters of both kinds. In every living creature we may feel assured that a host of long-lost characters lie ready to be evolved under proper conditions. How can we make intelligible and connect with other facts, this wonderful and common capacity of reversion,—this power of calling back to life long-lost characters?

## PART II.

I have now enumerated the chief facts which every one would desire to see connected by some intelligible bond. This can be done, if we make the following assumptions, and much may be advanced in favour of the chief one. The secondary assumptions can likewise be supported by various physiological considerations. It is universally admitted that the cells or units of the body increase by self-division or proliferation, retaining the same nature, and that they ultimately become converted into the various tissues and substances of the body. But besides this means of increase I assume that the units throw off minute granules which are dispersed throughout the whole system; that these, when supplied with proper nutriment, multiply by self-division, and are ultimately developed into units like those from which they were originally derived. These granules may be called gemmules. They are collected from all parts of the system to constitute the sexual elements, and their development in the next generation forms a new being; but they are likewise capable of transmission in a dormant state to future generations and may then be developed. Their development depends on their union with other partially developed or nascent cells which precede them in the regular course of growth. Why I use the term union, will be seen when we discuss the direct action of pollen on the tissues of the mother-plant. Gemmules are supposed to be thrown off by every unit, not only during the adult state, but during each stage of development of every organism; but not necessarily during the continued existence of the same unit. Lastly, I assume that the gemmules in their dormant state have a mutual affinity for each other, leading to their aggregation into buds or into the sexual elements. Hence, it is not the reproductive organs or buds which generate new organisms, but the units of which each individual is composed. These assumptions constitute the provisional hypothesis which I have called Pangenesis. Views in many respects similar have been propounded by various authors. (27/42. Mr. G.H. Lewes ('Fortnightly Review' November 1, 1868 page 506) remarks on the number of writers who have advanced nearly similar views. More than two thousand years ago Aristotle combated a view of this kind, which, as I hear from Dr. W. Ogle, was held by Hippocrates and others. Ray, in his 'Wisdom of God' (2nd edition 1692 page 68), says that "every part of the body seems to club and contribute to the seed." The "organic molecules" of Buffon ('Hist. Nat. Gen.' edition of 1749 tome 2 pages 54, 62, 329, 333, 420, 425) appear at first sight to be the same as the gemmules of my hypothesis, but they are essentially different. Bonnet ('Oeuvres d'Hist. Nat.' tome 5 part 1 1781 4to edition page 334) speaks of the limbs having germs adapted for the reparation of all possible losses; but whether these germs are supposed to be the same with those within buds and the sexual organs is not clear. Prof. Owen says ('Anatomy of Vertebrates' volume 3 1868 page 813) that he fails to see any fundamental difference between the views which he propounded in his 'Parthenogenesis' (1849 pages 5- 8), and which he now considers as erroneous, and my hypothesis of pangenesis: but a reviewer ('Journal of Anat. and Phys.' May 1869 page 441) shows how different they really are. I formerly thought that the "physiological units" of Herbert Spencer ('Principles of Biology' volume 1 chapters 4 and 8 1863-64) were the same as my gemmules, but I now know that this is not the case. Lastly, it appears from a review of the present work by Prof. Mantegazza ('Nuova Antologia, Maggio' 1868), that he (in his 'Elementi di Igiene' Ediz. 3 page 540) clearly foresaw the doctrine of pangenesis.)

Before proceeding to show, firstly, how far these assumptions are in themselves probable, and secondly, how far they connect and explain the various groups of facts with which we are concerned, it may be useful to give an illustration, as simple as possible, of the hypothesis. If one of the Protozoa be formed, as it appears under the microscope, of a small mass of homogeneous gelatinous matter, a minute particle or gemmule thrown off from any part and nourished under favourable circumstances would reproduce the whole; but if the upper and lower surfaces were to differ in texture from each other and from the central portion, then all three parts would have to throw off gemmules, which when aggregated by mutual affinity would form either buds or the sexual elements, and would ultimately be developed into a similar organism. Precisely the same view may be extended to one of the higher animals; although in this case many thousand gemmules must be thrown off from the various parts of the body at each stage of development; these gemmules being developed in union with pre-existing nascent cells in due order of succession.

Physiologists maintain, as we have seen, that each unit of the body, though to a large extent dependent on others, is likewise to a certain extent independent or autonomous, and has the power of increasing by self-division. I go one step further, and assume that each unit casts off free gemmules which are dispersed throughout the system, and are capable under proper conditions of being developed into similar units. Nor can this assumption be considered as gratuitous and improbable. It is manifest that the sexual elements and buds include formative matter of some kind, capable of development; and we now know from the production of graft-hybrids that similar matter is dispersed throughout the tissues of plants, and can combine with that of another and distinct plant, giving rise to a new being, intermediate in character. We know also that the male element can act directly on the partially developed tissues of the mother-plant, and on the future progeny of female animals. The formative matter which is thus dispersed throughout the tissues of plants, and which is capable of being developed into each unit or part, must be generated there by some means; and my chief assumption is that this matter consists of minute particles or gemmules cast off from each unit or cell. (27/43. Mr. Lowne has observed ('Journal of Queckett Microscopical Club' September 23, 1870) certain remarkable changes in the tissues of the larva of a fly, which makes him believe "it possible that organs and organisms are sometimes developed by the aggregation of excessively minute gemmules, such as those which Mr. Darwin's hypothesis demands.")

But I have further to assume that the gemmules in their undeveloped state are capable of largely multiplying themselves by self-division, like independent organisms. Delpino insists that to "admit of multiplication by fissiparity in corpuscles, analogous to seeds or buds…is repugnant to all analogy." But this seems a strange objection, as Thuret (27/44. 'Annales des Sc. Nat.' 3rd series Bot. tome 14 1850 page 244.) has seen the zoospore of an alga divide itself, and each half germinated. Haeckel divided the segmented ovum of a siphonophora into many pieces, and these were developed. Nor does the extreme minuteness of the gemmules, which can hardly differ much in nature from the lowest and simplest organisms, render it improbable that they should grow and multiply. A great authority, Dr. Beale (27/45. 'Disease Germs' page 20.), says "that minute yeast cells are capable of throwing off buds or gemmules, much less than the 1/100000 of an inch in diameter;" and these he thinks are "capable of subdivision practically ad infinitum."

A particle of small-pox matter, so minute as to be borne by the wind, must multiply itself many thousandfold in a person thus inoculated; and so with the contagious matter of scarlet fever. (27/46. See some very interesting papers on this subject by Dr. Beale in 'Medical Times and Gazette' September 9, 1865 pages 273, 330.) It has recently been ascertained (27/47. Third Report of the R. Comm. on the Cattle Plague as quoted in 'Gardener's Chronicle' 1866 page 446.) that a minute portion of the mucous discharge from an animal affected with rinderpest, if placed in the blood of a healthy ox, increases so fast that in a short space of time "the whole mass of blood, weighing many pounds, is infected, and every small particle of that blood contains enough poison to give, within less than forty-eight hours, the disease to another animal."

The retention of free and undeveloped gemmules in the same body from early youth to old age will appear improbable, but we should remember how long seeds lie dormant in the earth and buds in the bark of a tree. Their transmission from generation to generation will appear still more improbable; but here again we should remember that many rudimentary and useless organs have been transmitted during an indefinite number of generations.

We shall presently see how well the long-continued transmission of undeveloped gemmules explains many facts.

As each unit, or group of similar units, throughout the body, casts off its gemmules, and as all are contained within the smallest ovule, and within each spermatozoon or pollen-grain, and as some animals and plants produce an astonishing number of pollen-grains and ovules (27/48. Mr. F. Buckland found 6,867,840 eggs in a cod-fish ('Land and Water' 1868 page 62). An Ascaris produces about 64,000,000 eggs (Carpenter's 'Comp. Phys.' 1854 page 590). Mr. J. Scott, of the Royal Botanic Garden of Edinburgh, calculated, in the same manner as I have done for some British Orchids ('Fertilisation of Orchids' page 344), the number of seeds in a capsule of an Acropera and found the number to be 371,250. Now this plant produces several flowers on a raceme, and many racemes during a season. In an allied genus, Gongora, Mr. Scott has seen twenty capsules produced on a single raceme; ten such racemes on the Acropera would yield above seventy-four millions of seed.), the number and minuteness of the gemmules must be something inconceivable. But considering how minute the molecules are, and how many go to the formation of the smallest granule of any ordinary substance, this difficulty with respect to the gemmules is not insuperable. From the data arrived at by Sir W. Thomson, my son George finds that a cube of 1/10000 of an inch of glass or water must consist of between 16 million millions, and 131 thousand million million molecules. No doubt the molecules of which an organism is formed are larger, from being more complex, than those of an inorganic substance, and probably many molecules go to the formation of a gemmule; but when we bear in mind that a cube of 1/10000 of an inch is much smaller than any pollen-grain, ovule or bud, we can see what a vast number of gemmules one of these bodies might contain.

The gemmules derived from each part or organ must be thoroughly dispersed throughout the whole system. We know, for instance, that even a minute fragment of a leaf of a Begonia will reproduce the whole plant; and that if a fresh-water worm is chopped into small pieces, each will reproduce the whole animal. Considering also the minuteness of the gemmules and the permeability of all organic tissues, the thorough dispersion of the gemmules is not surprising. That matter may be readily transferred without the aid of vessels from part to part of the body, we have a good instance in a case recorded by Sir J. Paget of a lady, whose hair lost its colour at each successive attack of neuralgia and recovered it again in the course of a few days. With plants, however, and probably with compound animals, such as corals, the gemmules do not ordinarily spread from bud to bud, but are confined to the parts developed from each separate bud; and of this fact no explanation can be given.

The assumed elective affinity of each gemmule for that particular cell which precedes it in due order of development is supported by many analogies. In all ordinary cases of sexual reproduction, the male and female elements certainly have a mutual affinity for each other: thus, it is believed that about ten thousand species of Compositae exist, and there can be no doubt that if the pollen of all these species could be simultaneously or successively placed on the stigma of any one species, this one would elect with unerring certainty its own pollen. This elective capacity is all the more wonderful, as it must have been acquired since the many species of this great group of plants branched off from a common progenitor. On any view of the nature of sexual reproduction, the formative matter of each part contained within the ovules and the male element act on each other by some law of special affinity, so that corresponding parts affect one another; thus, a calf produced from a short-horned cow by a long-horned bull has its horns affected by the union of the two forms, and the offspring from two birds with differently coloured tails have their tails affected.

The various tissues of the body plainly show, as many physiologists have insisted (27/49. Paget 'Lectures on Pathology' page 27; Virchow 'Cellular Pathology' translated by Dr. Chance pages 123, 126, 294. Claude Bernard 'Des Tissus Vivants' pages 177, 210, 337; Muller 'Physiology' English translation page 290.), an affinity for special organic substances, whether natural or foreign to the body. We see this in the cells of the kidneys attracting urea from the blood; in curare affecting certain nerves; Lytta vesicatoria the kidneys; and the poisonous matter of various diseases, as small-pox, scarlet-fever, hooping-cough, glanders, and hydrophobia, affecting certain definite parts of the body. It has also been assumed that the development of each gemmule depends on its union with another cell or unit which has just commenced its development, and which precedes it in due order of growth. That the formative matter within the pollen of plants, which by our hypothesis consists of gemmules, can unite with and modify the partially developed cells of the mother-plant, we have clearly seen in the section devoted to this subject. As the tissues of plants are formed, as far as is known, only by the proliferation of pre-existing cells, we must conclude that the gemmules derived from the foreign pollen do not become developed into new and separate cells, but penetrate and modify the nascent cells of the mother-plant. This process may be compared with what takes place in the act of ordinary fertilisation, during which the contents of the pollen-tubes penetrate the closed embryonic sac within the ovule, and determine the development of the embryo. According to this view, the cells of the mother-plant may almost literally be said to be fertilised by the gemmules derived from the foreign pollen. In this case and in all others the proper gemmules must combine in due order with pre-existing nascent cells, owing to their elective affinities. A slight difference in nature between the gemmules and the nascent cells would be far from interfering with their mutual union and development, for we well know in the case of ordinary reproduction that such slight differentiation in the sexual elements favours in a marked manner their union and subsequent development, as well as the vigour of the offspring thus produced.

Thus far we have been able by the aid of our hypothesis to throw some obscure light on the problems which have come before us; but it must be confessed that many points remain altogether doubtful. Thus it is useless to speculate at what period of development each unit of the body casts off its gemmules, as the whole subject of the development of the various tissues is as yet far from clear. We do not know whether the gemmules are merely collected by some unknown means at certain seasons within the reproductive organs, or whether after being thus collected they rapidly multiply there, as the flow of blood to these organs at each breeding season seems to render probable. Nor do we know why the gemmules collect to form buds in certain definite places, leading to the symmetrical growth of trees and corals. We have no means of deciding whether the ordinary wear and tear of the tissues is made good by means of gemmules, or merely by the proliferation of pre-existing cells. If the gemmules are thus consumed, as seems probable from the intimate connection between the repair of waste, regrowth, and development, and more especially from the periodical changes which many male animals undergo in colour and structure, then some light would be thrown on the phenomena of old age, with its lessened power of reproduction and of the repair of injuries, and on the obscure subject of longevity. The fact of castrated animals, which do not cast off innumerable gemmules in the act of reproduction, not being longer-lived than perfect males, seems opposed to the belief that gemmules are consumed in the ordinary repair of wasted tissues; unless indeed the gemmules after being collected in small numbers within the reproductive organs are there largely multiplied. (27/50. Prof. Ray Lankester has discussed several of the points here referred to as bearing on pangenesis, in his interesting essay, 'On Comparative Longevity in Man and the Lower Animals' 1870 pages 33, 77, etc.)

That the same cells or units may live for a long period and continue multiplying without being modified by their union with free gemmules of any kind, is probable from such cases as that of the spur of a cock which grew to an enormous size when grafted into the ear of an ox. How far units are modified during their normal growth by absorbing peculiar nutriment from the surrounding tissues, independently of their union with gemmules of a distinct nature, is another doubtful point. (27/51. Dr. Ross refers to this subject in his 'Graft Theory of Disease' 1872 page 53.) We shall appreciate this difficulty by calling to mind what complex yet symmetrical growths the cells of plants yield when inoculated by the poison of a gall-insect. With animals various polypoid excrescences and tumours are generally admitted (27/52. Virchow 'Cellular Pathology' translated by Dr. Chance 1860 pages 60, 162, 245, 441, 454.) to be the direct product, through proliferation, of normal cells which have become abnormal. In the regular growth and repair of bones, the tissues undergo, as Virchow remarks (27/53. Ibid pages 412-426.), a whole series of permutations and substitutions. "The cartilage cells may be converted by a direct transformation into marrow-cells, and continue as such; or they may first be converted into osseous and then into medullary tissue; or lastly, they may first be converted into marrow and then into bone. So variable are the permutations of these tissues, in themselves so nearly allied, and yet in their external appearance so completely distinct." But as these tissues thus change their nature at any age, without any obvious change in their nutrition, we must suppose in accordance with our hypothesis that gemmules derived from one kind of tissue combine with the cells of another kind, and cause the successive modifications.

We have good reason to believe that several gemmules are requisite for the development of one and the same unit or cell; for we cannot otherwise understand the insufficiency of a single or even of two or three pollen-grains or spermatozoa. But we are far from knowing whether the gemmules of all the units are free and separate from one another, or whether some are from the first united into small aggregates. A feather, for instance, is a complex structure, and, as each separate part is liable to inherited variations, I conclude that each feather generates a large number of gemmules; but it is possible that these may be aggregated into a compound gemmule. The same remark applies to the petals of flowers, which are sometimes highly complex structures, with each ridge and hollow contrived for a special purpose, so that each part must have been separately modified, and the modifications transmitted; consequently, separate gemmules, according to our hypothesis, must have been thrown off from each cell or unit. But, as we sometimes see half an anther or a small portion of a filament becoming petali-form, or parts or mere stripes of the calyx assuming the colour and texture of the corolla, it is probable that with petals the gemmules of each cell are not aggregated together into a compound gemmule, but are free and separate. Even in so simple a case as that of a perfect cell, with its protoplasmic contents, nucleus, nucleolus, and walls, we do not know whether or not its development depends on a compound gemmule derived from each part. (27/54. See some good criticisms on this head by Delpino and by Mr. G.H. Lewes in the 'Fortnightly Review' November 1, 1868 page 509.)

Having now endeavoured to show that the several foregoing assumptions are to a certain extent supported by analogous facts, and having alluded to some of the most doubtful points, we will consider how far the hypothesis brings under a single point of view the various cases enumerated in the First Part. All the forms of reproduction graduate into one another and agree in their product; for it is impossible to distinguish between organisms produced from buds, from self-division, or from fertilised germs; such organisms are liable to variations of the same nature and to reversions of the same kind; and as, according to our hypothesis, all the forms of reproduction depend on the aggregation of gemmules derived from the whole body, we can understand this remarkable agreement. Parthenogenesis is no longer wonderful, and if we did not know that great good followed from the union of the sexual elements derived from two distinct individuals, the wonder would be that parthenogenesis did not occur much oftener than it does. On any ordinary theory of reproduction the formation of graft-hybrids, and the action of the male element on the tissues of the mother-plant, as well as on the future progeny of female animals, are great anomalies; but they are intelligible on our hypothesis. The reproductive organs do not actually create the sexual elements; they merely determine the aggregation and perhaps the multiplication of the gemmules in a special manner. These organs, however, together with their accessory parts, have high functions to perform. They adapt one or both elements for independent temporary existence, and for mutual union. The stigmatic secretion acts on the pollen of a plant of the same species in a wholly different manner to what it does on the pollen of one belonging to a distinct genus or family. The spermatophores of the Cephalopoda are wonderfully complex structures, which were formerly mistaken for parasitic worms; and the spermatozoa of some animals possess attributes which, if observed in an independent animal, would be put down to instinct guided by sense-organs,—as when the spermatozoa of an insect find their way into the minute micropyle of the egg.

The antagonism which has long been observed (27/55. Mr. Herbert Spencer ('Principles of Biology' volume 2 page 430) has fully discussed this antagonism.), with certain exceptions, between growth and the power of sexual reproduction (27/56. The male salmon is known to breed at a very early age. The Triton and Siredon, whilst retaining their larval branchiae, according to Filippi and Dumeril ('Annals and Mag. of Nat. Hist.' 3rd series 1866 page 157) are capable of reproduction. Ernst Haeckel has recently ('Monatsbericht Akad. Wiss. Berlin' February 2, 1865) observed the surprising case of a medusa, with its reproductive organs active, which produces by budding a widely different form of medusa; and this latter also has the power of sexual reproduction. Krohn has shown ('Annals and Mag. of Nat. Hist.' 3rd series volume 19 1862 page 6) that certain other medusae, whilst sexually mature, propagate by gemmae. See also Kolliker 'Morphologie und Entwickelungsgeschichte des Pennatulidenstammes' 1872 page 12.)—between the repair of injuries and gemmation—and with plants, between rapid increase by buds, rhizomes, etc., and the production of seed, is partly explained by the gemmules not existing in sufficient numbers for these processes to be carried on simultaneously.

Hardly any fact in physiology is more wonderful than the power of regrowth; for instance, that a snail should be able to reproduce its head, or a salamander its eyes, tail, and legs, exactly at the points where they have been cut off. Such cases are explained by the presence of gemmules derived from each part, and disseminated throughout the body. I have heard the process compared with that of the repair of the broken angles of a crystal by re- crystallisation; and the two processes have this much in common, that in the one case the polarity of the molecules is the efficient cause, and in the other the affinity of the gemmules for particular nascent cells. But we have here to encounter two objections which apply not only to the regrowth of a part, or of a bisected individual, but to fissiparous generation and budding. The first objection is that the part which is reproduced is in the same stage of development as that of the being which has been operated on or bisected; and in the case of buds, that the new beings thus produced are in the same stage as that of the budding parent. Thus a mature salamander, of which the tail has been cut off, does not reproduce a larval tail; and a crab does not reproduce a larval leg. In the case of budding it was shown in the first part of this chapter that the new being thus produced does not retrograde in development,—that is, does not pass through those earlier stages, which the fertilised germ has to pass through. Nevertheless, the organisms operated on or multiplying themselves by buds must, by our hypothesis, include innumerable gemmules derived from every part or unit of the earlier stages of development; and why do not such gemmules reproduce the amputated part or the whole body at a corresponding early stage of development?

The second objection, which has been insisted on by Delpino, is that the tissues, for instance, of a mature salamander or crab, of which a limb has been removed, are already differentiated and have passed through their whole course of development; and how can such tissues in accordance with our hypothesis attract and combine with the gemmules of the part which is to be reproduced? In answer to these two objections we must bear in mind the evidence which has been advanced, showing that at least in a large number of cases the power of regrowth is a localised faculty, acquired for the sake of repairing special injuries to which each particular creature is liable; and in the case of buds or fissiparous generation, for the sake of quickly multiplying the organism at a period of life when it can be supported in large numbers. These considerations lead us to believe that in all such cases a stock of nascent cells or of partially developed gemmules are retained for this special purpose either locally or throughout the body, ready to combine with the gemmules derived from the cells which come next in due succession. If this be admitted we have a sufficient answer to the above two objections. Anyhow, pangenesis seems to throw a considerable amount of light on the wonderful power of regrowth.

It follows, also, from the view just given, that the sexual elements differ from buds in not including nascent cells or gemmules in a somewhat advanced stage of development, so that only the gemmules belonging to the earliest stages are first developed. As young animals and those which stand low in the scale generally have a much greater capacity for regrowth than older and higher animals, it would also appear that they retain cells in a nascent state, or partially developed gemmules, more readily than do animals which have already passed through a long series of developmental changes. I may here add that although ovules can be detected in most or all female animals at an extremely early age, there is no reason to doubt that gemmules derived from parts modified during maturity can pass into the ovules.

With respect to hybridism, pangenesis agrees well with most of the ascertained facts. We must believe, as previously shown, that several gemmules are requisite for the development of each cell or unit. But from the occurrence of parthenogenesis, more especially from those cases in which an embryo is only partially formed, we may infer that the female element generally includes gemmules in nearly sufficient number for independent development, so that when united with the male element the gemmules are superabundant. Now, when two species or races are crossed reciprocally, the offspring do not commonly differ, and this shows that the sexual elements agree in power, in accordance with the view that both include the same gemmules. Hybrids and mongrels are also generally intermediate in character between the two parent-forms, yet occasionally they closely resemble one parent in one part and the other parent in another part, or even in their whole structure: nor is this difficult to understand on the admission that the gemmules in the fertilised germ are superabundant in number, and that those derived from one parent may have some advantage in number, affinity, or vigour over those derived from the other parent. Crossed forms sometimes exhibit the colour or other characters of either parent in stripes or blotches; and this occurs in the first generation, or through reversion in succeeding bud and seminal generations, of which fact several instances were given in the eleventh chapter. In these cases we must follow Naudin (27/57. See his excellent discussion on this subject in 'Nouvelles Archives du Museum' tome 1 page 151.) and admit that the "essence" or "element" of the two species,— terms which I should translate into the gemmules,—have an affinity for their own kind, and thus separate themselves into distinct stripes or blotches; and reasons were given, when discussing in the fifteenth chapter the incompatibility of certain characters to unite, for believing in such mutual affinity. When two forms are crossed, one is not rarely found to be prepotent in the transmission of its characters over the other; and this we can explain by again assuming that the one form has some advantage over the other in the number, vigour, or affinity of its gemmules. In some cases, however, certain characters are present in the one form and latent in the other; for instance, there is a latent tendency in all pigeons to become blue, and, when a blue pigeon is crossed with one of any other colour, the blue tint is generally prepotent. The explanation of this form of prepotency will be obvious when we come to the consideration of Reversion.

When two distinct species are crossed, it is notorious that they do not yield the full or proper number of offspring; and we can only say on this head that, as the development of each organism depends on such nicely-balanced affinities between a host of gemmules and nascent cells, we need not feel at all surprised that the commixture of gemmules derived from two distinct species should lead to partial or complete failure of development. With respect to the sterility of hybrids produced from the union of two distinct species, it was shown in the nineteenth chapter that this depends exclusively on the reproductive organs being specially affected; but why these organs should be thus affected we do not know, any more than why unnatural conditions of life, though compatible with health, should cause sterility; or why continued close interbreeding, or the illegitimate unions of heterostyled plants, induce the same result. The conclusion that the reproductive organs alone are affected, and not the whole organisation, agrees perfectly with the unimpaired or even increased capacity in hybrid plants for propagation by buds; for this implies, according to our hypothesis, that the cells of the hybrids throw off hybridised gemmules, which become aggregated into buds, but fail to become aggregated within the reproductive organs, so as to form the sexual elements. In a similar manner many plants, when placed under unnatural conditions, fail to produce seed, but can readily be propagated by buds. We shall presently see that pangenesis agrees well with the strong tendency to reversion exhibited by all crossed animals and plants.

Each organism reaches maturity through a longer or shorter course of growth and development: the former term being confined to mere increase of size, and development to changed structure. The changes may be small and insensibly slow, as when a child grows into a man, or many, abrupt, and slight, as in the metamorphoses of certain ephemerous insects, or, again, few and strongly- marked, as with most other insects. Each newly formed part may be moulded within a previously existing and corresponding part, and in this case it will appear, falsely as I believe, to be developed from the old part; or it may be formed within a distinct part of the body, as in the extreme cases of metagenesis. An eye, for instance, may be developed at a spot where no eye previously existed. We have also seen that allied organic beings in the course of their metamorphoses sometimes attain nearly the same structure after passing through widely different forms; or conversely, after passing through nearly the same early forms, arrive at widely different mature forms. In these cases it is very difficult to accept the common view that the first-formed cells or units possess the inherent power, independently of any external agency, of producing new structures wholly different in form, position, and function. But all these cases become plain on the hypothesis of pangenesis. The units, during each stage of development, throw off gemmules, which, multiplying, are transmitted to the offspring. In the offspring, as soon as any particular cell or unit becomes partially developed, it unites with (or, to speak metaphorically, is fertilised by) the gemmule of the next succeeding cell, and so onwards. But organisms have often been subjected to changed conditions of life at a certain stage of their development, and in consequence have been slightly modified; and the gemmules cast off from such modified parts will tend to reproduce parts modified in the same manner. This process may be repeated until the structure of the part becomes greatly changed at one particular stage of development, but this will not necessarily affect other parts, whether previously or subsequently formed. In this manner we can understand the remarkable independence of structure in the successive metamorphoses, and especially in the successive metageneses of many animals. In the case, however, of diseases which supervene during old age, subsequently to the ordinary period of procreation, and which, nevertheless, are sometimes inherited, as occurs with brain and heart complaints, we must suppose that the organs were affected at an early age and threw off at this period affected gemmules; but that the affection became visible or injurious only after the prolonged growth, in the strict sense of the word, of the part. In all the changes of structure which regularly supervene during old age, we probably see the effects of deteriorated growth, and not of true development.

The principle of the independent formation of each part, owing to the union of the proper gemmules with certain nascent cells, together with the superabundance of the gemmules derived from both parents, and the subsequent self-multiplication of the gemmules, throws light on a widely different group of facts, which on any ordinary view of development appears very strange. I allude to organs which are abnormally transposed or multiplied. For instance, a curious case has been recorded by Dr. Elliott Coues (27/58. 'Proc. Boston Soc. of Nat. Hist.' republished in 'Scientific Opinion' November 10, 1869 page 488.) of a monstrous chicken with a perfect additional RIGHT leg articulated to the LEFT side of the pelvis. Gold-fish often have supernumerary fins placed on various parts of their bodies. When the tail of a lizard is broken off, a double tail is sometimes reproduced; and when the foot of the salamander was divided longitudinally by Bonnet, additional digits were occasionally formed. Valentin injured the caudal extremity of an embryo, and three days afterwards it produced rudiments of a double pelvis and of double hind-limbs. (27/59. Todd

'Cyclop. of Anat. and Phys.' volume 4 1849-52 page 975.) When frogs, toads, etc., are born with their limbs doubled, as sometimes happens, the doubling, as Gervais remarks (27/60. 'Compte Rendus' November 14, 1865 page 800.), cannot be due to the complete fusion of two embryos, with the exception of the limbs, for the larvae are limbless. The same argument is applicable (27/61. As previously remarked by Quatrefages in his 'Metamorphoses de l'Homme' etc. 1862 page 129.) to certain insects produced with multiple legs or antennae, for these are metamorphosed from apodal or antennae-less larvae. Alphonse Milne-Edwards (27/62. Gunther 'Zoological Record' 1864 page 279.) has described the curious case of a crustacean in which one eye-peduncle supported, instead of a complete eye, only an imperfect cornea, and out of the centre of this a portion of an antenna was developed. A case has been recorded (27/63. Sedgwick 'Medico-Chirurg. Review' April 1863 page 454.) of a man who had during both dentitions a double tooth in place of the left second incisor, and he inherited this peculiarity from his paternal grandfather. Several cases are known (27/64. Isid. Geoffroy Saint-Hilaire 'Hist. des Anomalies' tome 1 1832 pages 435, 657; and tome 2 page 560.) of additional teeth having been developed in the orbit of the eye, and, more especially with horses, in the palate. Hairs occasionally appear in strange situations, as "within the substance of the brain." (27/65. Virchow 'Cellular Pathology' 1860 page 66.) Certain breeds of sheep bear a whole crowd of horns on their foreheads. As many as five spurs have been seen on both legs of certain Game-fowls. In the Polish fowl the male is ornamented with a topknot of hackles like those on his neck, whilst the female has a top-knot formed of common feathers. In feather-footed pigeons and fowls, feathers like those on the wing arise from the outer side of the legs and toes. Even the elemental parts of the same feather may be transposed; for in the Sebastopol goose, barbules are developed on the divided filaments of the shaft. Imperfect nails sometimes appear on the stumps of the amputated fingers of man (27/66. Muller 'Phys.' English Translation volume 1 1833 page 407. A case of this kind has lately been communicated to me.) and it is an interesting fact that with the snake-like Saurians, which present a series with more and more imperfect limbs, the terminations of the phalanges first disappear, "the nails becoming transferred to their proximal remnants, or even to parts which are not phalanges." (27/67. Dr. Furbringer 'Die Knochen etc. bei den schlangenahnlichen Sauriern' as reviewed in 'Journal of Anat. and Phys.' May 1870 page 286.)

Analogous cases are of such frequent occurrence with plants that they do not strike us with sufficient surprise. Supernumerary petals, stamens, and pistils, are often produced. I have seen a leaflet low down in the compound leaf of Vicia sativa replaced by a tendril; and a tendril possesses many peculiar properties, such as spontaneous movement and irritability. The calyx sometimes assumes, either wholly or by stripes, the colour and texture of the corolla. Stamens are so frequently converted into petals, more or less completely, that such cases are passed over as not deserving notice; but as petals have special functions to perform, namely, to protect the included organs, to attract insects, and in not a few cases to guide their entrance by well-adapted contrivances, we can hardly account for the conversion of stamens into petals merely by unnatural or superfluous nourishment. Again, the edge of a petal may occasionally be found including one of the highest products of the plant, namely, pollen; for instance, I have seen the pollen-mass of an Ophrys, which is a very complex structure, developed in the edge of an upper petal. The segments of the calyx of the common pea have been observed partially converted into carpels, including ovules, and with their tips converted into stigmas. Mr. Salter and Dr. Maxwell Masters have found pollen within the ovules of the passion-flower and of the rose. Buds may be developed in the most unnatural positions, as on the petal of a flower. Numerous analogous facts could be given. (27/68. Moquin-Tandon 'Teratologie Veg.' 1841 pages 218, 220, 353. For the case of the pea see 'Gardener's Chronicle' 1866 page 897. With respect to pollen within ovules see Dr. Masters in 'Science Review' October 1873 page 369. The Rev. J.M. Berkeley describes a bud developed on a petal of a Clarkia in 'Gardener's Chronicle' April 28, 1866.)

I do not know how physiologists look at such facts as the foregoing. According to the doctrine of pangenesis, the gemmules of the transposed organs become developed in the wrong place, from uniting with wrong cells or aggregates of cells during their nascent state; and this would follow from a slight modification in their elective affinities. Nor ought we to feel much surprise at the affinities of cells and gemmules varying, when we remember the many curious cases given in the seventeenth chapter, of plants which absolutely refuse to be fertilised by their own pollen, though abundantly fertile with that of any other individual of the same species, and in some cases only with that of a distinct species. It is manifest that the sexual elective affinities of such plants—to use the term employed by Gartner—have been modified. As the cells of adjoining or homologous parts will have nearly the same nature, they will be particularly liable to acquire by variation each other's elective affinities; and we can thus understand to a certain extent such cases as a crowd of horns on the heads of certain sheep, of several spurs on the legs of fowls, hackle-like feathers on the heads of the males of other fowls, and with the pigeon wing-like feathers on their legs and membrane between their toes, for the leg is the homologue of the wing. As all the organs of plants are homologous and spring from a common axis, it is natural that they should be eminently liable to transposition. It ought to be observed that when any compound part, such as an additional limb or an antenna, springs from a false position, it is only necessary that the few first gemmules should be wrongly attached; for these whilst developing would attract other gemmules in due succession, as in the regrowth of an amputated limb. When parts which are homologous and similar in structure, as the vertebrae of snakes or the stamens of polyandrous flowers, etc., are repeated many times in the same organism, closely allied gemmules must be extremely numerous, as well as the points to which they ought to become united; and, in accordance with the foregoing views, we can to a certain extent understand Isid. Geoffroy Saint-Hilaire's law, that parts, which are already multiple, are extremely liable to vary in number.

Variability often depends, as I have attempted to show, on the reproductive organs being injuriously affected by changed conditions; and in this case the gemmules derived from the various parts of the body are probably aggregated in an irregular manner, some superfluous and others deficient. Whether a superabundance of gemmules would lead to the increased size of any part cannot be told; but we can see that their partial deficiency, without necessarily leading to the entire abortion of the part, might cause considerable modifications; for in the same manner as plants, if their own pollen be excluded, are easily hybridised, so, in the case of cells, if the properly succeeding gemmules were absent, they would probably combine easily with other and allied gemmules, as we have just seen with transposed parts.

In variations caused by the direct action of changed conditions, of which several instances have been given, certain parts of the body are directly affected by the new conditions, and consequently throw off modified gemmules, which are transmitted to the offspring. On any ordinary view it is unintelligible how changed conditions, whether acting on the embryo, the young or the adult, can cause inherited modifications. It is equally or even more unintelligible on any ordinary view, how the effects of the long-continued use or disuse of a part, or of changed habits of body or mind, can be inherited. A more perplexing problem can hardly be proposed; but on our view we have only to suppose that certain cells become at last structurally modified; and that these throw off similarly modified gemmules. This may occur at any period of development, and the modification will be inherited at a corresponding period; for the modified gemmules will unite in all ordinary cases with the proper preceding cells, and will consequently be developed at the same period at which the modification first arose. With respect to mental habits or instincts, we are so profoundly ignorant of the relation between the brain and the power of thought that we do not know positively whether a fixed habit induces any change in the nervous system, though this seems highly probable; but when such habit or other mental attribute, or insanity, is inherited, we must believe that some actual modification is transmitted (27/69. See some remarks to this effect by Sir H. Holland in his 'Medical Notes' 1839 page 32.); and this implies, according to our hypothesis, that gemmules derived from modified nerve-cells are transmitted to the offspring.

It is generally necessary that an organism should be exposed during several generations to changed conditions or habits, in order that any modification thus acquired should appear in the offspring. This may be partly due to the changes not being at first marked enough to catch attention, but this explanation is insufficient; and I can account for the fact only by the assumption, which we shall see under the head of reversion is strongly supported, that gemmules derived from each unmodified unit or part are transmitted in large numbers to successive generations, and that the gemmules derived from the same unit after it has been modified go on multiplying under the same favourable conditions which first caused the modification, until at last they become sufficiently numerous to overpower and supplant the old gemmules.

A difficulty may be here noticed; we have seen that there is an important difference in the frequency, though not in the nature, of the variations in plants propagated by sexual and asexual generation. As far as variability depends on the imperfect action of the reproductive organs under changed conditions, we can at once see why plants propagated asexually should be far less variable than those propagated sexually. With respect to the direct action of changed conditions, we know that organisms produced from buds do not pass through the earlier phases of development; they will therefore not be exposed, at that period of life when structure is most readily modified, to the various causes inducing variability in the same manner as are embryos and young larval forms; but whether this is a sufficient explanation I know not.

With respect to variations due to reversion, there is a similar difference between plants propagated from buds and seeds. Many varieties can be propagated securely by buds, but generally or invariably revert to their parent-forms by seed. So, also, hybridised plants can be multiplied to any extent by buds, but are continually liable to reversion by seed,—that is, to the loss of their hybrid or intermediate character. I can offer no satisfactory explanation of these facts. Plants with variegated leaves, phloxes with striped flowers, barberries with seedless fruit, can all be securely propagated by buds taken from the stem or branches; but buds from the roots of these plants almost invariably lose their character and revert to their former condition. This latter fact is also inexplicable, unless buds developed from the roots are as distinct from those on the stem, as is one bud on the stem from another, and we know that these latter behave like independent organisms.

Finally, we see that on the hypothesis of pangenesis variability depends on at least two distinct groups of causes. Firstly, the deficiency, superabundance, and transposition of gemmules, and the redevelopment of those which have long been dormant; the gemmules themselves not having undergone any modification; and such changes will amply account for much fluctuating variability. Secondly, the direct action of changed conditions on the organisation, and of the increased use or disuse of parts; and in this case the gemmules from the modified units will be themselves modified, and, when sufficiently multiplied, will supplant the old gemmules and be developed into new structures.

Turning now to the laws of Inheritance. If we suppose a homogeneous gelatinous protozoon to vary and assume a reddish colour, a minute separated particle would naturally, as it grew to full size, retain the same colour; and we should have the simplest form of inheritance. (27/70. This is the view taken by Prof. Hackel in his 'Generelle Morphologie' b. 2 s. 171, who says: "Lediglich die partielle Identitat der specifisch constituirten Materie im elterlichen und im kindlichen Organismus, die Theilung dieser Materie bei der Fortpflanzung, ist die Ursache der Erblichkeit.") Precisely the same view may be extended to the infinitely numerous and diversified units of which the whole body of one of the higher animals is composed; the separated particles being our gemmules. We have already sufficiently discussed by implication, the important principle of inheritance at corresponding ages. Inheritance as limited by sex and by the season of the year (for instance with animals becoming white in winter) is intelligible if we may believe that the elective affinities of the units of the body are slightly different in the two sexes, especially at maturity, and in one or both sexes at different seasons, so that they unite with different gemmules. It should be remembered that, in the discussion on the abnormal transposition of organs, we have seen reason to believe that such elective affinities are readily modified. But I shall soon have to recur to sexual and seasonal inheritance. These several laws are therefore explicable to a large extent through pangenesis, and on no other hypothesis which has as yet been advanced.

But it appears at first sight a fatal objection to our hypothesis that a part or organ may be removed during several successive generations, and if the operation be not followed by disease, the lost part reappears in the offspring. Dogs and horses formerly had their tails docked during many generations without any inherited effect; although, as we have seen, there is some reason to believe that the tailless condition of certain sheep-dogs is due to such inheritance. Circumcision has been practised by the Jews from a remote period, and in most cases the effects of the operation are not visible in the offspring; though some maintain that an inherited effect does occasionally appear. If inheritance depends on the presence of disseminated gemmules derived from all the units of the body, why does not the amputation or mutilation of a part, especially if effected on both sexes, invariably affect the offspring? The answer in accordance with our hypothesis probably is that gemmules multiply and are transmitted during a long series of generations—as we see in the reappearance of zebrine stripes on the horse—in the reappearance of muscles and other structures in man which are proper to his lowly organised progenitors, and in many other such cases. Therefore the long-continued inheritance of a part which has been removed during many generations is no real anomaly, for gemmules formerly derived from the part are multiplied and transmitted from generation to generation.

We have as yet spoken only of the removal of parts, when not followed by morbid action: but when the operation is thus followed, it is certain that the deficiency is sometimes inherited. In a former chapter instances were given, as of a cow, the loss of whose horn was followed by suppuration, and her calves were destitute of a horn on the same side of their heads. But the evidence which admits of no doubt is that given by Brown-Sequard with respect to guinea-pigs, which after their sciatic nerves had been divided, gnawed off their own gangrenous toes, and the toes of their offspring were deficient in at least thirteen instances on the corresponding feet. The inheritance of the lost part in several of these cases is all the more remarkable as only one parent was affected; but we know that a congenital deficiency is often transmitted from one parent alone—for instance, the offspring of hornless cattle of either sex, when crossed with perfect animals, are often hornless. How, then, in accordance with our hypothesis can we account for mutilations being sometimes strongly inherited, if they are followed by diseased action? The answer probably is that all the gemmules of the mutilated or amputated part are gradually attracted to the diseased surface during the reparative process, and are there destroyed by the morbid action.

A few words must be added on the complete abortion of organs. When a part becomes diminished by disuse prolonged during many generations, the principle of economy of growth, together with intercrossing, will tend to reduce it still further as previously explained, but this will not account for the complete or almost complete obliteration of, for instance, a minute papilla of cellular tissue representing a pistil, or of a microscopically minute nodule of bone representing a tooth. In certain cases of suppression not yet completed, in which a rudiment occasionally reappears through reversion, dispersed gemmules derived from this part must, according to our view, still exist; we must therefore suppose that the cells, in union with which the rudiment was formerly developed, fail in their affinity for such gemmules, except in the occasional cases of reversion. But when the abortion is complete and final, the gemmules themselves no doubt perish; nor is this in any way improbable, for, though a vast number of active and long-dormant gemmules are nourished in each living creature, yet there must be some limit to their number; and it appears natural that gemmules derived from reduced and useless parts would be more liable to perish than those freshly derived from other parts which are still in full functional activity.

The last subject that need be discussed, namely, Reversion, rests on the principle that transmission and development, though generally acting in conjunction, are distinct powers; and the transmission of gemmules with their subsequent development shows us how this is possible. We plainly see the distinction in the many cases in which a grandfather transmits to his grandson, through his daughter, characters which she does not, or cannot, possess. But before proceeding, it will be advisable to say a few words about latent or dormant characters. Most, or perhaps all, of the secondary characters, which appertain to one sex, lie dormant in the other sex; that is, gemmules capable of development into the secondary male sexual characters are included within the female; and conversely female characters in the male: we have evidence of this in certain masculine characters, both corporeal and mental, appearing in the female, when her ovaria are diseased or when they fail to act from old age. In like manner female characters appear in castrated males, as in the shape of the horns of the ox, and in the absence of horns in castrated stags. Even a slight change in the conditions of life due to confinement sometimes suffices to prevent the development of masculine characters in male animals, although their reproductive organs are not permanently injured. In the many cases in which masculine characters are periodically renewed, these are latent at other seasons; inheritance as limited by sex and season being here combined. Again, masculine characters generally lie dormant in male animals until they arrive at the proper age for reproduction. The curious case formerly given of a Hen which assumed the masculine characters, not of her own breed but of a remote progenitor, illustrates the close connection between latent sexual characters and ordinary reversion.

With those animals and plants which habitually produce several forms, as with certain butterflies described by Mr. Wallace, in which three female forms and one male form co-exist, or, as with the trimorphic species of Lythrum and Oxalis, gemmules capable of reproducing these different forms must be latent in each individual.

Insects are occasionally produced with one side or one quarter of their bodies like that of the male, with the other half or three-quarters like that of the female. In such cases the two sides are sometimes wonderfully different in structure, and are separated from each other by a sharp line. As gemmules derived from every part are present in each individual of both sexes, it must be the elective affinities of the nascent cells which in these cases differ abnormally on the two sides of the body. Almost the same principle comes into play with those animals, for instance, certain gasteropods and Verruca amongst cirripedes, which normally have the two sides of the body constructed on a very different plan; and yet a nearly equal number of individuals have either side modified in the same remarkable manner.

Reversion, in the ordinary sense of the word, acts so incessantly, that it evidently forms an essential part of the general law of inheritance. It occurs with beings, however propagated, whether by buds or seminal generation, and sometimes may be observed with advancing age even in the same individual. The tendency to reversion is often induced by a change of conditions, and in the plainest manner by crossing. Crossed forms of the first generation are generally nearly intermediate in character between their two parents; but in the next generation the offspring commonly revert to one or both of their grandparents, and occasionally to more remote ancestors. How can we account for these facts? Each unit in a hybrid must throw off, according to the doctrine of pangenesis, an abundance of hybridised gemmules, for crossed plants can be readily and largely propagated by buds; but by the same hypothesis dormant gemmules derived from both pure parent-forms are likewise present; and as these gemmules retain their normal condition, they would, it is probable, be enabled to multiply largely during the lifetime of each hybrid. Consequently the sexual elements of a hybrid will include both pure and hybridised gemmules; and when two hybrids pair, the combination of pure gemmules derived from the one hybrid with the pure gemmules of the same parts derived from the other, would necessarily lead to complete reversion of character; and it is, perhaps, not too bold a supposition that unmodified and undeteriorated gemmules of the same nature would be especially apt to combine. Pure gemmules in combination with hybridised gemmules would lead to partial reversion. And lastly, hybridised gemmules derived from both parent-hybrids would simply reproduce the original hybrid form. (27/71. In these remarks I, in fact, follow Naudin, who speaks of the elements or essences of the two species which are crossed. See his excellent memoir in the 'Nouvelles Archives du Museum' tome 1 page 151.) All these cases and degrees of reversion incessantly occur.

It was shown in the fifteenth chapter that certain characters are antagonistic to each other or do not readily blend; hence, when two animals with antagonistic characters are crossed, it might well happen that a sufficiency of gemmules in the male alone for the reproduction of his peculiar characters, and in the female alone for the reproduction of her peculiar characters, would not be present; and in this case dormant gemmules derived from the same part in some remote progenitor might easily gain the ascendancy, and cause the reappearance of the long-lost character. For instance, when black and white pigeons, or black and white fowls, are crossed,—colours which do not readily blend,—blue plumage in the one case, evidently derived from the rock-pigeon, and red plumage in the other case, derived from the wild jungle-cock, occasionally reappear. With uncrossed breeds the same result follows, under conditions which favour the multiplication and development of certain dormant gemmules, as when animals become feral and revert to their pristine character. A certain number of gemmules being requisite for the development of each character, as is known to be the case from several spermatozoa or pollen-grains being necessary for fertilisation, and time favouring their multiplication, will perhaps account for the curious cases, insisted on by Mr. Sedgwick, of certain diseases which regularly appear in alternate generations. This likewise holds good, more or less strictly, with other weakly inherited modifications. Hence, as I have heard it remarked, certain diseases appear to gain strength by the intermission of a generation. The transmission of dormant gemmules during many successive generations is hardly in itself more improbable, as previously remarked, than the retention during many ages of rudimentary organs, or even only of a tendency to the production of a rudiment; but there is no reason to suppose that dormant gemmules can be transmitted and propagated for ever. Excessively minute and numerous as they are believed to be, an infinite number derived, during a long course of modification and descent, from each unit of each progenitor, could not be supported or nourished by the organism. But it does not seem improbable that certain gemmules, under favourable conditions, should be retained and go on multiplying for a much longer period than others. Finally, on the view here given, we certainly gain some insight into the wonderful fact that the child may depart from the type of both its parents, and resemble its grandparents, or ancestors removed by many hundreds of generations.

## CONCLUSION.

The hypothesis of Pangenesis, as applied to the several great classes of facts just discussed, no doubt is extremely complex, but so are the facts. The chief assumption is that all the units of the body, besides having the universally admitted power of growing by self-division, throw off minute gemmules which are dispersed through the system. Nor can this assumption be considered as too bold, for we know from the cases of graft-hybridisation that formative matter of some kind is present in the tissues of plants, which is capable of combining with that included in another individual, and of reproducing every unit of the whole organism. But we have further to assume that the gemmules grow, multiply, and aggregate themselves into buds and the sexual elements; their development depending on their union with other nascent cells or units. They are also believed to be capable of transmission in a dormant state, like seeds in the ground, to successive generations.

In a highly-organised animal, the gemmules thrown off from each different unit throughout the body must be inconceivably numerous and minute. Each unit of each part, as it changes during development, and we know that some insects undergo at least twenty metamorphoses, must throw off its gemmules. But the same cells may long continue to increase by self-division, and even become modified by absorbing peculiar nutriment, without necessarily throwing off modified gemmules. All organic beings, moreover, include many dormant gemmules derived from their grandparents and more remote progenitors, but not from all their progenitors. These almost infinitely numerous and minute gemmules are contained within each bud, ovule, spermatozoon, and pollen-grain. Such an admission will be declared impossible; but number and size are only relative difficulties. Independent organisms exist which are barely visible under the highest powers of the microscope, and their germs must be excessively minute. Particles of infectious matter, so small as to be wafted by the wind or to adhere to smooth paper, will multiply so rapidly as to infect within a short time the whole body of a large animal. We should also reflect on the admitted number and minuteness of the molecules composing a particle of ordinary matter. The difficulty, therefore, which at first appears insurmountable, of believing in the existence of gemmules so numerous and small as they must be according to our hypothesis, has no great weight.

The units of the body are generally admitted by physiologists to be autonomous. I go one step further and assume that they throw off reproductive gemmules. Thus an organism does not generate its kind as a whole, but each separate unit generates its kind. It has often been said by naturalists that each cell of a plant has the potential capacity of reproducing the whole plant; but it has this power only in virtue of containing gemmules derived from every part. When a cell or unit is from some cause modified, the gemmules derived from it will be in like manner modified. If our hypothesis be provisionally accepted, we must look at all the forms of asexual reproduction, whether occurring at maturity or during youth, as fundamentally the same, and dependent on the mutual aggregation and multiplication of the gemmules. The regrowth of an amputated limb and the healing of a wound is the same process partially carried out. Buds apparently include nascent cells, belonging to that stage of development at which the budding occurs, and these cells are ready to unite with the gemmules derived from the next succeeding cells. The sexual elements, on the other hand, do not include such nascent cells; and the male and female elements taken separately do not contain a sufficient number of gemmules for independent development, except in the cases of parthenogenesis. The development of each being, including all the forms of metamorphosis and metagenesis, depends on the presence of gemmules thrown off at each period of life, and on their development, at a corresponding period, in union with preceding cells. Such cells may be said to be fertilised by the gemmules which come next in due order of development. Thus the act of ordinary impregnation and the development of each part in each being are closely analogous processes. The child, strictly speaking, does not grow into the man, but includes germs which slowly and successively become developed and form the man. In the child, as well as in the adult, each part generates the same part. Inheritance must be looked at as merely a form of growth, like the self- division of a lowly-organised unicellular organism. Reversion depends on the transmission from the forefather to his descendants of dormant gemmules, which occasionally become developed under certain known or unknown conditions. Each animal and plant may be compared with a bed of soil full of seeds, some of which soon germinate, some lie dormant for a period, whilst others perish. When we hear it said that a man carries in his constitution the seeds of an inherited disease, there is much truth in the expression. No other attempt, as far as I am aware, has been made, imperfect as this confessedly is, to connect under one point of view these several grand classes of facts. An organic being is a microcosm—a little universe, formed of a host of self-propagating organisms, inconceivably minute and numerous as the stars in heaven.

# CHAPTER 2.XXVIII.

## CONCLUDING REMARKS.

## DOMESTICATION. NATURE AND CAUSES OF VARIABILITY. SELECTION. DIVERGENCE AND DISTINCTNESS OF CHARACTER. EXTINCTION OF RACES. CIRCUMSTANCES FAVOURABLE TO SELECTION BY MAN. ANTIQUITY OF CERTAIN RACES. THE QUESTION WHETHER EACH PARTICULAR VARIATION HAS BEEN SPECIALLY PREORDAINED.

As summaries have been added to nearly all the chapters, and as, in the chapter on pangenesis, various subjects, such as the forms of reproduction, inheritance, reversion, the causes and laws of variability, etc., have been recently discussed, I will here only make a few general remarks on the more important conclusions which may be deduced from the multifarious details given throughout this work.

Savages in all parts of the world easily succeed in taming wild animals; and those inhabiting any country or island, when first visited by man, would probably have been still more easily tamed. Complete subjugation generally depends on an animal being social in its habits, and on receiving man as the chief of the herd or family. In order that an animal should be domesticated it must be fertile under changed conditions of life, and this is far from being always the case. An animal would not have been worth the labour of domestication, at least during early times, unless of service to man. From these circumstances the number of domesticated animals has never been large. With respect to plants, I have shown in the ninth chapter how their varied uses were probably first discovered, and the early steps in their cultivation. Man could not have known, when he first domesticated an animal or plant, whether it would flourish and multiply when transported to other countries, therefore he could not have been thus influenced in his choice. We see that the close adaptation of the reindeer and camel to extremely cold and hot countries has not prevented their domestication. Still less could man have foreseen whether his animals and plants would vary in succeeding generations and thus give birth to new races; and the small capacity of variability in the goose has not prevented its domestication from a remote epoch.

With extremely few exceptions, all animals and plants which have been long domesticated have varied greatly. It matters not under what climate, or for what purpose they are kept, whether as food for man or beast, for draught or hunting, for clothing or mere pleasure,—under all these circumstances races have been produced which differ more from one another than do the forms which in a state of nature are ranked as different species. Why certain animals and plants have varied more under domestication than others we do not know, any more than why some are rendered more sterile than others under changed conditions of life. But we have to judge of the amount of variation which our domestic productions have undergone, chiefly by the number and amount of difference between the races which have been formed, and we can often clearly see why many and distinct races have not been formed, namely, because slight successive variations have not been steadily accumulated; and such variations will never be accumulated if an animal or plant be not closely observed, much valued, and kept in large numbers.

The fluctuating, and, as far as we can judge, never-ending variability of our domesticated productions,—the plasticity of almost their whole organisation,--is one of the most important lessons which we learn from the numerous details given in the earlier chapters of this work. Yet domesticated animals and plants can hardly have been exposed to greater changes in their conditions of life than have many natural species during the incessant geological, geographical, and climatal changes to which the world has been subject; but domesticated productions will often have been exposed to more sudden changes and to less continuously uniform conditions. As man has domesticated so many animals and plants belonging to widely different classes, and as he certainly did not choose with prophetic instinct those species which would vary most, we may infer that all natural species, if exposed to analogous conditions, would, on an average, vary to the same degree. Few men at the present day will maintain that animals and plants were created with a tendency to vary, which long remained dormant, in order that fanciers in after ages might rear, for instance, curious breeds of the fowl, pigeon, or canary-bird.

From several causes it is difficult to judge of the amount of modification which our domestic productions have undergone. In some cases the primitive parent-stock has become extinct; or it cannot be recognised with certainty, owing to its supposed descendants having been so much modified. In other cases two or more closely-allied forms, after being domesticated, have crossed; and then it is difficult to estimate how much of the character of the present descendants ought to be attributed to variation, and how much to the influence of the several parent-stocks. But the degree to which our domesticated breeds have been modified by the crossing of distinct species has probably been much exaggerated by some authors. A few individuals of one form would seldom permanently affect another form existing in greater numbers; for, without careful selection, the stain of the foreign blood would soon be obliterated, and during early and barbarous times, when our animals were first domesticated, such care would seldom have been taken.

There is good reason to believe in the case of the dog, ox, pig, and of some other animals, that several of our races are descended from distinct wild prototypes; nevertheless the belief in the multiple origin of our domesticated animals has been extended by some few naturalists and by many breeders to an unauthorised extent. Breeders refuse to look at the whole subject under a single point of view; I have heard it said by a man, who maintained that our fowls were descended from at least half-a-dozen aboriginal species, that the evidence of the common origin of pigeons, ducks and rabbits, was of no avail with respect to fowls. Breeders overlook the improbability of many species having been domesticated at an early and barbarous period. They do not consider the improbability of species having existed in a state of nature which, if they resembled our present domestic breeds, would have been highly abnormal in comparison with all their congeners. They maintain that certain species, which formerly existed, have become extinct, or are now unknown, although formerly known. The assumption of so much recent extinction is no difficulty in their eyes; for they do not judge of its probability by the facility or difficulty of the extinction of other closely-allied wild forms. Lastly, they often ignore the whole subject of geographical distribution as completely as if it were the result of chance.

Although from the reasons just assigned it is often difficult to judge accurately of the amount of change which our domesticated productions have undergone, yet this can be ascertained in the cases in which all the breeds are known to be descended from a single species,—as with the pigeon, duck, rabbit, and almost certainly with the fowl; and by the aid of analogy this can be judged of to a certain extent with domesticated animals descended from several wild stocks. It is impossible to read the details given in the earlier chapters and in many published works, or to visit our various exhibitions, without being deeply impressed with the extreme variability of our domesticated animals and cultivated plants. No part of the organisation escapes the tendency to vary. The variations generally affect parts of small vital or physiological importance, but so it is with the differences which exist between closely-allied species. In these unimportant characters there is often a greater difference between the breeds of the same species than between the natural species of the same genus, as Isidore Geoffroy has shown to be the case with size, and as is often the case with the colour, texture, form, etc., of the hair, feathers, horns, and other dermal appendages.

It has often been asserted that important parts never vary under domestication, but this is a complete error. Look at the skull of the pig in any one of the highly improved breeds, with the occipital condyles and other parts greatly modified; or look at that of the niata ox. Or, again, in the several breeds of the rabbit, observe the elongated skull, with the differently shaped occipital foramen, atlas, and other cervical vertebrae. The whole shape of the brain, together with the skull, has been modified in Polish fowls; in other breeds of the fowl the number of the vertebrae and the forms of the cervical vertebrae have been changed. In certain pigeons the shape of the lower jaw, the relative length of the tongue, the size of the nostrils and eyelids, the number and shape of the ribs, the form and size of the oesophagus, have all varied. In certain quadrupeds the length of the intestines has been much increased or diminished. With plants we see wonderful differences in the stones of various fruits. In the Cucurbitaceae several highly important characters have varied, such as the sessile position of the stigmas on the ovarium, the position of the carpels, and the projection of the ovarium out of the receptacle. But it would be useless to run through the many facts given in the earlier chapters.

It is notorious how greatly the mental disposition, tastes, habits, consensual movements, loquacity or silence, and tone of voice have varied and been inherited in our domesticated animals. The dog offers the most striking instance of changed mental attributes, and these differences cannot be accounted for by descent from distinct wild types.

New characters may appear and old ones disappear at any stage of development, being inherited at a corresponding stage. We see this in the difference between the eggs, the down on the chickens and the first plumage of the various breeds of the fowl; and still more plainly in the differences between the caterpillars and cocoons of the various breeds of the silk-moth. These facts, simple as they appear, throw light on the differences between the larval and adult states of allied natural species, and on the whole great subject of embryology. New characters first appearing late in life are apt to become attached exclusively to that sex in which they first arose, or they may be developed in a much higher degree in this than in the other sex; or again, after having become attached to one sex, they may be transferred to the opposite sex. These facts, and more especially the circumstance that new characters seem to be particularly liable, from some unknown cause, to become attached to the male sex, have an important bearing on the acquirement of secondary sexual characters by animals in a state of nature.

It has sometimes been said that our domestic races do not differ in constitutional peculiarities, but this cannot be maintained. In our improved cattle, pigs, etc., the period of maturity, including that of the second dentition, has been much hastened. The period of gestation varies much, and has been modified in a fixed manner in one or two cases. In some breeds of poultry and pigeons the period at which the down and the first plumage are acquired, differs. The number of moults through which the larvae of silk-moths pass, varies. The tendency to fatten, to yield much milk, to produce many young or eggs at a birth or during life, differs in different breeds. We find different degrees of adaptation to climate, and different tendencies to certain diseases, to the attacks of parasites, and to the action of certain vegetable poisons. With plants, adaptation to certain soils, the power of resisting frost, the period of flowering and fruiting, the duration of life, the period of shedding the leaves or of retaining them throughout the winter, the proportion and nature of certain chemical compounds in the tissues or seeds, all vary.

There is, however, one important constitutional difference between domestic races and species; I refer to the sterility which almost invariably follows, in a greater or less degree, when species are crossed, and to the perfect fertility of the most distinct domestic races, with the exception of a very few plants, when similarly crossed. It is certainly a most remarkable fact that many closely-allied species, which in appearance differ extremely little, should yield when crossed only a few more or less sterile offspring, or none at all; whilst domestic races which differ conspicuously from each other are, when united, remarkably fertile, and yield perfectly fertile offspring. But this fact is not in reality so inexplicable as it at first appears. In the first place, it was clearly shown in the nineteenth chapter that the sterility of crossed species does not depend chiefly on differences in their external structure or general constitution, but on differences in the reproductive system, analogous to those which cause the lessened fertility of the illegitimate unions of dimorphic and trimorphic plants. In the second place, the Pallasian doctrine, that species after having been long domesticated lose their natural tendency to sterility when crossed, has been shown to be highly probable or almost certain. We cannot avoid this conclusion when we reflect on the parentage and present fertility of the several breeds of the dog, of the Indian or humped and European cattle, and of the two chief kinds of pigs. Hence it would be unreasonable to expect that races formed under domestication should acquire sterility when crossed, whilst at the same time we admit that domestication eliminates the normal sterility of crossed species. Why with closely-allied species their reproductive systems should almost invariably have been modified in so peculiar a manner as to be mutually incapable of acting on each other—though in unequal degrees in the two sexes, as shown by the difference in fertility between reciprocal crosses of the same species—we do not know, but may with much probability infer the cause to be as follows. Most natural species have been habituated to nearly uniform conditions of life for an incomparably longer time than have domestic races; and we positively know that changed conditions exert an especial and powerful influence on the reproductive system. Hence this difference may well account for the difference in the power of reproduction between domestic races when crossed and species when crossed. It is probably in chief part owing to the same cause that domestic races can be suddenly transported from one climate to another, or placed under widely different conditions, and yet retain in most cases their fertility unimpaired; whilst a multitude of species subjected to lesser changes are rendered incapable of breeding.

The offspring of crossed domestic races and of crossed species resemble each other in most respects, with the one important exception of fertility; they often partake in the same unequal degree of the characters of their parents, one of which is often prepotent over the other; and they are liable to reversion of the same kind. By successive crosses one species may be made to absorb completely another, and so it notoriously is with races. The latter resemble species in many other ways. They sometimes inherit their newly-acquired characters almost or even quite as firmly as species. The conditions leading to variability and the laws governing its nature appear to be the same in both. Varieties can be classed in groups under groups, like species under genera, and these under families and orders; and the classification may be either artificial,—that is, founded on any arbitrary character,—or natural. With varieties a natural classification is certainly founded, and with species is apparently founded, on community of descent, together with the amount of modification which the forms have undergone. The characters by which domestic varieties differ from one another are more variable than those distinguishing species, though hardly more so than with certain polymorphic species; but this greater degree of variability is not surprising, as varieties have generally been exposed within recent times to fluctuating conditions of life, and are much more liable to have been crossed; they are also in many cases still undergoing, or have recently undergone, modification by man's methodical or unconscious selection.

Domestic varieties as a general rule certainly differ from one another in less important parts than do species; and when important differences occur, they are seldom firmly fixed; but this fact is intelligible, if we consider man's method of selection. In the living animal or plant he cannot observe internal modifications in the more important organs; nor does he regard them as long as they are compatible with health and life. What does the breeder care about any slight change in the molar teeth of his pigs, or for an additional molar tooth in the dog; or for any change in the intestinal canal or other internal organ? The breeder cares for the flesh of his cattle being well marbled with fat, and for an accumulation of fat within the abdomen of his sheep, and this he has effected. What would the floriculturist care for any change in the structure of the ovarium or of the ovules? As important internal organs are certainly liable to numerous slight variations, and as these would probably be transmitted, for many strange monstrosities are inherited, man could undoubtedly effect a certain amount of change in these organs. When he has produced any modification in an important part, he has generally done so unintentionally, in correlation with some other conspicuous part. For instance, he has given ridges and protuberances to the skulls of fowls, by attending to the form of the comb, or to the plume of feathers on the head. By attending to the external form of the pouter-pigeon, he has enormously increased the size of the oesophagus, and has added to the number of the ribs, and given them greater breadth. With the carrier-pigeon, by increasing through steady selection the wattles on the upper mandible, he has greatly modified the form of the lower mandible; and so in many other cases. Natural species, on the other hand, have been modified exclusively for their own good, to fit them for infinitely diversified conditions of life, to avoid enemies of all kinds, and to struggle against a host of competitors. Hence, under such complex conditions, it would often happen that modifications of the most varied kinds, in important as well as in unimportant parts, would be advantageous or even necessary; and they would slowly but surely be acquired through the survival of the fittest. Still more important is the fact that various indirect modifications would likewise arise through the law of correlated variation.

Domestic breeds often have an abnormal or semi-monstrous character, as amongst dogs, the Italian greyhound, bulldog, Blenheim spaniel, and bloodhound,—some breeds of cattle and pigs,—several breeds of the fowl,—and the chief breeds of the pigeon. In such abnormal breeds, parts which differ but slightly or not at all in the allied natural species, have been greatly modified. This may be accounted for by man's often selecting, especially at first, conspicuous and semi-monstrous deviations of structure. We should, however, be cautious in deciding what deviations ought to be called monstrous: there can hardly be a doubt that, if the brush of horse-like hair on the breast of the turkey-cock had first appeared in the domesticated bird, it would have been considered as a monstrosity; the great plume of feathers on the head of the Polish cock has been thus designated, though plumes are common on the heads of many kinds of birds; we might call the wattle or corrugated skin round the base of the beak of the English carrier-pigeon a monstrosity, but we do not thus speak of the globular fleshy excrescence at the base of the beak of the Carpophaga oceanica.

Some authors have drawn a wide distinction between artificial and natural breeds; although in extreme cases the distinction is plain, in many other cases it is arbitrary; the difference depending chiefly on the kind of selection which has been applied. Artificial breeds are those which have been intentionally improved by man; they frequently have an unnatural appearance, and are especially liable to lose their characters through reversion and continued variability. The so-called natural breeds, on the other hand, are those which are found in semi-civilised countries, and which formerly inhabited separate districts in nearly all the European kingdoms. They have been rarely acted on by man's intentional selection; more frequently by unconscious selection, and partly by natural selection, for animals kept in semi-civilised countries have to provide largely for their own wants. Such natural breeds will also have been directly acted on by the differences, though slight, in the surrounding conditions.

There is a much more important distinction between our several breeds, namely, in some having originated from a strongly-marked or semi-monstrous deviation of structure, which, however, may subsequently have been augmented by selection; whilst others have been formed in so slow and insensible a manner, that if we could see their early progenitors we should hardly be able to say when or how the breed first arose. From the history of the racehorse, greyhound, gamecock, etc., and from their general appearance, we may feel nearly confident that they were formed by a slow process of improvement; and we know that this has been the case with the carrier-pigeon, as well as with some other pigeons. On the other hand, it is certain that the ancon and mauchamp breeds of sheep, and almost certain that the niata cattle, turnspit, and pug-dogs, jumper and frizzled fowls, short-faced tumbler pigeons, hook- billed ducks, etc., suddenly appeared in nearly the same state as we now see them. So it has been with many cultivated plants. The frequency of these cases is likely to lead to the false belief that natural species have often originated in the same abrupt manner. But we have no evidence of the appearance, or at least of the continued procreation, under nature, of abrupt modifications of structure; and various general reasons could be assigned against such a belief.

On the other hand, we have abundant evidence of the constant occurrence under nature of slight individual differences of the most diversified kinds; and we are thus led to conclude that species have generally originated by the natural selection of extremely slight differences. This process may be strictly compared with the slow and gradual improvement of the racehorse, greyhound, and gamecock. As every detail of structure in each species has to be closely adapted to its habits of life, it will rarely happen that one part alone will be modified; but, as was formerly shown, the co-adapted modifications need not be absolutely simultaneous. Many variations, however, are from the first connected by the law of correlation. Hence it follows that even closely-allied species rarely or never differ from one another by one character alone; and the same remark is to a certain extent applicable to domestic races; for these, if they differ much, generally differ in many respects.

Some naturalists boldly insist (28/1. Godron 'De l'Espece' 1859 tome 2 page 44 etc.) that species are absolutely distinct productions, never passing by intermediate links into one another; whilst they maintain that domestic varieties can always be connected either with one another or with their parent-forms. But if we could always find the links between the several breeds of the dog, horse, cattle, sheep, pigs, etc., there would not have been such incessant doubts whether they were descended from one or several species. The greyhound genus, if such a term may be used, cannot be closely connected with any other breed, unless, perhaps, we go back to the ancient Egyptian monuments. Our English bulldog also forms a very distinct breed.

In all these cases crossed breeds must of course be excluded, for distinct natural species can thus be likewise connected. By what links can the Cochin fowl be closely united with others? By searching for breeds still preserved in distant lands, and by going back to historical records, tumbler-pigeons, carriers, and barbs can be closely connected with the parent rock-pigeon; but we cannot thus connect the turbit or the pouter. The degree of distinctness between the various domestic breeds depends on the amount of modification which they have undergone, and more especially on the neglect and final extinction of intermediate and less-valued forms.

It has often been argued that no light is thrown on the changes which natural species are believed to undergo from the admitted changes of domestic races, as the latter are said to be mere temporary productions, always reverting, as soon as they become feral, to their pristine form. This argument has been well combated by Mr. Wallace (28/2. 'Journal Proc. Linn. Soc.' 1858 volume 3 page 60.) and full details were given in the thirteenth chapter, showing that the tendency to reversion in feral animals and plants has been greatly exaggerated, though no doubt it exists to a certain extent. It would be opposed to all the principles inculcated in this work, if domestic animals, when exposed to new conditions and compelled to struggle for their own wants against a host of foreign competitors, were not modified in the course of time. It should also be remembered that many characters lie latent in all organic beings, ready to be evolved under fitting conditions; and in breeds modified within recent times, the tendency to reversion is particularly strong. But the antiquity of some of our breeds clearly proves that they remain nearly constant as long as their conditions of life remain the same.

It has been boldly maintained by some authors that the amount of variation to which our domestic productions are liable is strictly limited; but this is an assertion resting on little evidence. Whether or not the amount of change in any particular direction is limited, the tendency to general variability is, as far as we can judge, unlimited. Cattle, sheep, and pigs have varied under domestication from the remotest period, as shown by the researches of Rutimeyer and others; yet these animals have been improved to an unparalleled degree, within quite recent times, and this implies continued variability of structure. Wheat, as we know from the remains found in the Swiss lake- dwellings, is one of the most anciently cultivated plants, yet at the present day new and better varieties frequently arise. It may be that an ox will never be produced of larger size and finer proportions, or a racehorse fleeter, than our present animals, or a gooseberry larger than the London variety; but he would be a bold man who would assert that the extreme limit in these respects has been finally attained. With flowers and fruit it has repeatedly been asserted that perfection has been reached, but the standard has soon been excelled. A breed of pigeons may never be produced with a beak shorter than that of the present short-faced tumbler, or with one longer than that of the English carrier, for these birds have weak constitutions and are bad breeders; but shortness and length of beak are the points which have been steadily improved during the last 150 years, and some of the best judges deny that the goal has yet been reached. From reasons which could be assigned, it is probable that parts which have now reached their maximum development, might, after remaining constant during a long period, vary again in the direction of increase under new conditions of life. But there must be, as Mr. Wallace has remarked with much truth (28/3. 'The Quarterly Journal of Science' October 1867 page 486.), a limit to change in certain directions both with natural and domestic productions; for instance, there must be a limit to the fleetness of any terrestrial animal, as this will be determined by the friction to be overcome, the weight to be carried, and the power of contraction in the muscular fibres. The English racehorse may have reached this limit; but it already surpasses in fleetness its own wild progenitor and all other equine species. The short-faced tumbler-pigeon has a beak shorter, and the carrier a beak longer, relatively to the size of their bodies, than that of any natural species of the family. Our apples, pears and gooseberries bear larger fruit than those of any natural species of the same genera; and so in many other cases.

It is not surprising, seeing the great difference between many domestic breeds, that some few naturalists have concluded that each is descended from a distinct aboriginal stock, more especially as the principle of selection has been ignored, and the high antiquity of man, as a breeder of animals, has only recently become known. Most naturalists, however, freely admit that our various breeds, however dissimilar, are descended from a single stock, although they do not know much about the art of breeding, cannot show the connecting links, nor say where and when the breeds arose. Yet these same naturalists declare, with an air of philosophical caution, that they will never admit that one natural species has given birth to another until they behold all the transitional steps. Fanciers use exactly the same language with respect to domestic breeds; thus, an author of an excellent treatise on pigeons says he will never allow that the carrier and fantail are the descendants of the wild rock-pigeon, until the transitions have "actually been observed, and can be repeated whenever man chooses to set about the task." No doubt it is difficult to realise that slight changes added up during long centuries can produce such great results; but he who wishes to understand the origin of domestic breeds or of natural species must overcome this difficulty.

The causes which excite and the laws which govern variability have been discussed so lately, that I need here only enumerate the leading points. As domesticated organisms are much more liable to slight deviations of structure and to monstrosities than species living under their natural conditions, and as widely-ranging species generally vary more than those which inhabit restricted areas, we may infer that variability mainly depends on changed conditions of life. We must not overlook the effects of the unequal combination of the characters derived from both parents, or reversion to former progenitors. Changed conditions have an especial tendency to render the reproductive organs more or less impotent, as shown in the chapter devoted to this subject; and these organs consequently often fail to transmit faithfully the parental characters. Changed conditions also act directly and definitely on the organisation, so that all or nearly all the individuals of the same species thus exposed become modified in the same manner; but why this or that part is especially affected we can seldom or ever say. In most cases, however, a change in the conditions seems to act indefinitely, causing diversified variations in nearly the same manner as exposure to cold or the absorption of the same poison affects different individuals in different ways. We have reason to suspect that an habitual excess of highly-nutritious food, or an excess relatively to the wear and tear of the organisation from exercise, is a powerful exciting cause of variability. When we see the symmetrical and complex outgrowths, caused by a minute drop of the poison of a gall-insect, we may believe that slight changes in the chemical nature of the sap or blood would lead to extraordinary modifications of structure.

The increased use of a muscle with its various attached parts, and the increased activity of a gland or other organ, lead to their increased development. Disuse has a contrary effect. With domesticated productions, although their organs sometimes become rudimentary through abortion, we have no reason to suppose that this has ever followed solely from disuse. With natural species, on the contrary, many organs appear to have been rendered rudimentary through disuse, aided by the principle of the economy of growth together with intercrossing. Complete abortion can be accounted for only by the hypothesis given in the last chapter, namely, the final destruction of the germs or gemmules of useless parts. This difference between species and domestic varieties may be partly accounted for by disuse having acted on the latter for an insufficient length of time, and partly from their exemption from any severe struggle for existence entailing rigid economy in the development of each part, to which all species under nature are subjected. Nevertheless the law of compensation or balancement, which likewise depends on the economy of growth, apparently has affected to a certain extent our domesticated productions.

As almost every part of the organisation becomes highly variable under domestication, and as variations are easily selected both consciously and unconsciously, it is very difficult to distinguish between the effects of the selection of indefinite variations and the direct action of the conditions of life. For instance, it is possible that the feet of our water-dogs and of the

American dogs which have to travel much over the snow, may have become partially webbed from the stimulus of widely extending their toes; but it is more probable that the webbing, like the membrane between the toes of certain pigeons, spontaneously appeared and was afterwards increased by the best swimmers and the best snow-travellers being preserved during many generations. A fancier who wished to decrease the size of his bantams or tumbler-pigeons would never think of starving them, but would select the smallest individuals which spontaneously appeared. Quadrupeds are sometimes born destitute of hair and hairless breeds have been formed, but there is no reason to believe that this is caused by a hot climate. Within the tropics heat often causes sheep to lose their fleeces; on the other hand, wet and cold act as a direct stimulus to the growth of hair; but who will pretend to decide how far the thick fur of arctic animals, or their white colour, is due to the direct action of a severe climate, and how far to the preservation of the best-protected individuals during a long succession of generations?

Of all the laws governing variability, that of correlation is one of the most important. In many cases of slight deviations of structure as well as of grave monstrosities, we cannot even conjecture what is the nature of the bond of connexion. But between homologous parts—between the fore and hind limbs— between the hair, hoofs, horns, and teeth—which are closely similar during their early development and which are exposed to similar conditions, we can see that they would be eminently liable to be modified in the same manner. Homologous parts, from having the same nature, are apt to blend together, and, when many exist, to vary in number.

Although every variation is either directly or indirectly caused by some change in the surrounding conditions, we must never forget that the nature of the organisation which is acted on, is by far the more important factor in the result. We see this in different organisms, which when placed under similar conditions vary in a different manner, whilst closely-allied organisms under dissimilar conditions often vary in nearly the same manner. We see this, in the same modification frequently reappearing in the same variety at long intervals of time, and likewise in the several striking cases given of analogous or parallel variations. Although some of these latter cases are due to reversion, others cannot thus be accounted for.

From the indirect action of changed conditions on the organisation, owing to the reproductive organs being thus affected—from the direct action of such conditions, and these will cause the individuals of the same species either to vary in the same manner, or differently in accordance with slight differences in their constitution—from the effects of the increased or decreased use of parts—and from correlation,—the variability of our domesticated productions is complicated to an extreme degree. The whole organisation becomes slightly plastic. Although each modification must have its own exciting cause, and though each is subjected to law, yet we can so rarely trace the precise relation between cause and effect, that we are tempted to speak of variations as if they arose spontaneously. We may even call them accidental, but this must be only in the sense in which we say that a fragment of rock dropped from a height owes its shape to accident.

It may be worth while briefly to consider the result of the exposure to unnatural conditions of a large number of animals of the same species and allowed to cross freely with no selection of any kind, and afterwards to consider the result when selection is brought into play. Let us suppose that 500 wild rock-pigeons were confined in their native land in an aviary and fed in the same manner as pigeons usually are; and that they were not allowed to increase in number. As pigeons propagate so rapidly, I suppose that a thousand or fifteen hundred birds would have to be annually killed. After several generations had been thus reared, we may feel sure that some of the young birds would vary, and the variations would tend to be inherited; for at the present day slight deviations of structure often occur and are inherited. It would be tedious even to enumerate the multitude of points which still go on varying or have recently varied. Many variations would occur in correlation with one another, as the length of the wing and tail feathers—the number of the primary wing-feathers, as well as the number and breadth of the ribs, in correlation with the size and form of the body—the number of the scutellae with the size of the feet—the length of the tongue with the length of the beak—the size of the nostrils and eyelids and the form of lower jaw in correlation with the development of wattle—the nakedness of the young with the future colour of the plumage—the size of the feet with that of the beak, and other such points. Lastly, as our birds are supposed to be confined in an aviary, they would use their wings and legs but little, and certain parts of the skeleton, such as the sternum, scapulae and feet, would in consequence become slightly reduced in size.

As in our assumed case many birds have to be indiscriminately killed every year, the chances are against any new variety surviving long enough to breed. And as the variations which arise are of an extremely diversified nature, the chances are very great against two birds pairing which have varied in the same manner; nevertheless, a varying bird even when not thus paired would occasionally transmit its character to its young; and these would not only be exposed to the same conditions which first caused the variation in question to appear, but would in addition inherit from their modified parent a tendency again to vary in the same manner. So that, if the conditions decidedly tended to induce some particular variation, all the birds might in the course of time become similarly modified. But a far commoner result would be, that one bird would vary in one way and another bird in another way; one would be born with a beak a little longer, and another with a shorter beak; one would gain some black feathers, another some white or red feathers. And as these birds would be continually intercrossing, the final result would be a body of individuals differing from each other in many ways, but only slightly; yet more than did the original rock-pigeons. But there would not be the least tendency towards the formation of several distinct breeds.

If two separate lots of pigeons were treated in the manner just described, one in England and the other in a tropical country, the two lots being supplied with different kinds of food, would they after many generations differ? When we reflect on the cases given in the twenty-third chapter, and on such facts as the difference in former times between the breeds of cattle, sheep, etc., in almost every district of Europe, we are strongly inclined to admit that the two lots would be differently modified through the influence of climate and food. But the evidence on the definite action of changed conditions is in most cases insufficient; and, with respect to pigeons, I have had the opportunity of examining a large collection of domesticated kinds, sent to me by Sir W. Elliot from India, and they varied in a remarkably similar manner with our European birds.

If two distinct breeds were mingled together in equal numbers, there is reason to suspect that they would to a certain extent prefer pairing with their own kind; but they would often intercross. From the greater vigour and fertility of the crossed offspring, the whole body would by this means become interblended sooner than would otherwise have occurred. From certain breeds being prepotent over others, it does not follow that the interblended progeny would be strictly intermediate in character. I have, also, proved that the act of crossing in itself gives a strong tendency to reversion, so that the crossed offspring would tend to revert to the state of the aboriginal rock-pigeon; and in the course of time they would probably be not much more heterogeneous in character than in our first case, when birds of the same breed were confined together.

I have just said that the crossed offspring would gain in vigour and fertility. From the facts given in the seventeenth chapter there can be no doubt of this fact; and there can be little doubt, though the evidence on this head is not so easily acquired, that long-continued close interbreeding leads to evil results. With hermaphrodites of all kinds, if the sexual elements of the same individual habitually acted on each other, the closest possible interbreeding would be perpetual. But we should bear in mind that the structure of all hermaphrodite animals, as far as I can learn, permits and frequently necessitates a cross with a distinct individual. With hermaphrodite plants we incessantly meet with elaborate and perfect contrivances for this same end. It is no exaggeration to assert that, if the use of the talons and tusks of a carnivorous animal, or of the plumes and hooks on a seed, may be safely inferred from their structure, we may with equal safety infer that many flowers are constructed for the express purpose of ensuring a cross with a distinct plant. From these various considerations, not to mention the result of a long series of experiments which I have tried, the conclusion arrived at in the chapter just referred to—namely, that great good of some kind is derived from the sexual concourse of distinct individuals—must be admitted.

To return to our illustration: we have hitherto assumed that the birds were kept down to the same number by indiscriminate slaughter; but if the least choice be permitted in their preservation, the whole result will be changed. Should the owner observe any slight variation in one of his birds, and wish to obtain a breed thus characterised, he would succeed in a surprisingly short time by careful selection. As any part which has once varied generally goes on varying in the same direction, it is easy, by continually preserving the most strongly marked individuals, to increase the amount of difference up to a high, predetermined standard of excellence. This is methodical selection.

If the owner of the aviary, without any thought of making a new breed, simply admired, for instance, short-beaked more than long-beaked birds, he would, when he had to reduce the number, generally kill the latter; and there can be no doubt that he would thus in the course of time sensibly modify his stock. It is improbable, if two men were to keep pigeons and act in this manner, that they would prefer exactly the same characters; they would, as we know, often prefer directly opposite characters, and the two lots would ultimately come to differ. This has actually occurred with strains or families of cattle, sheep, and pigeons, which have been long kept and carefully attended to by different breeders, without any wish on their part to form new and distinct sub-breeds. This unconscious kind of selection will more especially come into action with animals which are highly serviceable to man; for every one tries to get the best dogs, horses, cows, or sheep, without thinking about their future progeny, yet these animals would transmit more or less surely their good qualities to their offspring. Nor is any one so careless as to breed from his worst animals. Even savages, when compelled from extreme want to kill some of their animals, would destroy the worst and preserve the best. With animals kept for use and not for mere amusement, different fashions prevail in different districts, leading to the preservation, and consequently to the transmission, of all sorts of trifling peculiarities of character. The same process will have been pursued with our fruit-trees and vegetables, for the best will always have been the most largely cultivated, and will occasionally have yielded seedlings better than their parents.

The different strains, just alluded to, which have been actually produced by breeders without any wish on their part to obtain such a result, afford excellent evidence of the power of unconscious selection. This form of selection has probably led to far more important results than methodical selection, and is likewise more important under a theoretical point of view from closely resembling natural selection. For during this process the best or most valued individuals are not separated and prevented from crossing with others of the same breed, but are simply preferred and preserved; yet this inevitably leads to their gradual modification and improvement; so that finally they prevail, to the exclusion of the old parent-form.

With our domesticated animals natural selection checks the production of races with any injurious deviation of structure. In the case of animals which, from being kept by savages or semi-civilised people, have to provide largely for their own wants under different circumstances, natural selection will have played a more important part. Hence it probably is that they often closely resemble natural species.

As there is no limit to man's desire to possess animals and plants more and more useful in any respect, and as the fancier always wishes, owing to fashions running into extremes, to produce each character more and more strongly pronounced, there is, through the prolonged action of methodical and unconscious selection, a constant tendency in every breed to become more and more different from its parent-stock; and when several breeds have been produced and are valued for different qualities, to differ more and more from each other. This leads to Divergence of Character. As improved sub-varieties and races are slowly formed, the older and less improved breeds are neglected and decrease in number. When few individuals of any breed exist within the same locality, close interbreeding, by lessening their vigour and fertility, aids in their final extinction. Thus the intermediate links are lost, and the remaining breeds gain in Distinctness of Character.

In the chapters on the Pigeon, it was proved by historical evidence and by the existence of connecting sub-varieties in distant lands that several breeds have steadily diverged in character, and that many old and intermediate sub-breeds have been lost. Other cases could be adduced of the extinction of domestic breeds, as of the Irish wolf-dog, the old English hound, and of two breeds in France, one of which was formerly highly valued. (28/4. M. Rufz de Lavison in 'Bull. Soc. Imp. d'Acclimat.' December 1862 page 1009.) Mr. Pickering remarks (28/5. 'Races of Man' 1850 page 315.) that "the sheep figured on the most ancient Egyptian monuments is unknown at the present day; and at least one variety of the bullock, formerly known in Egypt, has in like manner become extinct." So it has been with some animals and with several plants cultivated by the ancient inhabitants of Europe during the neolithic period. In Peru, Von Tschudi (28/6. 'Travels in Peru' English translation page 177.) found in certain tombs, apparently prior to the dynasty of the Incas, two kinds of maize not now known in the country. With our flowers and culinary vegetables, the production of new varieties and their extinction has incessantly recurred. At the present time improved breeds sometimes displace older breeds at an extraordinarily rapid rate; as has recently occurred throughout England with pigs. The Longhorn cattle in their native home were "suddenly swept away as if by some murderous pestilence," by the introduction of Shorthorns. (28/7. Youatt on 'Cattle' 1834 page 200. On Pigs see 'Gardener's Chronicle' 1854 page 410.)

What grand results have followed from the long-continued action of methodical and unconscious selection, regulated to a certain extent by natural selection, we see on every side of us. Compare the many animals and plants which are displayed at our exhibitions with their parent-forms when these are known, or consult old historical records with respect to their former state. Most of our domesticated animals have given rise to numerous and distinct races, but those which cannot be easily subjected to selection must be excepted—such as cats, the cochineal insect, and the hive-bee. In accordance with what we know of the process of selection, the formation of our many races has been slow and gradual. The man who first observed and preserved a pigeon with its oesophagus a little enlarged, its beak a little longer, or its tail a little more expanded than usual, never dreamed that he had made the first step in the creation of a pouter, carrier, and fantail-pigeon. Man can create not only anomalous breeds, but others having their whole structure admirably co-ordinated for certain purposes, such as the racehorse and dray-horse, or the greyhound and bulldog. It is by no means necessary that each small change of structure throughout the body, leading towards excellence, should simultaneously arise and be selected. Although man seldom attends to differences in organs which are important under a physiological point of view, yet he has so profoundly modified some breeds, that assuredly, if found wild, they would be ranked as distinct genera.

The best proof of what selection has effected is perhaps afforded by the fact that whatever part or quality in any animal, and more especially in any plant, is most valued by man, that part or quality differs most in the several races. This result is well seen by comparing the amount of difference between the fruits produced by the several varieties of fruit-trees, between the flowers of our flower-garden plants, between the seeds, roots, or leaves of our culinary and agricultural plants, in comparison with the other and not valued parts of the same varieties. Striking evidence of a different kind is afforded by the fact ascertained by Oswald Heer (28/8. 'Die Pflanzen der Pfahlbauten' 1865.) namely, that the seeds of a large number of plants,—wheat, barley, oats, peas, beans, lentils, poppies,—cultivated for their seed by the ancient Lake-inhabitants of Switzerland, were all smaller than the seeds of our existing varieties. Rutimeyer has shown that the sheep and cattle which were kept by the earlier Lake-inhabitants were likewise smaller than our present breeds. In the middens of Denmark, the earliest dog of which the remains have been found was the weakest; this was succeeded during the Bronze age by a stronger kind, and this again during the Iron age by one still stronger. The sheep of Denmark during the Bronze period had extraordinarily slender limbs, and the horse was smaller than our present animal. (28/9. Morlot 'Soc. Vaud. des Scien. Nat.' Mars 1860 page 298.) No doubt in most of these cases the new and larger breeds were introduced from foreign lands by the immigration of new hordes of men. But it is not probable that each larger breed, which in the course of time has supplanted a previous and smaller breed, was the descendant of a distinct and larger species; it is far more probable that the domestic races of our various animals were gradually improved in different parts of the great Europaeo-Asiatic continent, and thence spread to other countries. This fact of the gradual increase in size of our domestic animals is all the more striking as certain wild or half-wild animals, such as red-deer, aurochs, park-cattle, and boars (28/10. Rutimeyer 'Die Fauna der Pfahlbauten' 1861 s. 30.) have within nearly the same period decreased in size.

The conditions favourable to selection by man are,—the closest attention to every character,—long-continued perseverance,—facility in matching or separating animals,—and especially a large number being kept, so that the inferior individuals may be freely rejected or destroyed, and the better ones preserved. When many are kept there will also be a greater chance of the occurrence of well-marked deviations of structure. Length of time is all-important; for as each character, in order to become strongly pronounced, has to be augmented by the selection of successive variations of the same kind, this can be effected only during a long series of generations. Length of time will, also, allow any new feature to become fixed by the continued rejection of those individuals which revert or vary, and by the preservation of those which still inherit the new character. Hence, although some few animals have varied rapidly in certain respects under new conditions of life, as dogs in India and sheep in the West Indies, yet all the animals and plants which have produced strongly marked races were domesticated at an extremely remote epoch, often before the dawn of history. As a consequence of this, no record has been preserved of the origin of our chief domestic breeds. Even at the present day new strains or sub-breeds are formed so slowly that their first appearance passes unnoticed. A man attends to some particular character, or merely matches his animals with unusual care, and after a time a slight difference is perceived by his neighbours;—the difference goes on being augmented by unconscious and methodical selection, until at last a new sub-breed is formed, receives a local name, and spreads; but by this time its history is almost forgotten. When the new breed has spread widely, it gives rise to new strains and sub-breeds, and the best of these succeed and spread, supplanting other and older breeds; and so always onwards in the march of improvement.

When a well-marked breed has once been established, if not supplanted by still further improved sub-breeds, and if not exposed to greatly changed conditions of life inducing further variability or reversion to long-lost characters, it may apparently last for an enormous period. We may infer that this is the case from the high antiquity of certain races; but some caution is necessary on this head, for the same variation may appear independently after long intervals of time, or in distant places. We may safely assume that this has occurred with the turnspit-dog, of which one is figured on the ancient Egyptian monuments—with the solid-hoofed swine (28/11. Godron 'De l'Espece' tome 1 1859 page 368.) mentioned by Aristotle—with five-toed fowls described by Columella—and certainly with the nectarine. The dogs represented on the Egyptian monuments, about 2000 B.C., show us that some of the chief breeds then existed, but it is extremely doubtful whether any are identically the same with our present breeds. A great mastiff sculptured on an Assyrian tomb, 640 B.C., is said to be the same with the dog still imported from Thibet into the same region. The true greyhound existed during the Roman classical period. Coming down to a later period, we have seen that, though most of the chief breeds of the pigeon existed between two and three centuries ago, they have not all retained exactly the same character to the present day; but this has occurred in certain cases in which no improvement was desired, for instance, in the case of the Spot and Indian ground-tumbler.

De Candolle (28/12. 'Geographie Botan.' 1855 page 989.) has fully discussed the antiquity of various races of plants; he states that the black seeded poppy was known in the time of Homer, the white-seeded sesamum by the ancient Egyptians, and almonds with sweet and bitter kernels by the Hebrews; but it does not seem improbable that some of these varieties may have been lost and reappeared. One variety of barley and apparently one of wheat, both of which were cultivated at an immensely remote period by the Lake-inhabitants of Switzerland, still exist. It is said (28/13. Pickering 'Races of Man' 1850 page 318.) that "specimens of a small variety of gourd which is still common in the market of Lima were exhumed from an ancient cemetery in Peru." De Candolle remarks that, in the books and drawings of the sixteenth century, the principal races of the cabbage, turnip, and gourd can be recognised: this might have been expected at so late a period, but whether any of these plants are absolutely identical with our present sub-varieties is not certain. It is, however, said that the Brussels sprout, a variety which in some places is liable to degeneration, has remained genuine for more than four centuries in the district where it is believed to have originated. (28/14. 'Journal of a Horticultural Tour' by a Deputation of the Caledonian Hist. Soc. 1823 page 293.)

In accordance with the views maintained by me in this work and elsewhere, not only the various domestic races, but the most distinct genera and orders within the same great class—for instance, mammals, birds, reptiles, and fishes—are all the descendants of one common progenitor, and we must admit that the whole vast amount of difference between these forms has primarily arisen from simple variability. To consider the subject under this point of view is enough to strike one dumb with amazement. But our amazement ought to be lessened when we reflect that beings almost infinite in number, during an almost infinite lapse of time, have often had their whole organisation rendered in some degree plastic, and that each slight modification of structure which was in any way beneficial under excessively complex conditions of life has been preserved, whilst each which was in any way injurious has been rigorously destroyed. And the long-continued accumulation of beneficial variations will infallibly have led to structures as diversified, as beautifully adapted for various purposes and as excellently co-ordinated, as we see in the animals and plants around us. Hence I have spoken of selection as the paramount power, whether applied by man to the formation of domestic breeds, or by nature to the production of species. I may recur to the metaphor given in a former chapter: if an architect were to rear a noble and commodious edifice, without the use of cut stone, by selecting from the fragments at the base of a precipice wedge-formed stones for his arches, elongated stones for his lintels, and flat stones for his roof, we should admire his skill and regard him as the paramount power. Now, the fragments of stone, though indispensable to the architect, bear to the edifice built by him the same relation which the fluctuating variations of organic beings bear to the varied and admirable structures ultimately acquired by their modified descendants.

Some authors have declared that natural selection explains nothing, unless the precise cause of each slight individual difference be made clear. If it were explained to a savage utterly ignorant of the art of building, how the edifice had been raised stone upon stone, and why wedge-formed fragments were used for the arches, flat stones for the roof, etc.; and if the use of each part and of the whole building were pointed out, it would be unreasonable if he declared that nothing had been made clear to him, because the precise cause of the shape of each fragment could not be told. But this is a nearly parallel case with the objection that selection explains nothing, because we know not the cause of each individual difference in the structure of each being.

The shape of the fragments of stone at the base of our precipice may be called accidental, but this is not strictly correct; for the shape of each depends on a long sequence of events, all obeying natural laws; on the nature of the rock, on the lines of deposition or cleavage, on the form of the mountain, which depends on its upheaval and subsequent denudation, and lastly on the storm or earthquake which throws down the fragments. But in regard to the use to which the fragments may be put, their shape may be strictly said to be accidental. And here we are led to face a great difficulty, in alluding to which I am aware that I am travelling beyond my proper province. An omniscient Creator must have foreseen every consequence which results from the laws imposed by Him. But can it be reasonably maintained that the Creator intentionally ordered, if we use the words in any ordinary sense, that certain fragments of rock should assume certain shapes so that the builder might erect his edifice? If the various laws which have determined the shape of each fragment were not predetermined for the builder's sake, can it be maintained with any greater probability that He specially ordained for the sake of the breeder each of the innumerable variations in our domestic animals and plants;—many of these variations being of no service to man, and not beneficial, far more often injurious, to the creatures themselves? Did He ordain that the crop and tail-feathers of the pigeon should vary in order that the fancier might make his grotesque pouter and fantail breeds? Did He cause the frame and mental qualities of the dog to vary in order that a breed might be formed of indomitable ferocity, with jaws fitted to pin down the bull for man's brutal sport? But if we give up the principle in one case,—if we do not admit that the variations of the primeval dog were intentionally guided in order that the greyhound, for instance, that perfect image of symmetry and vigour, might be formed,—no shadow of reason can be assigned for the belief that variations, alike in nature and the result of the same general laws, which have been the groundwork through natural selection of the formation of the most perfectly adapted animals in the world, man included, were intentionally and specially guided. However much we may wish it, we can hardly follow Professor Asa Gray in his belief "that variation has been led along certain beneficial lines," like a stream "along definite and useful lines of irrigation." If we assume that each particular variation was from the beginning of all time preordained, then that plasticity of organisation, which leads to many injurious deviations of structure, as well as the redundant power of reproduction which inevitably leads to a struggle for existence, and, as a consequence, to the natural selection or survival of the fittest, must appear to us superfluous laws of nature. On the other hand, an omnipotent and omniscient Creator ordains everything and foresees everything. Thus we are brought face to face with a difficulty as insoluble as is that of free will and predestination.

## END OF VOLUME II.